Jean-Pierre Gazeau

Coherent States in Quantum Physics

Related Titles

Schleich, W. P., Walther, H. (eds.)

Elements of Quantum Information

2007
ISBN 978-3-527-40725-5

Glauber, R. J.

Quantum Theory of Optical Coherence
Selected Papers and Lectures

2007
ISBN 978-3-527-40687-6

Vogel, W., Welsch, D.-G.

Quantum Optics

2006
ISBN 978-3-527-40507-7

Aharonov, Y., Rohrlich, D.

Quantum Paradoxes
Quantum Theory for the Perplexes

2005
ISBN 978-3-527-40391-2

Schleich, W. P.

Quantum Optics in Phase Space

2001
ISBN 978-3-527-29435-0

Jean-Pierre Gazeau

Coherent States in Quantum Physics

WILEY-VCH Verlag GmbH & Co. KGaA

The Author

Prof. Jean-Pierre Gazeau
Astroparticules et Cosmologie
Université Paris Diderot
Paris, France
gazeau@apc.univ-paris7.fr

Cover Picture
With permission of Guy Ropard,
Université de Rennes 1, France

Library of Congress Card No.: applied for
British Library Cataloguing-in-Publication Data: A catalogue record for this book is available from the British Library.
Bibliographic information published by the Deutsche Nationalbibliothek
The Deutsche Nationalbibliothek lists this publication in the Deutsche Nationalbibliografie; detailed bibliographic data are available on the Internet at <http://dnb.d-nb.de>.

Printed in the Federal Republic of Germany
Printed on acid-free paper

Typesetting le-tex publishing services GmbH, Leipzig
Printing betz-druck GmbH, Darmstadt
Binding Litges & Dopf GmbH, Heppenheim

ISBN 978-3-527-40709-5

Contents

Coherent States in Quantum Physics. Jean-Pierre Gazeau

ISBN: 978-3-527-40709-5

Preface

This book originated from a series of advanced lectures on coherent states in physics delivered in Strasbourg, Louvain-la Neuve, Paris, Rio de Janeiro, Rabat, and Bialystok, over the period from 1997 to 2008. In writing this book, I have attempted to maintain a cohesive self-contained content.

Let me first give some insights into the notion of a coherent state in physics. Within the context of classical mechanics, a physical system is described by states which are points of its phase space (and more generally densities). In quantum mechanics, the system is described by states which are vectors (up to a phase) in a Hilbert space (and more generally by density operators).

There exist superpositions of quantum states which have many features (properties or dynamical behaviors) analogous to those of their classical counterparts: they are the so-called coherent states, already studied by Schrödinger in 1926 and rediscovered by Klauder, Glauber, and Sudarshan at the beginning of the 1960s.

The phrase "coherent states" was proposed by Glauber in 1963 in the context of quantum optics. Indeed, these states are superpositions of Fock states of the quantized electromagnetic field that, up to a complex factor, are not modified by the action of photon annihilation operators. They describe a reservoir with an undetermined number of photons, a situation that can be viewed as formally close to the classical description in which the concept of a photon is absent.

The purpose of these lecture notes is to explain the notion of coherent states and of their various generalizations, since Schrödinger up to some of the most recent conceptual advances and applications in different domains of physics and signal analysis. The guideline of the book is based on a unifying method of construction of coherent states, of minimal complexity. This method has a substantially probabilistic content and allows one to establish a simple and natural link between practically all families of coherent states proposed until now. This approach embodies the originality of the book in regard to well-established procedures derived essentially from group theory (e.g., coherent state family viewed as the orbit under the action of a group representation) or algebraic constraints (e.g., coherent states viewed as eigenvectors of some lowering operator), and comprehensively presented in previous treatises, reviews, an extensive collection of important papers, and proceedings.

Coherent States in Quantum Physics. Jean-Pierre Gazeau

ISBN: 978-3-527-40709-5

A working knowledge of basic quantum mechanics and related mathematical formalisms, e.g., Hilbert spaces and operators, is required to understand the contents of this book. Nevertheless, I have attempted to recall necessary definitions throughout the chapters and the appendices.

The book is divided into two parts.

- The first part introduces the reader progressively to the most familiar coherent states, their origin, their construction (for which we adopt an original and unifying procedure), and their application/relevance to various (although selected) domains of physics.
- The second part, mostly based on recent original results, is devoted to the question of quantization of various sets (including traditional phase spaces of classical mechanics) through coherent states.

Acknowledgements My thanks go to Barbara Heller (Illinois Institute of Technology, Chicago), Ligia M.C.S. Rodrigues (Centro Brasileiro de Pesquisas Fisicas, Rio de Janeiro), and Nicolas Treps (Laboratoire Kastler Brossel, Université Pierre et Marie Curie – Paris 6) for reading my manuscript and offering valuable advice, suggestions, and corrections. They also go to my main coworkers who contributed to various extents to this book – Syad Twareque Ali (Concordia University, Montreal), Jean-Pierre Antoine (Université Catholique de Louvain), Nicolae Cotfas (University of Bucharest), Eric Huguet (Université Paris Diderot – Paris 7), John Klauder (University of Florida, Gainesville), Pascal Monceau (Université d'Ivry), Jihad Mourad (Université Paris Diderot – Paris 7), Karol Penson (Université Pierre et Marie Curie – Paris 6), Włodzimierz Piechocki (Sołtan Institute for Nuclear Studies, Warsaw), and Jacques Renaud (Université Paris-Est) – and my PhD or former PhD students Lenin Arcadio García de León Rumazo, Mónica Suárez Esteban, Julien Quéva, Petr Siegl, and Ahmed Youssef.

Paris, February 2009 *Jean Pierre Gazeau*

Part One Coherent States

1
Introduction

1.1
The Motivations

Coherent states were first studied by Schrödinger in 1926 [1] and were rediscovered by Klauder [2–4], Glauber [5–7], and Sudarshan [8] at the beginning of the 1960s. The term "coherent" itself originates in the terminology in use in quantum optics (e.g., coherent radiation, sources emitting coherently). Since then, coherent states and their various generalizations have disseminated throughout quantum physics and related mathematical methods, for example, nuclear, atomic, and condensed matter physics, quantum field theory, quantization and dequantization problems, path integrals approaches, and, more recently, quantum information through the questions of entanglement or quantum measurement.

The purpose of this book is to explain the notion of coherent states and of their various generalizations, since Schrödinger up to the most recent conceptual advances and applications in different domains of physics, with some incursions into signal analysis. This presentation, illustrated by various selected examples, does not have the pretension to be exhaustive, of course. Its main feature is a unifying method of construction of coherent states, of minimal complexity and of probabilistic nature. The procedure followed allows one to establish a simple and natural link between practically all families of coherent states proposed until now. It embodies the originality of the book in regard to well-established constructions derived essentially from group theory (e.g., coherent state family viewed as the orbit under the action of a group representation) or algebraic constraints (e.g., coherent states viewed as eigenvectors of some lowering operator), and comprehensively presented in previous treatises [10, 11], reviews [9, 12–14], an extensive collection of important papers [15], and proceedings [16].

As early as 1926, at the very beginning of quantum mechanics, Schrödinger [1] was interested in studying quantum states, which mimic their classical counterparts through the time evolution of the position operator:

$$Q(t) = e^{\frac{i}{\hbar}Ht}\, Q\, e^{-\frac{i}{\hbar}Ht}\,. \tag{1.1}$$

In this relation, $H = P^2/2m + V(Q)$ is the quantum Hamiltonian of the system. Schrödinger understood classical behavior to mean that the average or expected

Coherent States in Quantum Physics. Jean-Pierre Gazeau

ISBN: 978-3-527-40709-5

value of the position operator,

$$\overline{q}(t) = \langle \text{coherent state}|Q(t)|\text{coherent state}\rangle ,$$

in the desired state, would obey the classical equation of motion:

$$m\ddot{\overline{q}}(t) + \overline{\frac{\partial V}{\partial q}} = 0 . \tag{1.2}$$

Schrödinger was originally concerned with the harmonic oscillator, $V(q) = \frac{1}{2}m^2\omega^2 q^2$. The states parameterized by the complex number $z = |z|e^{i\varphi}$, and denoted by $|z\rangle$, are defined in a way such that one recovers the familiar sinusoidal solution

$$\langle z|Q(t)|z\rangle = 2Q_o|z|\cos(\omega t - \varphi) , \tag{1.3}$$

where $Q_o = (\hbar/2m\omega)^{1/2}$ is a fundamental quantum length built from the universal constant $\hbar$ and the constants m and ω characterizing the quantum harmonic oscillator under consideration.

In this way, states $|z\rangle$ mediate a "smooth" transition from classical to quantum mechanics. But one should not be misled: coherent states are rigorously quantum states (witness the constant $\hbar$ appearing in the definition of Q_o), yet they allow for a classical "reading" in a host of quantum situations. This unique qualification results from a set of properties satisfied by these Schrödinger–Klauder–Glauber coherent states, also called *canonical coherent states* or *standard coherent states.*

The most important among them are the following:

(CS1) The states $|z\rangle$ saturate the Heisenberg inequality:

$$\langle \Delta Q\rangle_z \, \langle \Delta P\rangle_z = \tfrac{1}{2}\hbar , \tag{1.4}$$

where $\langle \Delta Q\rangle_z := [\langle z|Q^2|z\rangle - \langle z|Q|z\rangle^2]^{1/2}$.

(CS2) The states $|z\rangle$ are eigenvectors of the annihilation operator, with eigenvalue z:

$$a|z\rangle = z|z\rangle, \quad z \in \mathbb{C} , \tag{1.5}$$

where $a = (2m\hbar\omega)^{-1/2}(m\omega Q + iP)$.

(CS3) The states $|z\rangle$ are obtained from the ground state $|0\rangle$ of the harmonic oscillator by a unitary action of the Weyl–Heisenberg group. The latter is a key Lie group in quantum mechanics, whose Lie algebra is generated by $\{Q, P, I\}$, with $[Q, P] = i\hbar I_d$ (which implies $[a, a^\dagger] = I$):

$$|z\rangle = e^{(za^\dagger - \overline{z}a)}|0\rangle . \tag{1.6}$$

(CS4) The coherent states $\{|z\rangle\}$ constitute an overcomplete family of vectors in the Hilbert space of the states of the harmonic oscillator. This property is encoded in the following resolution of the identity or unity:

$$I_d = \frac{1}{\pi}\int_{\mathbb{C}} d\,\mathrm{Re}\,z \; d\,\mathrm{Im}\,z \; |z\rangle\langle z| . \tag{1.7}$$

These four properties are, to various extents, the basis of the many generalizations of the canonical notion of coherent states, illustrated by the family $\{|z\rangle\}$. Property (CS4) is in fact, both historically and conceptually, the one that survives. As far as physical applications are concerned, this property has gradually emerged as the one most fundamental for the analysis, or decomposition, of states in the Hilbert space of the problem, or of operators acting on this space. Thus, property (CS4) will be a sort of *motto* for the present volume, like it was in the previous, more mathematically oriented, book by Ali, Antoine, and the author [11]. We shall explain in much detail this point of view in the following pages, but we can say very schematically, that given a measure space (X, ν) and a Hilbert space $\mathcal{H}$, a family of coherent states $\{|x\rangle \mid x \in X\}$ must satisfy the operator identity

$$\int_X |x\rangle\langle x| \, \nu(dx) = I_d \,. \tag{1.8}$$

Here, the integration is carried out on projectors and has to be interpreted in a *weak sense*, that is, in terms of expectation values in arbitrary states $|\psi\rangle$. Hence, the equation in (1.8) is understood as

$$\langle\psi| \int_X x\rangle\langle x| \, \nu(dx) \, |\psi\rangle = \int_X |\langle x|\psi\rangle|^2 \, \nu(dx) = |\psi|^2 \,. \tag{1.9}$$

In the ultimate analysis, what is desired is to make the family $\{|x\rangle\}$ operational through the identity (1.8). This means being able to use it as a "frame", through which one reads the information contained in an arbitrary state in $\mathcal{H}$, or in an operator on $\mathcal{H}$, or in a setup involving both operators and states, such as an evolution equation on $\mathcal{H}$. At this point one can say that (1.8) realizes a "quantization" of the "classical" space (X, ν) and the measurable functions on it through the operator-valued maps:

$$x \mapsto |x\rangle\langle x| \,, \tag{1.10}$$

$$f \mapsto A_f \stackrel{\text{def}}{=} \int_X f(x) \, |x\rangle\langle x| \, \nu(dx) \,. \tag{1.11}$$

The second part of this volume contains a series of examples of this quantization procedure.

As already stressed in [11], the family $\{|x\rangle\}$ allows a "classical reading" of operators A acting on $\mathcal{H}$ through their expected values in coherent states, $\langle x|A|x\rangle$ ("lower symbols"). In this sense, a family of coherent states provides the opportunity to study quantum reality through a framework formally similar to classical reality. It was precisely this symbolic formulation that enabled Glauber and others to treat a quantized boson or fermion field like a classical field, particularly for computing correlation functions or other quantities of statistical physics, such as partition functions and derived quantities. In particular, one can follow the dynamical evolution of a system in a "classical" way, elegantly going back to the study of classical "trajectories" in the space X.

The formalisms of quantum mechanics and signal analysis are similar in many aspects, particularly if one considers the identities (1.8) and (1.9). In signal analysis, $\mathcal{H}$ is a Hilbert space of finite energy signals, (X, ν) a space of parameters, suitably chosen for emphasizing certain aspects of the signal that may interest us in particular situations, and (1.8) and (1.9) bear the name of "conservation of energy". Every signal contains "noise", but the nature and the amount of noise is different for different signals. In this context, choosing $(X, \nu, \{|x\rangle\})$ amounts to selecting a part of the signal that we wish to isolate and interpret, while eliminating or, at least, strongly damping a noise that has (once and for all) been regarded as unessential . Here too we have in effect chosen a frame. Perfect illustrations of the deep analogy between quantum mechanics and signal processing are *Gabor analysis* and *wavelet analysis*. These analyses yield a time–frequency ("Gaboret") or a time-scale (wavelet) representation of the signal. The built-in scaling operation makes it a very efficient tool for analyzing singularities in a signal, a function, an image, and so on – that is, the portion of the signal that contains the most significant information. Now, not surprisingly, Gaborets and wavelets can be viewed as coherent states from a group-theoretical viewpoint. The first ones are associated with the Weyl–Heisenberg group, whereas the latter are associated with the affine group of the appropriate dimension, consisting of translations, dilations, and also rotations if we deal with dimensions higher than one.

Let us now give an overview of the content and organization of the book.

Part One. Coherent States

The first part of the book is devoted to the construction and the description of different families of coherent states, with the chapters organized as follows.

Chapter 2. The Standard Coherent States: the Basics

In the second chapter, we present the basics of the Schrödinger–Glauber–Klauder–Sudarshan or "standard" coherent states $|z\rangle \equiv |q, p\rangle$ introduced as a specific superposition of all energy eigenstates of the one-dimensional harmonic oscillator. We do this through four representations of this system, namely, "position", "momentum", "Fock" or "number", and "analytical" or "Fock–Bargmann". We then describe the specific role coherent states play in quantum mechanics and in quantum optics, for which those objects are precisely the coherent states of a radiation quantum field.

Chapter 3. The Standard Coherent States: the (Elementary) Mathematics

In the third chapter, we focus on the main elementary mathematical features of the standard coherent states, particularly that essential property of being a continuous *frame*, resolving the unit operator in an "overcomplete" fashion in the space of quantum states, and also their relation to the Weyl–Heisenberg group. Appendix B is devoted to Lie algebra, Lie groups, and their representations on a very basic level to help the nonspecialist become familiar with such notions. Next, we state the probabilistic content of the coherent states and describe their links with three important quantum distributions, namely, the "P", "Q" distribution and the Wigner distribution. Appendix A is devoted to probabilities and will also help the reader grasp these essential aspects. Finally, we indicate the way in which coherent states naturally occur in the Feynman path integral formulation of quantum mechanics. In more mathematical language, we tentatively explain in intelligible terms the coherent state properties such as (CS1)–(CS4) and others characterizing on a mathematical level the standard coherent states.

Chapter 4. Coherent States in Quantum Information
Chapter 4 gives an account of a recent experimental evidence of a feedback-mediated quantum measurement aimed at discriminating between optical coherent states under photodetection. The description of the experiment and of its theoretical motivations is aimed at counterbalancing the abstract character of the mathematical formalism presented in the previous two chapters.

Chapter 5. Coherent States: a General Construction
In Chapter 5 we go back to the formalism by presenting a general method of construction of coherent states, starting from some observations on the structure of coherent states as superpositions of number states. Given a set X, equipped with a measure ν and the resulting Hilbert space $L^2(X,\nu)$ of square-integrable functions on X, we explain how the choice of an orthonormal system of functions in $L^2(X,\nu)$, precisely $\{\phi_j(x) \mid j \in \text{index set } \mathcal{J}\}$, $\int_X \overline{\phi_j(x)}\phi_{j'}(x)\,\nu(dx) = \delta_{jj'}$, carrying a probabilistic content, $\sum_{j\in\mathcal{J}} |\phi_j(x)|^2 = 1$, determines the family of coherent states $|x\rangle = \sum_j \overline{\phi_j(x)}|\phi_j\rangle$. The relation to the underlying existence of a reproducing kernel space will be clarified.

This coherent state construction is the main guideline ruling the content of the subsequent chapters concerning each family of coherent states examined (in a generalized sense). As an elementary illustration of the method, we present the coherent states for the quantum motion of a particle on the circle.

Chapter 6. Spin Coherent States
Chapter 6 is devoted to the second most known family of coherent states, namely, the so-called spin or Bloch or atomic coherent states. The way of obtaining them follows the previous construction. Once they have been made explicit, we describe their main properties: that is, we depict and comment on the sequence of properties like we did in the third chapter, the link with $SU(2)$ representations, their classical aspects, and so on.

Chapter 7. Selected Pieces of Applications of Standard and Bloch Coherent States
In Chapter 7 we proceed to a (small, but instructive) panorama of applications of the standard coherent states and spin coherent states in some problems encountered in physics, quantum physics, statistical physics, and so on. The selected pa-

pers that are presented as examples, despite their ancient publication, were chosen by virtue of their high pedagogical and illustrative content.

Application to the Driven Oscillator This is a simple and very pedagogical model for which the Weyl–Heisenberg displacement operator defining standard coherent states is identified with the S matrix connecting ingoing and outgoing states of a driven oscillator.

Application in Statistical Physics: Superradiance This is another nice example of application of the coherent state formalism. The object pertains to atomic physics: two-level atoms in resonant interaction with a radiation field (Dicke model and superradiance).

Application to Quantum Magnetism We explain how the spin coherent states can be used to solve exactly or approximately the Schrödinger equation for some systems, such as a spin interacting with a variable magnetic field.

Classical and Thermodynamical Limits Coherent states are useful in thermodynamics. For instance, we establish a representation of the partition function for systems of quantum spins in terms of coherent states. After introducing the so-called Berezin–Lieb inequalities, we show how that coherent state representation makes crossed studies of classical and thermodynamical limits easier.

Chapter 8. $SU(1,1)$, $SL(2,\mathbb{R})$, and $Sp(2,\mathbb{R})$ Coherent States

Chapter 8 is devoted to the third most known family of coherent states, namely, the $SU(1,1)$ Perelomov and Barut–Girardello coherent states. Again, the way of obtaining them follows the construction presented in Chapter 5. We then describe the main properties of these coherent states: probabilistic interpretation, link with $SU(1,1)$ representations, classical aspects, and so on. We also show the relationship between wavelet analysis and the coherent states that emerge from the unitary irreducible representations of the affine group of the real line viewed as a subgroup of $SL(2,\mathbb{R}) \sim SU(1,1)$.

Chapter 9. $SU(1,1)$ Coherent States and the Infinite Square Well

In Chapter 9 we describe a direct illustration of the $SU(1,1)$ Barut–Girardello coherent states, namely, the example of a particle trapped in an infinite square well and also in Pöschl–Teller potentials of the trigonometric type.

Chapter 10. $SU(1,1)$ Coherent States and Squeezed States in Quantum Optics

Chapter 10 is an introduction to the squeezed coherent states by insisting on their relations with the unitary irreducible representations of the symplectic groups $Sp(2,\mathbb{R}) \simeq SU(1,1)$ and their importance in quantum optics (reduction of the uncertainty on one of the two noncommuting observables present in the measurements of the electromagnetic field).

Chapter 11. Fermionic Coherent States
In Chapter 11 we present the so-called fermionic coherent states and their utilization in the study of many-fermion systems (e.g., the Hartree–Fock–Bogoliubov approach).

Part Two. Coherent State Quantization

This second part is devoted to what we call "coherent state quantization". This procedure of quantization of a measure space is quite straightforward and can be applied to many physical situations, such as motions in different geometries (line, circle, interval, torus, etc.) as well as to various geometries themselves (interval, circle, sphere, hyperboloid, etc.), to give a noncommutative or "fuzzy" version for them.

Chapter 12. Coherent State Quantization: The Klauder–Berezin Approach
We explain in Chapter 12 the way in which standard coherent states allow a natural quantization of a large class of functions and distributions, including tempered distributions, on the complex plane viewed as the phase space of the particle motion on the line. We show how they offer a classical-like representation of the evolution of quantum observables. They also help to set Heisenberg inequalities concerning the "phase operator" and the number operator for the oscillator Fock states. By restricting the formalism to the finite dimension, we present new quantum inequalities concerning the respective spectra of "position" and "momentum" matrices that result from such a coherent state quantization scheme for the motion on the line.

Chapter 13. Coherent State or Frame Quantization
In Chapter 13 we extend the procedure of standard coherent state quantization to any measure space labeling a total family of vectors solving the identity in some Hilbert space. We thus advocate the idea that, to a certain extent, quantization pertains to a larger discipline than just being restricted to specific domains of physics such as mechanics or field theory. We also develop the notion of lower and upper symbols resulting from such a quantization scheme, and we discuss the probabilistic content of the construction.

Chapter 14. Elementary Examples of Coherent State Quantization
The examples which are presented in Chapter 14 are, although elementary, rather unusual. In particular, we start with measure sets that are not necessarily phase spaces. Such sets are far from having any physical meaning in the common sense.

Finite Set We first consider a two-dimensional quantization of a N-element set that leads, for $N \geq 4$, to a Pauli algebra of observables.

Unit Interval We study two-dimensional (and higher-dimensional) quantizations of the unit segment.

Unit Circle We apply the same quantization procedure to the unit circle in the plane. As an interesting byproduct of this "fuzzy circle", we give an expression for the phase or angle operator, and we discuss its relevance in comparison with various phase operators proposed by many authors.

Chapter 15. Motions on Simple Geometries
Two examples of coherent state quantization of classical motions taking place in simple geometries are presented in Chapter 15.

Motion on the Circle Quantization of the motion of a particle on the circle (like the quantization of polar coordinates in a plane) is an old question with so far no really satisfactory answers. Many questions concerning this subject have been addressed, more specifically devoted to the problem of angular localization and related Heisenberg inequalities. We apply our scheme of coherent state quantization to this particular problem.

Motion on the Hyperboloid Viewed as a $1 + 1$ *de Sitter Space-Time* To a certain extent, the motion of a massive particle on a $1 + 1$ de Sitter background, which means a one-sheeted hyperboloid embedded in a 2+1 Minkowski space, has characteristics similar to those of the phase space for the motion on the circle. Hence, the same type of coherent state is used to perform the quantization.

Motion in an Interval We revisit the quantum motion in an infinite square well with our coherent state approach by exploiting the fact that the quantization problem is similar, to a certain extent, to the quantization of the motion on the circle S^1. However, the boundary conditions are different, and this leads us to introduce vector coherent states to carry out the quantization.

Motion on a Discrete Set of Points We end this series of examples by the consideration of a problem inspired by modern quantum geometry, where geometrical entities are treated as quantum observables, as they have to be in order for them to be promoted to the status of objects and not to be simply considered as a substantial arena in which physical objects "live".

Chapter 16. Motion on the Torus
Chapter 16 is devoted to the coherent states associated with the discrete Weyl–Heisenberg group and to their utilization for the quantization of the chaotic motion on the torus.

Chapter 17. Fuzzy Geometries: Sphere and Hyperboloid
In Chapter 17, we end this series of examples of coherent state quantization with the application of the procedure to familiar geometries, yielding a noncommutative or "fuzzy" structure for these objects.

Fuzzy Sphere This is an extension to the sphere S^2 of the quantization of the unit circle. It is a nice illustration of noncommutative geometry (approached in a rather pedestrian way). We show explicitly how the coherent state quantization of the ordinary sphere leads to its fuzzy geometry. The continuous limit at infinite spins restores commutativity.

Fuzzy Hyperboloid We then describe the construction of the two-dimensional fuzzy de Sitter hyperboloids by using a coherent state quantization.

Chapter 18. Conclusion and Outlook
In this last chapter we give some final remarks and suggestions for future developments of the formalism presented.

2
The Standard Coherent States: the Basics

2.1
Schrödinger Definition

The coherent states, as they were found by Schrödinger [1, 17], are denoted by $|z\rangle$ in Dirac ket notation, where $z = |z|\, e^{i\varphi}$ is a complex parameter. They are states for which the mean values are the classical sinusoidal solutions of a one-dimensional harmonic oscillator with mass m and frequency ω:

$$\langle z|Q(t)|z\rangle = 2l_c|z|\, \cos(\omega t - \varphi)\,. \tag{2.1}$$

The various symbols that are involved in this definition are as follows:

- the characteristic length $l_c = \sqrt{\frac{\hbar}{2m\omega}}$,
- the Hilbert space $\mathcal{H}$ of quantum states for an object which classically would be viewed as a point particle of mass m, moving on the real line, and subjected to a harmonic potential with constant $k = m\omega^2$,
- $H = \frac{P^2}{2m} + \frac{1}{2}m\omega^2 Q^2$ is the Hamiltonian,
- operators "position" Q and "momentum" P are self-adjoint in the Hilbert space $\mathcal{H}$ of quantum states,
- their commutation rule is canonical, that is,

$$[Q, P] = i\hbar I_d\,, \tag{2.2}$$

- the time evolution of the position operator is defined as $Q(t) = e^{\frac{i}{\hbar}Ht}Qe^{-\frac{i}{\hbar}Ht}$.

In the sequel, we present the different ways to construct these specific states and their basic properties. We also explain the *raison d'être* of the adjective *coherent*.

2.2
Four Representations of Quantum States

The formalism of quantum mechanics allows different representations of quantum states: "position," "momentum," "energy" or "number" or Fock representation, and "phase space" or "analytical" or Fock–Bargmann representation.

Coherent States in Quantum Physics. Jean-Pierre Gazeau

ISBN: 978-3-527-40709-5

2.2.1 Position Representation

The original Schrödinger approach was carried out in the position representation. Operator Q is a multiplication operator acting in the space $\mathcal{H}$ of wave functions $\Psi(x,t)$ of the quantum entity.

$$Q\,\Psi(x,t) = x\,\Psi(x,t)\,, \quad P\,\Psi(x,t) = -i\hbar\frac{\partial}{\partial x}\Psi(x,t)\,. \tag{2.3}$$

The quantity $\mathcal{P}(S) = \int_S |\Psi(x,t)|^2\,dx$ is interpreted as the probability that, at the instant t, the object considered lies within the set $S \subset \mathbb{R}$, in the sense that a classical localization experiment would find it in S with probability $\mathcal{P}(S)$. Consistently, we have the normalization $1 = \int_{\mathbb{R}} |\Psi(x,t)|^2\,dx < \infty$, and so, at a given time t, $\mathcal{H} \cong L^2(\mathbb{R})$.

Time evolution of the wave function is ruled by the Schrödinger equation

$$H\Psi(x,t) = i\hbar\frac{\partial}{\partial t}\Psi(x,t) \tag{2.4}$$

or equivalently $\Psi(x,t) = e^{-\frac{i}{\hbar}H(t-t_0)}\,\Psi(x,t_0)$.

Stationary solutions read as $\Psi(x,t) = e^{-\frac{i}{\hbar}E_n t}\psi_n(x)$, where the energy eigenvalues are equally distributed on the positive line, $E_n = \hbar\omega\left(n + \frac{1}{2}\right)$, $n = 0,1,2,\ldots$. To each eigenvalue corresponds the normalized eigenstate ψ_n, $H\psi_n = E_n\psi_n$,

$$\begin{aligned} \psi_n(x) &= \sqrt[4]{\frac{1}{2\pi l_c^2}}\frac{1}{\sqrt{2^n n!}}\,e^{-\frac{x^2}{4l_c^2}}\,\mathrm{H}_n\left(\frac{x}{\sqrt{2}l_c}\right)\,, \\ \|\psi_n\|^2 &= \int_{-\infty}^{+\infty}|\psi_n(x)|^2\,dx = 1\,. \end{aligned} \tag{2.5}$$

Here, H_n denotes the Hermite polynomial of degree n [18], with n nodes. The functions $\{\psi_n, n \in \mathbb{N}\}$ form an orthonormal basis of the Hilbert space $\mathcal{H} = L^2(\mathbb{R})$:

$$\delta_{mn} = \langle\psi_m|\psi_n\rangle \overset{\text{def}}{=} \int_{-\infty}^{+\infty}\overline{\psi_m(x)}\psi_n(x)\,dx\,, \tag{2.6}$$

$$\forall\psi \in \mathcal{H},\ \psi = \sum_{n\in\mathbb{N}} c_n\psi_n\,, \quad c_n = \langle\psi_n|\psi\rangle\,. \tag{2.7}$$

Note that the characteristic length is the standard deviation of the position in the ground state, $n = 0$, $l_c = \sqrt{\frac{\hbar}{2m\omega}} = \sqrt{\langle\psi_0|x^2|\psi_0\rangle}$.

2.2.2 Momentum Representation

In momentum representation, it is the turn of operator P to be realized as a multiplication operator on $\mathcal{H} = L^2(\mathbb{R})$:

$$P\widehat{\Psi}(p,t) = p\widehat{\Psi}(p,t)\,, \quad Q\widehat{\Psi}(p,t) = i\hbar\frac{\partial}{\partial p}\widehat{\Psi}(p,t)\,. \tag{2.8}$$

The function $\widehat{\Psi}(p,t)$ (or $\widehat{\psi}(p)$), its pure momentum part in the stationary case, is the Fourier transform of $\Psi(x,t)$ at fixed time t (or $\psi(x)$):

$$\widehat{\Psi}(p,t) = \frac{1}{\sqrt{2\pi\hbar}} \int_{-\infty}^{+\infty} e^{-\frac{i}{\hbar}px}\,\Psi(x,t)\,dx\,, \tag{2.9}$$

$$\Psi(x,t) = \frac{1}{\sqrt{2\pi\hbar}} \int_{-\infty}^{+\infty} e^{\frac{i}{\hbar}px}\,\widehat{\Psi}(p,t)\,dp\,. \tag{2.10}$$

Energy eigenstates $\widehat{\psi}_n(p)$ in the momentum representation are similar to their Fourier counterpart $\psi_n(x)$:

$$\widehat{\psi}_n(p) = \sqrt[4]{\frac{1}{2\pi p_c^2}}\frac{1}{\sqrt{2^n n!}}\, e^{-\frac{p^2}{4p_c^2}}\,\mathrm{H}_n\left(\frac{p}{\sqrt{2}\,p_c}\right)\,, \tag{2.11}$$

where the characteristic momentum $p_c = \sqrt{\frac{\hbar m\omega}{2}}$ is the standard deviation of the momentum operator in the ground state.

2.2.3 Number or Fock Representation

The space $\mathcal{H}$ of states is viewed here on a more abstract level as the closure of the linear span of the kets $|\psi_n\rangle \equiv |n\rangle$, $n \in \mathbb{N}$, that is, the "standard" model of all separable Hilbert spaces, namely, the space $\ell^2(\mathbb{N})$ of square-summable sequences.

Operators Q and P are now realized via the *annihilation* operator a and its adjoint $a^\dagger$, the *creation* operator, defined by

$$a = \frac{1}{\sqrt{2\hbar m\omega}}(m\omega Q + iP)\,, \quad a^\dagger = \frac{1}{\sqrt{2\hbar m\omega}}(m\omega Q - iP)\,, \tag{2.12}$$

that is,

$$Q = l_c\,(a + a^\dagger)\,, \quad P = -i\;p_c(a - a^\dagger)\,. \tag{2.13}$$

The respective actions of a and $a^\dagger$ on the number basis read as

$$a|n\rangle = \sqrt{n}|n-1\rangle\,, \quad a^\dagger|n\rangle = \sqrt{n+1}|n+1\rangle\,, \tag{2.14}$$

together with the action of a on the ground or "vacuum" state $a|0\rangle = 0$. In this context, the Hamiltonian takes the simple form

$$H = \frac{1}{2}\hbar\omega(a^\dagger a + aa^\dagger) = \hbar\omega\left(N + \frac{1}{2}\right)\,, \tag{2.15}$$

where $N = a^\dagger a$ is the "number" operator, diagonal in the basis $\{|n\rangle, n \in \mathbb{N}\}$, with spectrum $\mathbb{N}$: $N|n\rangle = n|n\rangle$.

2.2.4
A Little (Lie) Algebraic Observation

From the *canonical* commutation rules

$$[Q, P] = i\hbar I_d \Rightarrow [a, a^\dagger] = I_d\,, \tag{2.16}$$

we infer that $\{Q, P, iI_d\}$ or alternatively $\{a, a^\dagger, I_d\}$ span a Lie algebra (see Appendix B) on the real numbers. It is the Weyl–Heisenberg algebra, denoted $\mathfrak{w}$, which plays a central role in the construction of coherent states. Let us adjoin to $\mathfrak{w}$ the number operator N that obeys

$$[a, N] = a\,, \quad [a^\dagger, N] = -a^\dagger\,. \tag{2.17}$$

We obtain a Lie algebra isomorphic to a *central extension* of the Euclidean group of the plane (rotations and translations). Next, the set of generators $\{Q^2, P^2, \frac{1}{2}(QP + PQ)\}$ or alternatively $\{N, a^2, a^{\dagger 2}\}$ spans the lowest-dimensional *symplectic* Lie algebra, denoted by $\mathfrak{sp}(2,\mathbb{R}) \simeq \mathfrak{sl}(2,\mathbb{R}) \simeq \mathfrak{su}(1,1)$. This algebra is involved in the construction of the so-called pure squeezed states. Finally, the union of these linear and quadratic generators,

$$\{Q, P, iI_d, Q^2, P^2, \tfrac{1}{2}(QP + PQ)\}\,,$$

or alternatively $\{a, a^\dagger, I_d, N, a^2, a^{\dagger 2}\}$, spans a six-dimensional Lie algebra, denoted by $\mathfrak{h}_6$, which is involved in the construction of the general squeezed states (see Chapter 10).

2.2.5
Analytical or Fock–Bargmann Representation

Starting from the position representation, let us apply the integral transform

$$f(z) = \int_{\mathbb{R}} \mathcal{K}(x, z)\psi(x)\, dx\,, \tag{2.18}$$

on $\psi \in \mathcal{H}$. In (2.18), z is element of the complex plane $\mathbb{C}$ with physical dimension a square-rooted action, and the integral kernel is defined as a generating function for the Hermite polynomials [18]:

$$\begin{aligned} \mathcal{K}(x, z) &= \sum_{n=0}^{+\infty} \psi_n(x) \frac{\left(z/\sqrt{\hbar}\right)^n}{\sqrt{n!}} \\ &= \sqrt[4]{\frac{1}{2\pi l_c^2}} \exp\left[\frac{1}{\hbar}\left(\frac{z^2}{2} - \left(\sqrt{\frac{m\omega}{2}}x - z\right)^2\right)\right]. \end{aligned} \tag{2.19}$$

Viewed through the transform (2.18), the eigenstate $\psi_n(x)$ is simply proportional to the nth power of z:

$$f_n(z) = \int_{\mathbb{R}} \mathcal{K}(x,z)\psi_n(x)\,dx = \frac{1}{\sqrt{n!}}\left(\frac{z}{\sqrt{\hbar}}\right)^n \equiv {}_s\langle \overline{z}|n\rangle\,. \tag{2.20}$$

The notation ${}_s\langle \overline{z}|$ will be explained soon. The inverse transformation for (2.18) reads as follows:

$$\psi(x) = \int_{\mathbb{C}} \overline{\mathcal{K}(x,z)}\, f(z)\,\mu_s(dz)\,. \tag{2.21}$$

Here $\mu_s(dz)$ is the Gaussian measure on the plane:

$$\mu_s(dz) = \frac{1}{\pi\hbar}e^{-\frac{|z|^2}{\hbar}}\,dx\,dy = \frac{i}{2\pi\hbar}e^{-\frac{|z|^2}{\hbar}}\,dz \wedge d\overline{z}\,, \tag{2.22}$$

with $z = x + iy$. The transform (2.18) maps the Hilbert space $L^2(\mathbb{R})$ onto the space $\mathcal{FB}$ of entire analytical functions that are square-integrable with respect to $\mu_s(dz)$:

$$f(z) = \sum_{n=0}^{+\infty} \alpha_n z^n \text{ converges absolutely for all } z \in \mathbb{C}\,,$$

that is, its convergence radius is infinite, and

$$\|f\|^2_{\mathcal{FB}} \stackrel{\text{def}}{=} \int_{\mathbb{C}} |f(z)|^2\,\mu_s(dz) < \infty\,. \tag{2.23}$$

The Hilbert space $\mathcal{FB}$ is known as a *Fock–Bargmann Hilbert* space. It is equipped with the scalar product

$$\langle f_1|f_2\rangle = \int_{\mathbb{C}} \overline{f_1(z)}\,f_2(z)\,\mu_s(dz) = \hbar\sum_{n=0}^{+\infty} n!\,\overline{\alpha_{1n}}\,\alpha_{2n}\,. \tag{2.24}$$

From (2.20) a natural orthonormal basis of $\mathcal{FB}$ is immediately found to be

$$f_n(z) = \frac{1}{\sqrt{n!}}\left(\frac{z}{\sqrt{\hbar}}\right)^n\,. \tag{2.25}$$

2.2.6 Operators in Fock–Bargmann Representation

The annihilation operator a is represented as a derivation, whereas its adjoint is a multiplication operator:

$$a f(z) = \sqrt{\hbar}\frac{d}{dz}f(z)\,, \quad a^\dagger f(z) = \frac{z}{\sqrt{\hbar}}f(z)\,. \tag{2.26}$$

In consequence, the number operator N realizes as a dilatation (Euler), $N = z\frac{d}{dz}$, and the Hamiltonian becomes a first order differential operator:

$$H = \hbar\omega\left(z\frac{d}{dz} + \frac{1}{2}\right). \tag{2.27}$$

Position and momentum then assume a quasi-symmetric form:

$$Q = l_c\left(\sqrt{\hbar}\frac{d}{dz} + \frac{1}{\sqrt{\hbar}}z\right), \quad P = -i\,p_c\left(\sqrt{\hbar}\frac{d}{dz} - \frac{1}{\sqrt{\hbar}}z\right). \tag{2.28}$$

2.3
Schrödinger Coherent States

Equipped with the basic quantum mechanical material presented in the previous section, we are now in the position to describe the coherent states appearing in (2.1). We first note that in position (like in momentum) representation, the ground state,

$$\psi_0(x) = \sqrt[4]{\frac{1}{2\pi l_c^2}}\, e^{-\frac{x^2}{4l_c^2}}, \tag{2.29}$$

is a Gaussian centered at the origin. Then, let us ask the question: what quantum states could keep this kind of Gaussian localization in other points of the real line

$$|\psi_{(q)}(x)|^2 \propto e^{-\text{const.}(x-q)^2}, \quad q \in \mathbb{R}\,? \tag{2.30}$$

In our Fock Hilbertian framework, the question amounts to finding the expansion coefficients b_n such that

$$|\psi_{(q)}\rangle = \sum_{n=0}^{+\infty} b_n |n\rangle. \tag{2.31}$$

The answer is immediate after having a look at the Bergman kernel $\mathcal{K}(x,z)$.

2.3.1
Bergman Kernel as a Coherent State

Let us first simplify our notation by putting from now on $\hbar = 1, m = 1, \omega = 1 \Rightarrow l_c = \frac{1}{\sqrt{2}} = p_c$. Consider again the expansion (2.19) of the kernel $\mathcal{K}(x,z)$:

$$\mathcal{K}(x,z) = \frac{1}{\sqrt[4]{\pi}}\exp\left[\left(\frac{z^2}{2} - \left(\frac{1}{\sqrt{2}}x - z\right)^2\right)\right] = \sum_{n=0}^{+\infty}\frac{z^n}{\sqrt{n!}}\psi_n(x), \tag{2.32}$$

where we have noted that $\psi_n = \overline{\psi_n}$. Let us put $z = \frac{1}{\sqrt{2}}(q + i\,p)$ and adopt the notation

$$\mathcal{K}(x,z) = \frac{1}{\sqrt[4]{\pi}} e^{\frac{|z|^2}{2}} \overbrace{e^{ix\,p} e^{-i\frac{qp}{2}}}^{\text{phase}} e^{-\frac{1}{2}(x-q)^2} \equiv \langle\delta_x|z\rangle_s \equiv \langle\delta_x|q,p\rangle_s. \tag{2.33}$$

These are the Schrödinger or nonnormalized coherent states in position representation (the index "s" is for "Schrödinger"). In Fock representation they read as

$$|z\rangle_s = |q, p\rangle_s = \sum_{n=0}^{+\infty} \frac{z^n}{\sqrt{n!}} |n\rangle \,. \tag{2.34}$$

We have obtained a continuous family of states, labeled by *all* points of the complex plane, and elements of the Hilbert space $\mathcal{H}$ with as orthonormal basis the set of kets $|n\rangle$, $n \in \mathbb{N}$.

2.3.2
A First Fundamental Property

For a given value of the labeling parameter z, the coherent state $|z\rangle_s$ is an eigenvector of the annihilation operator a, with eigenvalue z,

$$a|z\rangle_s = \sum_{n=0}^{+\infty} \frac{z^n}{\sqrt{n!}} a|n\rangle = \sum_{n=1}^{+\infty} \frac{z^n}{\sqrt{n!}} \sqrt{n}|n-1\rangle = z|z\rangle_s \,. \tag{2.35}$$

It thus follows from this equation that

(i) $f(a)|z\rangle_s = f(z)|z\rangle_s$ for an analytical function of z (with appropriate conditions),

(ii) and ${}_s\langle z|a^\dagger = \bar{z}\,{}_s\langle z|$.

2.3.3
Schrödinger Coherent States in the Two Other Representations

In momentum representation with variable k, coherent states $|z\rangle_s \equiv |q, p\rangle_s$ are essentially Gaussians centered at k:

$$\langle \delta_k|z\rangle_s = \frac{1}{\sqrt[4]{\pi}} e^{\frac{|z|^2}{2}} \overbrace{e^{-ixk} e^{-i\frac{qp}{2}}}^{\text{phase}} e^{-\frac{1}{2}(k-p)^2} \,. \tag{2.36}$$

In Fock–Bargmann representation with variable $\zeta = \frac{1}{\sqrt{2}}(x + ik) \in \mathbb{C}$, one gets

$$\langle \zeta|z\rangle_s = \sum_{n=0}^{+\infty} \frac{z^n}{\sqrt{n!}} \langle \zeta|n\rangle = \sum_{n=0}^{+\infty} \frac{z^n}{\sqrt{n!}} \frac{\bar{\zeta}^n}{\sqrt{n!}} = e^{z\bar{\zeta}} = e^{\frac{1}{2}(xq+kp)} e^{\frac{i}{2}(xp-qk)} \,. \tag{2.37}$$

One thus obtains a Gaussian depending on $z \cdot \zeta$ multiplied by a phase factor involving the form $\Im z\bar{\zeta} = \frac{1}{2}(xp - qk) \stackrel{\text{def}}{=} \zeta \wedge z$. After multiplication by Gaussian factors present in the measures $\mu_s(dz)$ and $\mu_s(d\zeta)$, one gets a Gaussian localization in the complex plane:

$$e^{-\frac{|\zeta|^2}{2}} \langle \zeta|z\rangle_s e^{-\frac{|z|^2}{2}} = e^{i(\zeta\wedge z)} e^{-\frac{|z-\zeta|^2}{2}} \,. \tag{2.38}$$

One should notice that the Schrödinger coherent states are not normalized:

$$\underset{s}{\langle z|z\rangle} = e^{|z|^2}\,. \tag{2.39}$$

2.4
Glauber–Klauder–Sudarshan or Standard Coherent States

In view of the last remark on the coherent state normalization, we now turn to the normalized or *standard* coherent states, those ones which were precisely introduced by Glauber[1] [5–7], Klauder, [3, 4], and Sudarshan [8]. They are obtained from the Schrödinger coherent states by including in the expression of the latter the Gaussian factor $e^{-\frac{|z|^2}{2}}$. They are denoted by

$$|z\rangle = e^{-\frac{|z|^2}{2}} \sum_{n=0}^{+\infty} \frac{z^n}{\sqrt{n!}}|n\rangle\,. \tag{2.40}$$

Therefore, the overlap between two such states follows a Gaussian law modulated by a "symplectic" phase factor:

$$\langle \zeta|z\rangle = e^{i(\zeta\wedge z)}e^{-\frac{|z-\zeta|^2}{2}}\,. \tag{2.41}$$

One should notice that the probability transition

$$|\langle n|z\rangle|^2 = e^{-|z|^2}\frac{|z|^{2n}}{n!} \tag{2.42}$$

is a Poisson distribution with parameter $|z|^2$. We will come back to this important point in the next chapter.

2.5
Why the Adjective *Coherent*?

Let us compare the two eigenvalue equations:

$$a|z\rangle = z|z\rangle\,, \quad a|n\rangle = \sqrt{n}|n-1\rangle\,. \tag{2.43}$$

Hence, *an infinite superposition of number states $|n\rangle$, each of the latter describing a determinate number of elementary quanta, describes a state which is left unmodified (up to a factor) under the action of the operator annihilating an elementary quantum. The factor is equal to the parameter z labeling the coherent state considered.*

1) In quantum optics the tradition has been to use the first letters of the Greek alphabet to denote the coherent state parameter: $|\alpha\rangle$, $|\beta\rangle$, ...

More generally, we have $f(a)|z\rangle = f(z)|z\rangle$ for an analytical function f. This is precisely the idea developed by Glauber [5, 6]. Indeed, an electromagnetic field in a box can be assimilated to a countably infinite assembly of harmonic oscillators. This results from a simple Fourier analysis of the Maxwell equations. The quantization of these classical harmonic oscillators yields a Fock space $\mathcal{F}$ spanned by all possible tensor products of number eigenstates $\bigotimes_k |n_k\rangle \equiv |n_1, n_2, \ldots, n_k, \ldots\rangle$, where "$k$" is a shortening for labeling the mode (including the photon polarization)

$$k \equiv \begin{cases} \vec{k} & \text{wave vector}\,, \\ \omega_k = \|\vec{k}\|c & \text{frequency}\,, \\ \lambda = 1, 2 & \text{helicity}\,, \end{cases} \tag{2.44}$$

and n_k is the number of photons in mode "k". The Fourier expansion of the quantum vector potential reads as

$$\vec{A}(\vec{r}, t) = c \sum_k \sqrt{\frac{\hbar}{2\omega_k}} \left(a_k \vec{u}_k(\vec{r}) e^{-i\omega_k t} + a_k^\dagger \vec{u}_k^*(\vec{r}) e^{i\omega_k t} \right) . \tag{2.45}$$

As an operator, it acts (up to a gauge) on the Fock space $\mathcal{F}$ via a_k and $a_k^\dagger$ defined by

$$a_{k_0} \prod_k |n_k\rangle = \sqrt{n_{k_0}} |n_{k_0} - 1\rangle \prod_{k \neq k_0} |n_k\rangle\,, \tag{2.46}$$

and obeying the canonical commutation rules

$$[a_k, a_{k'}] = 0 = [a_k^\dagger, a_{k'}^\dagger]\,, \quad [a_k, a_{k'}^\dagger] = \delta_{kk'}\, I_d\,. \tag{2.47}$$

Let us now give more insight into the modes, observables, and Hamiltonian. On the level of the mode functions $\vec{u}_k$, the Maxwell equations read as

$$\Delta \vec{u}_k(\vec{r}) + \frac{\omega_k^2}{c^2} \vec{u}_k(\vec{r}) = \vec{0}\,. \tag{2.48}$$

When confined to a cubic box C_L with size L, these functions form an orthonormal basis

$$\int_{C_L} \overline{\vec{u}_k(\vec{r})} \cdot \vec{u}_l(\vec{r})\, d^3\vec{r} = \delta_{kl}\,,$$

with obvious discretization constraints on "k". By choosing the gauge $\nabla \cdot \vec{u}_k(\vec{r}) = 0$, their expression is

$$\vec{u}_k(\vec{r}) = L^{-3/2} \widehat{e}^{(\lambda)} e^{i\vec{k}\cdot\vec{r}}\,, \quad \lambda = 1 \text{ or } 2\,, \quad \vec{k} \cdot \widehat{e}^{(\lambda)} = 0\,, \tag{2.49}$$

where the $\widehat{e}^{(\lambda)}$ stand for polarization vectors. The respective expressions of the electric and magnetic field operators are easily derived from the vector potential:

$$\vec{E} = -\frac{1}{c}\frac{\partial \vec{A}}{\partial t}\,, \quad \vec{B} = \vec{\nabla} \times \vec{A}\,.$$

Finally, the electromagnetic field Hamiltonian is given by

$$H = \frac{1}{2}\int \left(\|\vec{E}\|^2 + \|\vec{B}\|^2\right) d^3\vec{r} = \frac{1}{2}\sum_k \hbar\omega_k \left(a_k^\dagger a_k + a_k a_k^\dagger\right) .$$

Let us now decompose the electric field operator into positive and negative frequencies.

$$\vec{E} = \vec{E}^{(+)} + \vec{E}^{(-)}, \quad \vec{E}^{(-)} = \vec{E}^{(+)^\dagger} ,$$

$$\vec{E}^{(+)}(\vec{r}, t) = i\sum_k \sqrt{\frac{\hbar\omega_k}{2}} a_k \vec{u}_k(\vec{r}) e^{-i\omega_k t} . \tag{2.50}$$

We then consider the field described by the density (matrix) operator

$$\varrho = \sum_{n_k} c_{n_k} \prod_k |n_k\rangle\langle n_k| , \quad c_{n_k} \geq 0 , \ \mathrm{tr}\,\varrho = 1 , \tag{2.51}$$

and the derived sequence of correlation functions $G^{(n)}$. The Euclidean tensor components for the simplest one read as

$$G_{ij}^{(1)}(\vec{r}, t; \vec{r}', t') = \mathrm{tr}\left\{\varrho \vec{E}_i^{(-)}(\vec{r}, t)\vec{E}_j^{(+)}(\vec{r}', t')\right\} , \quad i, j = 1, 2, 3 . \tag{2.52}$$

They measure the correlation of the field state at different space-time points. A *coherent state* or *coherent radiation* $|\text{c.r.}\rangle$ for the electromagnetic field is then defined by

$$|\text{c.r.}\rangle = \prod_k |\alpha_k\rangle , \tag{2.53}$$

where $|\alpha_k\rangle$ is precisely the Glauber–Klauder–Sudarshan coherent state (2.40) for the "k" mode:

$$|\alpha_k\rangle = e^{-\frac{|\alpha_k|^2}{2}} \sum_{n_k} \frac{(\alpha_k)^{n_k}}{\sqrt{n_k!}} |n_k\rangle , \quad a_k|\alpha_k\rangle = \alpha_k|\alpha_k\rangle , \tag{2.54}$$

with $\alpha_k \in \mathbb{C}$. The particular status of the state $|\text{c.r.}\rangle$ is well understood through the action of the positive frequency electric field operator

$$\vec{E}^{(+)}(\vec{r}, t)|\text{c.r.}\rangle = \vec{\mathcal{E}}^{(+)}(\vec{r}, t)|\text{c.r.}\rangle . \tag{2.55}$$

The expression $\vec{\mathcal{E}}^{(+)}(\vec{r}, t)$ which shows up is precisely the classical field expression solution to the Maxwell equations.

$$\vec{\mathcal{E}}^{(+)}(\vec{r}, t) = i\sum_k \sqrt{\frac{\hbar\omega_k}{2}} \alpha_k \vec{u}_k(\vec{r}) e^{-i\omega_k t} . \tag{2.56}$$

Now, if the density operator is chosen as a pure coherent state, that is,

$$\varrho = |\text{c.r.}\rangle\langle\text{c.r.}| , \tag{2.57}$$

then the components (2.52) of the first-order correlation function factorize into independent terms:

$$G_{ij}^{(1)}(\vec{r},t;\vec{r}',t') = \overline{\vec{\mathcal{E}}_i^{(-)}(\vec{r},t)\vec{\mathcal{E}}_j^{(+)}(\vec{r}',t')}\,. \tag{2.58}$$

An electromagnetic field operator is said to be "fully coherent" in the Glauber sense if all of its correlation functions factorize like in (2.58). Nevertheless, one should notice that such a definition does not imply monochromaticity.

3
The Standard Coherent States: the (Elementary) Mathematics

3.1
Introduction

In this third chapter we develop, on an elementary level, the mathematical formalism of the standard coherent states: Hilbertian properties, resolution of the unity, Weyl–Heisenberg group. We also describe some probabilistic aspects of the coherent states and their essential role in the existence of four important quantum distributions, namely, the "$\mathcal{R}$", "$\mathcal{Q}$", and "$\mathcal{P}$" distributions and the Wigner distribution. Finally, we indicate the way in which coherent states naturally occur in the Feynman path integral formulation of quantum mechanics. In more mathematical language, we tentatively explain in intelligible terms the properties characterizing the standard coherent states,

$$|z\rangle = e^{-|z|^2/2} \sum_{n=0}^{\infty} \frac{z^n}{\sqrt{n!}} |n\rangle . \tag{3.1}$$

Let us list these properties that give, on their own, a strong status of uniqueness to the coherent states.

P$_0$ *The map* $\mathbb{C} \ni z \to |z\rangle \in L^2(\mathbb{R})$ *is continuous.*

P$_1$ $|z\rangle$ *is an eigenvector of the annihilation operator:* $a|z\rangle = z|z\rangle$.

P$_2$ *The coherent state family resolves the unity* $\frac{1}{\pi}\int_{\mathbb{C}} |z\rangle\langle z|\, d^2z = I$.

P$_3$ *The coherent states saturate the Heisenberg inequality:* $\Delta Q\, \Delta P = 1/2$.

P$_4$ *The coherent state family is temporally stable:* $e^{-iHt}|z\rangle = e^{-it/2}|e^{-i\omega t}z\rangle$, *where* H *is the harmonic oscillator Hamiltonian.*

P$_5$ *The mean value (or "lower symbol") of the Hamiltonian mimics the classical energy-action relation:* $\check{H}(z) \equiv \langle z|H|z\rangle = \omega|z|^2 + \frac{1}{2}$.

P$_6$ *The coherent state family is the orbit of the ground state under the action of the Weyl–Heisenberg displacement operator:* $|z\rangle = e^{(za^\dagger - \bar{z}a)}|0\rangle \equiv D(z)|0\rangle$.

P$_7$ *The Weyl–Heisenberg covariance follows from the above:*

$$\mathcal{U}(s,\zeta)|z\rangle = e^{i(s+\Im(\zeta\bar{z}))}|z+\zeta\rangle .$$

P$_8$ *The coherent states provide a straightforward quantization scheme:*

$$\textit{Classical state } z \to |z\rangle\langle z| \textit{ quantum state} .$$

Coherent States in Quantum Physics. Jean-Pierre Gazeau

ISBN: 978-3-527-40709-5

These properties cover a wide spectrum, starting from the "wave-packet" expression (3.1) together with properties P_3 and P_4, through an algebraic side (P_1), a group representation side (P_6 and P_7), a functional analysis side (P_2), and ending with the ubiquitous problem of the relationship between classical and quantum worlds (P_5 and P_8).

3.2
Properties in the Hilbertian Framework

3.2.1
A "Continuity" from the Classical Complex Plane to Quantum States

With the standard coherent states, we are in the presence of a "continuous" family of quantum states $\left|z = \frac{q+ip}{\sqrt{2}}\right\rangle$ or $|z\rangle_{s}$ in the Hilbert space $\mathcal{H}$, labeled by *all* points of the complex plane $\mathbb{C}$. We first notice that the scalar product or *overlap* between two elements of the coherent state family is given by

$$\langle z_1|z_2\rangle_{s} = e^{\bar{z}_1 z_2}\,, \quad \langle z_1|z_2\rangle = e^{i(z_1\wedge z_2)}\, e^{-\frac{|z_1-z_2|^2}{2}}\,. \tag{3.2}$$

The latter complex-valued expression never vanishes, and has a Gaussian decreasing at infinity. Now, the continuity of the map $\mathbb{C} \ni z \mapsto |z\rangle \in \mathcal{H}$, understood in terms of the respective metric topologies of $\mathbb{C}$ and $\mathcal{H}$, results from the above overlap:

$$\begin{aligned} \| |z\rangle - |z'\rangle \|^2 &= 2(1 - \Re\langle z|z'\rangle) \\ &= 2\left(1 - e^{-\frac{|z-z'|^2}{2}} \cos\left(\frac{qp' - pq'}{2}\right)\right) \underset{z'\to z}{\to} 0\,. \end{aligned} \tag{3.3}$$

3.2.2
"Coherent" Resolution of the Unity

One of the most important features of the coherent states is that not only do they form a total family in $\mathcal{H}$, that is, $\mathcal{H}$ is the closure of the linear span of the family, but they also *resolve the identity operator* on it.

$$\begin{aligned} \int_{\mathbb{C}} |z\rangle\langle z| \frac{d^2 z}{\pi} &= \sum_{nn'} |n\rangle\langle n'| \underbrace{\int_{\mathbb{C}} \frac{z^n}{\sqrt{n!}} \frac{\bar{z}^{n'}}{\sqrt{n'!}} e^{-|z|^2} \frac{d^2 z}{\pi}}_{\delta_{nn'}} \\ &= \sum_{n} |n\rangle\langle n| = I_d \left(= \int_{\mathbb{C}_s} |z\rangle_{s}\langle z| \mu_s(dz) \right)\,. \end{aligned} \tag{3.4}$$

The identity between operators,

$$\int_{\mathbb{C}} |z\rangle\langle z| \frac{d^2 z}{\pi} = I_d = \int_{\mathbb{C}_s} |z\rangle_{s}\langle z| \mu_s(dz)\,,$$

has to be mathematically understood in the so-called *weak* sense, that is,

$$\text{for all } \phi_1, \phi_2 \in \mathcal{H}, \quad \int_{\mathbb{C}} \langle \phi_1 | z \rangle \langle z | \phi_2 \rangle \frac{d^2 z}{\pi} = \langle \phi_1 | \phi_2 \rangle . \tag{3.5}$$

Actually, (3.4) also holds as a strong operator identity.

The resolution of the unity by the coherent states will hold as a guideline in the multiple generalizations presented throughout this book. The abstract nature of its content should not obscure the great significance of the formula. Before giving more mathematical insights into the latter, let us present the following elementary example that will help the reader grasp (3.4).

3.2.3 The Interplay Between the Circle (as a Set of Parameters) and the Plane (as a Euclidean Space)

Everyone is familiar with the orthonormal basis (or frame) of the Euclidean plane $\mathbb{R}^2$ defined by the two vectors (in Dirac ket notations) $|0\rangle$ and $\left|\frac{\pi}{2}\right\rangle$, where $|\theta\rangle$ denotes the unit vector with polar angle $\theta \in [0, 2\pi)$. This frame is such that

$$\langle 0 | 0 \rangle = 1 = \left\langle \frac{\pi}{2} \middle| \frac{\pi}{2} \right\rangle , \quad \left\langle 0 \middle| \frac{\pi}{2} \right\rangle = 0 ,$$

and such that the sum of their corresponding orthogonal projectors *resolves the unity*

$$I_d = |0\rangle\langle 0| + \left|\frac{\pi}{2}\right\rangle \left\langle \frac{\pi}{2}\right| . \tag{3.6}$$

This is a trivial reinterpretation of the matrix identity:

$$\begin{pmatrix} 1 & 0 \\ 0 & 1 \end{pmatrix} = \begin{pmatrix} 1 & 0 \\ 0 & 0 \end{pmatrix} + \begin{pmatrix} 0 & 0 \\ 0 & 1 \end{pmatrix} . \tag{3.7}$$

To the unit vector

$$|\theta\rangle = \cos\theta |0\rangle + \sin\theta \left|\frac{\pi}{2}\right\rangle , \tag{3.8}$$

there corresponds the orthogonal projector P_θ given by

$$P_\theta = |\theta\rangle\langle\theta| = \begin{pmatrix} \cos\theta \\ \sin\theta \end{pmatrix} \begin{pmatrix} \cos\theta & \sin\theta \end{pmatrix} = \begin{pmatrix} \cos^2\theta & \cos\theta\sin\theta \\ \cos\theta\sin\theta & \sin^2\theta \end{pmatrix} . \tag{3.9}$$

The θ-dependent superposition (3.8) can also be viewed as a *coherent state* superposition. Indeed, integrating the matrix elements of (3.9) over all angles and dividing by π leads to a *continuous* analog of (3.6)

$$\frac{1}{\pi} \int_0^{2\pi} d\theta \, |\theta\rangle\langle\theta| = I_d . \tag{3.10}$$

Thus, we have obtained a continuous frame for the plane, that is to say the continuous set of unit vectors forming the unit circle, for describing, with an extreme redundancy the Euclidean plane. The operator relation (3.10) is equally understood through its action on a vector $|v\rangle = \|v\| \, |\phi\rangle$ with polar coordinates $\|v\|, \phi$. By virtue of $\langle\theta|\theta'\rangle = \cos(\theta - \theta')$, we have

$$|v\rangle = \frac{\|v\|}{\pi} \int_0^{2\pi} d\theta \, \cos(\phi - \theta)|\theta\rangle , \tag{3.11}$$

a relation that illustrates the *overcompleteness* of the family $\{|\theta\rangle\}$. The vectors of this family are not linearly independent, and their mutual "overlappings" are given by the scalar products $\langle\phi|\theta\rangle = \cos(\phi - \theta)$. Moreover, the map $\theta \mapsto P_\theta$ furnishes a noncommutative version of the unit circle since

$$P_\theta P_{\theta'} - P_{\theta'} P_\theta = \sin(\theta - \theta') \begin{pmatrix} 0 & -1 \\ 1 & 0 \end{pmatrix} . \tag{3.12}$$

More generally, we find in this example the notion of a *positive-operator-valued measure* (POVM) on the unit circle, which means that to any measurable set $\Delta \subset [0, 2\pi)$ there corresponds the positive 2×2 matrix:

$$\begin{aligned} \Delta \mapsto P(\Delta) &\stackrel{\text{def}}{=} \frac{1}{\pi} \int_\Delta d\theta \, |\theta\rangle\langle\theta| \\ &= \begin{pmatrix} \frac{1}{\pi} \int_\Delta d\theta \, \cos^2\theta & \frac{1}{\pi} \int_\Delta d\theta \, \cos\theta \sin\theta \\ \frac{1}{\pi} \int_\Delta d\theta \, \cos\theta \sin\theta & \frac{1}{\pi} \int_\Delta d\theta \, \sin^2\theta \end{pmatrix} . \end{aligned} \tag{3.13}$$

This matrix is obviously positive since, for any nonzero vector $|v\rangle$ in the plane, we have $\langle v|P(\Delta)|v\rangle = \frac{1}{\pi} \int_\Delta d\theta \, |\langle v|\theta\rangle|^2 > 0$. POVMs are important partners of resolutions of the unity, as will be seen at different places in this book.

3.2.4 Analytical Bridge

The resolution of the identity (3.4) provided by the coherent states is the key for transforming any abstract or concrete realization of the quantum states into the Fock–Bargmann analytical one. Let

$$|\phi\rangle = \sum_n \varphi_n |n\rangle \tag{3.14}$$

be a vector in the Hilbert space $\mathcal{H}$ of quantum states in some of its realizations. Its scalar product with a Schrödinger coherent state (chosen preferably to be a standard coherent state to avoid the Gaussian weights) reads as the power series

$${}_{s}\langle\bar{z}|\phi\rangle = \sum_{n=0}^{+\infty} \varphi_n \frac{z^n}{\sqrt{n!}} \stackrel{\text{def}}{=} \Phi_s(z) , \tag{3.15}$$

with infinite convergence radius. The series (3.15) defines an entire analytical function, $\Phi_s(z)$, which is *square-integrable* with respect to the Fock–Bargmann measure $\mu_s(dz)$. Hence, this function is an element of the Fock–Bargmann space $\mathcal{FB}$ and we have established a linear map from $\mathcal{H}$ into $\mathcal{FB}$. This map is an isometry:

$$
\begin{aligned}
\|\Phi_s\|^2_{\mathcal{FB}} &= \int_{\mathbb{C}} |\Phi_s(z)|^2\, \mu_s(dz) \\
&= \int_{\mathbb{C}} |\underset{s}{\langle \bar z|}\phi\rangle|^2\, \mu_s(dz) \\
&= \int_{\mathbb{C}} \langle\phi\underset{s}{|\bar z\rangle}\underset{s}{\langle \bar z|}\phi\rangle\, \mu_s(dz) = \|\phi\|^2_{\mathcal{H}}\,,
\end{aligned}
\tag{3.16}
$$

where we have used the fact that $\mu_s(dz) = \mu_s(d\bar z)$ and the resolution of the identity (3.4).

3.2.5
Overcompleteness and Reproducing Properties

The coherent states form an *overcomplete* family of states in the sense that

(i) it is *total*, which is equivalent to stating that if there exists $\phi \in \mathcal{H}$ such that $\langle\phi\underset{s}{|z\rangle} = 0$ for all $z \in \mathbb{C}$, then $\phi = 0$. Now $\langle\phi\underset{s}{|z\rangle} = 0$ implies $\overline{\Phi_s(\bar z)} = 0$ for all $z \in \mathbb{C}$. So, by analyticity, $\overline{\varphi_n} = 0$ for all n and finally $\phi = 0$,

(ii) at least two of them in the family are not linearly independent.

Thus, the coherent states do not form a Hilbertian basis of $\mathcal{H}$, but they form a dense family in the Hilbert space $\mathcal{H}$ *and* they resolve the unity.
An immediate consequence of the latter is their reproducing action on the elements of $\mathcal{FB}$ that emerge from the map $\mathcal{H} \ni \phi \leftrightarrow \Phi_s \in \mathcal{FB}$:

$$
\begin{aligned}
\Phi_s(z) &= \underset{s}{\langle \bar z|}\phi\rangle = \underset{s}{\langle \bar z|}\int_{\mathbb{C}} \underset{s}{|\bar z'\rangle}\underset{s}{\langle \bar z'|}\phi\rangle\, \mu_s(dz') \\
&= \int_{\mathbb{C}} \underset{s}{\langle \bar z|\bar z'\rangle}\,\Phi_s(z')\, \mu_s(dz')\,.
\end{aligned}
\tag{3.17}
$$

Hence, the scalar product $\underset{s}{\langle \bar z|\bar z'\rangle} = e^{z\bar z'} = \underset{s}{\langle z'|z\rangle}$ plays the role of the *reproducing kernel* in the Fock–Bargmann space $\mathcal{FB}$. The latter *is* a *reproducing kernel space*. Note that this object indicates to what extent the coherent states are linearly dependent:

$$
\underset{s}{|z\rangle} = \int_{\mathbb{C}} e^{z\bar z'}\underset{s}{|z'\rangle}\, \mu_s(dz')\,,
\tag{3.18}
$$

$$
|z\rangle = \int_{\mathbb{C}} e^{-iz\wedge z'} e^{-\frac{1}{2}|z-z'|^2}|z'\rangle\,\frac{d^2z'}{\pi}\,.
\tag{3.19}
$$

This precisely shows that a coherent state is, by itself, a kind of average over all the coherent state family weighted with a Gaussian distribution (up to a phase).

3.3
Coherent States in the Quantum Mechanical Context

3.3.1
Symbols

As shown in 3.2.5, the coherent states form an overcomplete system (often abusively termed overcomplete "basis") or frame that allows one to analyze quantum states or observables from a "coherent state" point of view. Indeed, when we decompose a state $|\phi\rangle$ in $\mathcal{H}$ as

$$|\phi\rangle = \int_{\mathbb{C}} \langle z_s|\phi\rangle \, |z_s\rangle \, \mu_s(dz) = \int_{\mathbb{C}} \Phi_s(\bar{z}) \, |z_s\rangle \, \mu_s(dz) \,, \tag{3.20}$$

the continuous expansion component $\Phi_s(\bar{z})$ is an entire (anti-) holomorphic function called *symbol*. On the same footing, the nonanalytical function

$$\Phi(z) = e^{-\frac{|z|^2}{2}} \Phi_s(z) = \langle \bar{z}|\phi\rangle \tag{3.21}$$

is the symbol of $|\phi\rangle$ in its decomposition over the family of standard coherent states:

$$|\phi\rangle = \int_{\mathbb{C}} \langle z|\phi\rangle \, |z\rangle \, \frac{d^2 z}{\pi} = \int_{\mathbb{C}} \Phi(\bar{z}) \, |z\rangle \, \frac{d^2 z}{\pi} \,. \tag{3.22}$$

Note that the following upper bound, readily derived from the Cauchy–Schwarz inequality,

$$|\Phi(z)| = |\langle \bar{z}|\phi\rangle| \leq \sqrt{\langle z|z\rangle} \|\phi\| = \|\phi\| \,, \tag{3.23}$$

implies the continuity of the map $\phi \mapsto \Phi$ for the uniform convergence topology in the target space.

3.3.2
Lower Symbols

The mean value of a quantum observable in standard coherent states is called the "lower" [19] or "contravariant" [20] symbol of this operator. For instance, for the creation and annihilation operators ("ladder operators"), we have

$$\langle z|a|z\rangle = z \,, \quad \langle z|a^\dagger|z\rangle = \bar{z} \,. \tag{3.24}$$

For the number operator $N = a^\dagger a$, we have

$$\langle z|N|z\rangle = |z|^2 \,. \tag{3.25}$$

Hence, although a coherent state is an infinite superposition of Fock states $|n\rangle$, one gets a finite mean value of N that can be arbitrarily small at $z \to 0$.

For the canonical operator position $Q = \frac{1}{\sqrt{2}}(a+a^\dagger)$ and momentum $P = \frac{1}{\sqrt{2}i}(a-a^\dagger)$ and with the classical parameterization $z = \frac{1}{\sqrt{2}}(q + i\,p)$, we have

$$\langle z|Q|z\rangle = q\,, \quad \langle z|P|z\rangle = p\,. \tag{3.26}$$

These formulas give coherent states a quite "classical face", although they are rigorously quantal! The physical meaning of the variable x in the Schrödinger position representation of the wave is that of a sharp position. Unlike x in the Schrödinger representation, the variables p and q represent *mean values* in the coherent states. It is for this reason that we can specify both values simultaneously at the same time, something that could not be done if instead they both had represented sharp eigenvalues.

3.3.3 Heisenberg Inequalities

Let us now calculate the mean quadratic values of the position and of the momentum.

$$\langle z|Q^2|z\rangle = q^2 + \tfrac{1}{2}\,, \quad \langle z|P^2|z\rangle = p^2 + \tfrac{1}{2}\,. \tag{3.27}$$

It follows for the quantum harmonic oscillator Hamiltonian $H = \frac{1}{2}\left(P^2 + Q^2\right)$ that

$$\langle z|H|z\rangle = \tfrac{1}{2}(q^2 + p^2) + \tfrac{1}{2}\,. \tag{3.28}$$

The "absolutely quantal face" of the coherent states reappears here. Indeed, the additional term $\frac{1}{2}$, called "symplectic correction" or "vacuum energy" according to the context, has no classical counterpart.

Finally, from the respective variances of position and momentum,

$$(\Delta Q)^2 \equiv \langle z|Q^2 - \langle Q\rangle_z^2|z\rangle = \tfrac{1}{2} = (\Delta P)^2\,,$$

we infer the saturation of the uncertainty relation or, more properly, Heisenberg inequality:

$$\Delta Q\,\Delta P = \frac{1}{2}\left(\equiv \frac{\hbar}{2}\right)\,. \tag{3.29}$$

This is one of the most important features exhibited by coherent states: their quantal face is the closest possible to its classical counterpart.

3.3.4
Time Evolution and Phase Space

From the eigenvalue equation $H|n\rangle = \left(n + \frac{1}{2}\right)|n\rangle$ for the Hamiltonian of the oscillator, one deduces the time evolution of a coherent state:

$$\begin{aligned} e^{-iHt}|z\rangle &= e^{-\frac{|z|^2}{2}} \sum_{n=0}^{+\infty} \frac{z^n}{\sqrt{n!}} e^{-iHt}|n\rangle \\ &= e^{-i\frac{t}{2}} e^{-\frac{|z|^2}{2}} \sum_{n=0}^{+\infty} \frac{\left(e^{-it}z\right)^n}{\sqrt{n!}}|n\rangle \\ &= \underbrace{e^{-i\frac{t}{2}}}_{\text{phase}} |\underbrace{e^{-it}z}_{\text{rotation of } z}\rangle\,. \end{aligned} \tag{3.30}$$

We thus ascertain the *temporal stability* of the family $\{|z\rangle,\ z \in \mathbb{C}\}$ or $\{|z\rangle_s,\ z \in \mathbb{C}\}$ since a quantum state is defined up to a phase factor. There follows the illuminating form of this time evolution in position representation:

$$\begin{aligned} \langle \delta_x|z\rangle &= \frac{1}{\sqrt[4]{\pi}} e^{ixp} e^{-i\frac{qp}{2}} e^{-\frac{1}{2}(x-q)^2}\,, \\ \langle \delta_x|e^{-iHt}|z\rangle &= \frac{1}{\sqrt[4]{\pi}} e^{ixp'} e^{-i\frac{q'p'}{2}} e^{-\frac{1}{2}(x-q')^2}\,, \end{aligned} \tag{3.31}$$

where

$$q' = q\cos t + p\sin t\,, \quad p' = -q\sin t + p\cos t\,.$$

Hence, at given $z = (q, p)$, the time evolution of the Gaussian localization of the "particle" is a harmonic motion taking place between $-|z|$ and $|z|$. We exactly recover the classical behavior of the harmonic oscillator such as it is described by its phase diagram in which, at constant energy,

$$2E = p^2 + q^2 = 2|z|^2 = p'^2 + q'^2\,,$$

the phase trajectory is a circle (or, depending on the units utilized, an ellipse). The complex number $z = \frac{q+ip}{\sqrt{2}}$ can unambiguously be identified as a phase space point (a "classical state") and the Liouville measure or 2-form $\vec{\omega} = -idz \wedge d\bar{z} = dq \wedge dp$ on the phase space is naturally present in the definition of the Fock–Bargmann space. Given an initial state $z = z_0$, say, at $t = 0$, its time evolution is simply given by $z = z(t) = z_0\, e^{it}$. The alternative phase space formalism "angle–action" becomes apparent here. The angle is precisely the argument $\theta = t + \theta_0$ of z, whereas its canonical conjugate is the action

$$I_0 = \frac{1}{2\pi} \int_{|z'| \le |z_0|} \vec{\omega}' = |z_0|^2\,. \tag{3.32}$$

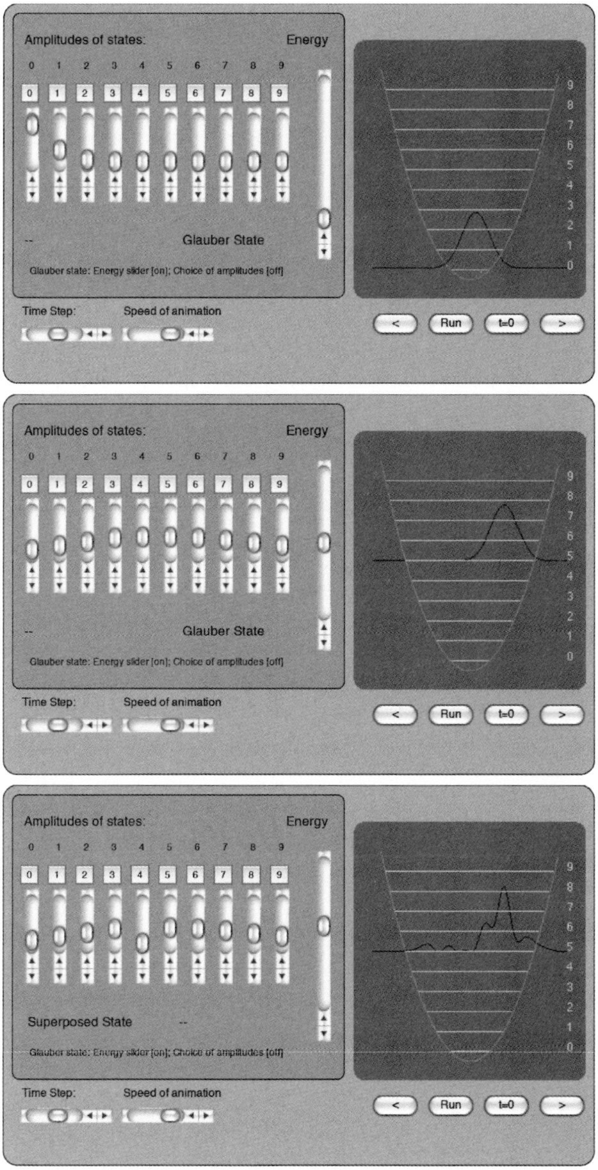

Fig. 3.1 This is a java animation developed by L. Kocbach from an original applet written by O. Psencik. It is found at http://web.ift.uib.no/AMOS/MOV/HO/. This interactive toy shows the shape of Glauber (i.e. standard) coherent states and their time evolution. It offers the possibility to select the energy $\sim |z|^2$: the vacuum $z = 0$ is shown on the top, and an example of CS with intermediate $z \neq 0$ is shown in the middle figure. The shape of a perfect Poisson law is easily recognized for these two cases. One can see in the third figure (bottom) to what extent a small pertubation of the Poisson distribution destroys the "coherent" Gaussian shape of CS states.

One notices that this physical quantity can be viewed as the average value of the number operator in the coherent state $|z_0\rangle$, $I_0 = \langle z_0|N|z_0\rangle$ (lower symbol).

At this point it is worth observing, as Schrödinger did, the close relationship between classical and quantum solutions to the oscillator problem. To be more concrete, let us reintroduce physical dimensions, namely, mass m and frequency ω. The solution to the Newton equation $\ddot{x} + \omega^2 x = 0$ is given by $x(t) = |A| \cos(\omega t - \varphi)$ with *amplitude* $= |A| \equiv \sqrt{2I_0}$ and *phase* $\varphi = \theta_0$ as initial conditions. Let us rewrite this solution as the real part of the complex $Ae^{-i\omega t}$, with $A = |A|e^{i\varphi} \equiv \sqrt{2I_0}e^{i\theta_0}$:

$$x(t) = \frac{1}{2}(Ae^{-i\omega t} + \bar{A}e^{i\omega t}) . \tag{3.33}$$

Let us now go to the quantum side by adopting the time-dependent (*Heisenberg*) point of view for position and momentum operators. Their respective time evolutions

$$Q(t) = e^{\frac{i}{\hbar}Ht}\, Q\, e^{-\frac{i}{\hbar}Ht} , \quad P(t) = e^{\frac{i}{\hbar}Ht}\, P\, e^{-\frac{i}{\hbar}Ht} , \tag{3.34}$$

produced by the Hamiltonian $H = \frac{P^2}{2m} + \frac{1}{2}m\omega^2 Q^2$, obey

$$\dot{Q}(t) = -\frac{i}{\hbar}[Q(t), H] = P(t)/m , \tag{3.35}$$

$$\dot{P}(t) = -m\omega^2 Q(t) . \tag{3.36}$$

From these two equations one easily derives the analog of the classical Newton equation:

$$\ddot{Q}(t) + \omega^2 Q(t) = 0 . \tag{3.37}$$

The same holds for $P(t)$, and, as well, for the time-dependent lowering and raising operators defined consistently as (2.12)

$$a(t) = \frac{1}{\sqrt{2\hbar m\omega}}(m\omega Q(t) + iP(t)) , \quad a^\dagger(t) = \frac{1}{\sqrt{2\hbar m\omega}}(m\omega Q(t) - iP(t)) . \tag{3.38}$$

Note that, in the Heisenberg representation, these operators act on quantum states at the initial time, since we have for any operator, say, $\mathcal{O}$, and any states, say, ψ_1, ψ_2, the identity

$$\langle\psi_1(t)| \underbrace{\mathcal{O}}_{\text{Schrödinger}} |\psi_2(t)\rangle = \langle\psi_1(0)| \underbrace{e^{\frac{i}{\hbar}Ht}\mathcal{O}e^{-\frac{i}{\hbar}Ht}}_{\text{Heisenberg } \mathcal{O}(t)} |\psi_2(0)\rangle . \tag{3.39}$$

Now, from (3.35), (3.36) and (3.38) we derive the time-evolution equation for $a(t)$ and $a^\dagger(t)$:

$$\dot{a}(t) = -i\omega a(t) , \quad \dot{a}^\dagger(t) = i\omega a^\dagger(t) . \tag{3.40}$$

They are easily solved with respective initial conditions $a(0) = a$, $a^\dagger(0) = a^\dagger$:

$$a(t) = a\, e^{-i\omega t} , \quad a^\dagger(0) = a^\dagger\, e^{i\omega t} . \tag{3.41}$$

Hence, the solution to (3.37) parallels exactly the classical solution (3.33) as

$$Q(t) = l_c \left(a\, e^{-i\omega t} + a^\dagger\, e^{i\omega t}\right) , \tag{3.42}$$

where the quantum characteristic length $l_c = \sqrt{\frac{\hbar}{2m\omega}}$ was introduced in Section 2.1.

Since the mean value of $a(t)$ in a coherent state $|z\rangle$ is given by $\langle z|a(t)|z\rangle = e^{-i\omega t}\langle z|a|z\rangle = ze^{-i\omega t}$, we can now understand Schrödinger's interest in these states as pointed out at the beginning of this book:

$$\langle z|Q(t)|z\rangle = 2l_c|z|\cos(\omega t - \varphi) . \tag{3.43}$$

3.4
Properties in the Group-Theoretical Context

3.4.1
The Vacuum as a Transported Probe...

From

$$|z\rangle = e^{-\frac{|z|^2}{2}} \sum_{n=0}^{+\infty} \frac{z^n}{\sqrt{n!}}|n\rangle$$

and

$$|n\rangle = \frac{(a^\dagger)^n}{\sqrt{n!}}|0\rangle ,$$

we obtain the alternative expression for the coherent states

$$|z\rangle = e^{-\frac{|z|^2}{2}} \sum_{n=0}^{+\infty} \frac{(za^\dagger)^n}{\sqrt{n!}}|0\rangle = e^{-\frac{|z|^2}{2}}\, e^{za^\dagger}|0\rangle . \tag{3.44}$$

One should notice here the presence of the "exponentiated" action of an alternative version, denoted by $\mathfrak{w}_m$, of the Weyl–Heisenberg Lie algebra

$$\mathfrak{w}_m = \text{linear span of } \{iQ, iP, iI_d\} . \tag{3.45}$$

A generic element of $\mathfrak{w}_m$ is written as

$$\mathfrak{w}_m \ni X = isI_d + i(pQ - qP) = isI_d + (za^\dagger - \bar{z}a) , \quad s \in \mathbb{R} , \tag{3.46}$$

with the notation $z = \frac{q+ip}{\sqrt{2}}$, $Q = \frac{a+ia^\dagger}{\sqrt{2}}$, and $P = \frac{a-ia^\dagger}{\sqrt{2}i}$. The operator X is anti-self-adjoint in $\mathcal{H}$ and is the infinitesimal generator of the *unitary* operator:

$$e^X = e^{is}e^{i(pQ-qP)} = e^{is}e^{za^\dagger - \bar{z}a} \overset{\text{def}}{=} e^{is}D(z) . \tag{3.47}$$

3.4.2
Under the Action of...

The operator-valued map $z \mapsto D(z) = e^{za^\dagger - \bar{z}a} = e^{i(pQ-qP)}$ that appears in (3.47) is a *unitary representation*, up to a (crucial!) phase factor, of the group of translations of the complex plane. Indeed, let us apply to the product $D(z_1)D(z_2) \equiv e^{A_1}e^{A_2}$, where $A_i = z_i a^\dagger - \bar{z}_i a$ and $[A_1, A_2] = 2i\Im(z_1\bar{z}_2) = -2iz_1 \wedge z_2$, the Weyl formula, $e^A e^B = e^{\left(\frac{1}{2}[A,B]\right)} e^{(A+B)}$, which is valid for any pair of operators that commute with their commutator, $[A,[A,B]] = 0 = [B,[A,B]]$. One gets the composition rule

$$D(z_1)D(z_2) = e^{-iz_1 \wedge z_2} D(z_1 + z_2) . \tag{3.48}$$

The unitarity of the operator $D(z)$, $D(z)^{-1} = D(z)^\dagger = e^{\bar{z}a - za^\dagger} = D(-z)$ is then easy to check:

$$D(z)D(-z) = e^{-iz\wedge z} D(z-z) = I_d . \tag{3.49}$$

More generally,

$$D(z_n)D(z_{n-1}) \cdots D(z_1) = e^{i\delta} D(z_1 + z_2 + \cdots + z_n) . \tag{3.50}$$

The phase $\delta = -\sum_{j<k} z_j \wedge z_k$ that appears in this expression has a topological (or symplectic) meaning: it is equal to the oriented area of the polygon $\mathcal{A}_\Gamma$ delimited by the path Γ with vertices $z_1,\ z_1+z_2,\ \ldots,\ z_1+z_2+\cdots+z_n$, as shown in Figure 3.2. In canonical coordinates $z_i = \frac{1}{\sqrt{2}}(q_i + p_i)$,

$$\delta = \sum_{j<k} \frac{1}{2}(q_j p_k - q_k p_j) , \tag{3.51}$$

and this represents a discrete version of the Stokes formula $\int_\Gamma p\, dq = \int_{\mathcal{A}_\Gamma} dq \wedge dp$.

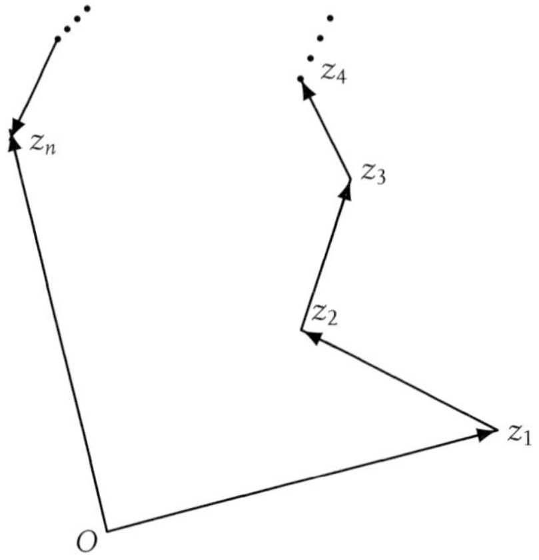

Fig. 3.2 An example of piecewise linear path in phase space.

3.4.3
... the *D*-Function

The operators X, through the map $X \mapsto e^X = e^{is}D(z)$, thus define an *irreducible unitary representation* of the Weyl–Heisenberg algebra $\mathfrak{w}$. All other representations, except an unimportant case, are of the same type: there is *uniqueness* of the realization of the canonical commutation rules (Stone–von Neumann theorem [21]). The phase factor e^{is} that appears here commutes with all other operators in the representation but nevertheless plays a crucial role in quantum mechanics, since it encodes its noncommutativity, as will be seen below.

The D-"function" (it is actually an operator, also named *displacement* operator) $D(z) = \exp(za^\dagger - \bar{z}a)$ is essential in determining the properties and applications of the coherent states. Using the Weyl formula, one can also write

$$D(z) = e^{-\frac{|z|^2}{2}} e^{za^\dagger} e^{-\bar{z}a} = e^{-\frac{i}{2}qp} e^{ipQ} e^{-iqP} . \tag{3.52}$$

With this formulation and from $a|0\rangle = 0 \Rightarrow e^{-\bar{z}a}|0\rangle = |0\rangle$, one gets a new definition of the standard coherent states, precisely the one that emerges from group theory [10, 11]:

$$|z\rangle = D(z)|0\rangle . \tag{3.53}$$

Hence, the family of coherent states is to be viewed as the orbit of the Fock vacuum under the action of operators $D(z)$, $z \in \mathbb{C}$.

Let us show how we derive from (3.53) the coherent states in the position representation. From

$$|z\rangle = D(z)|0\rangle , \quad \zeta_z(x) \stackrel{\text{def}}{=} \langle\delta_x|z\rangle = \langle D(-z)\delta_x|0\rangle , \tag{3.54}$$

and from

$$\begin{aligned} D(-z)\delta_x(y) &= e^{-\frac{i}{2}qp} e^{-ipQ} e^{iqP} \delta_x(y) = e^{-\frac{i}{2}qp} e^{-ipQ} \delta_x(y+q) \\ &= e^{\frac{i}{2}qp} e^{-ipx} \delta_{x-q}(y) , \end{aligned} \tag{3.55}$$

one finds the following expression for the coherent states:

$$\zeta_z(x) = \frac{1}{\sqrt[4]{\pi}} e^{-\frac{i}{2}qp} e^{ipx} e^{-\frac{(x-q)^2}{2}} ,$$

to be compared with (2.33).

3.4.4
Symplectic Phase and the Weyl–Heisenberg Group

In fact, the coherent state family should be viewed as the orbit of *any* particular coherent state since

$$D(z')|z\rangle = D(z')D(z)|0\rangle = e^{iz\wedge z'} D(z'+z)|0\rangle = e^{iz\wedge z'}|z+z'\rangle . \tag{3.56}$$

One notices that the presence of the phase factor in (3.56) prevents us from viewing the displacement operator $D(z)$ as the representation of a simple translation. Indeed, it appears as the subtle mark of noncommutativity of two successive displacements:

$$D(z_1)D(z_2) = e^{2iz_2 \wedge z_1} D(z_2)D(z_1)\,. \tag{3.57}$$

This equation *is* the integrated version of the canonical commutation rules $[Q, P] = iI_d$. In this regard, one speaks of *projective representation* $z \mapsto D(z)$ of the Abelian group $\mathbb{C}$, since the composition of two operators $D(z_1)$, $D(z_2)$ produces a phase factor, with phase $\Im z_1\bar{z}_2 = -z_1 \wedge z_2$. In other words and to insist on this important feature of the formalism, the unavoidable appearance of this phase compels us to work with a wider set than the complex numbers. This set is precisely the (Lie) Weyl–Heisenberg group $W \simeq \mathbb{R} \times \mathbb{C}$, the Lie algebra of which is $\mathfrak{w}_m$

$$\begin{aligned} W \ni g &= (s, z) = (s, q, p) \\ &\downdownarrows \\ \mathfrak{W}_m \ni X &= isI_d + za^\dagger - \bar{z}a\,. \end{aligned} \tag{3.58}$$

The group law is given by

$$\begin{aligned} (s_1, z_1)(s_2, z_2) &= (s_1 + s_2 - z_1 \wedge z_2, z_1 + z_2) \Leftrightarrow \\ (s_1, q_1, p_1)(s_2, q_2, p_2) &= \left(s_1 + s_2 - \tfrac{1}{2}(q_1 p_2 - q_2 p_1), q_1 + q_2, p_1 + p_2\right)\,. \end{aligned} \tag{3.59}$$

The neutral element is $(0, 0, 0)$ and the inverse is $(s, z)^{-1} = (-s, -z)$.

3.4.5
Coherent States as Tools in Signal Analysis

An intriguing question arises from the group-theoretical interpretation of the coherent states: what about transporting a state different from the vacuum? Concretely, let us make $D(z)$ act on an arbitrarily chosen state $|\psi\rangle \in \mathcal{H}$:

$$|z\rangle_\psi \stackrel{\text{def}}{=} D(z)|\psi\rangle\,. \tag{3.60}$$

For instance, in position representation, these states read as

$$\langle\delta_x|z\rangle_\psi \stackrel{\text{def}}{=} \psi_z(x) = e^{-\frac{i}{2}qp} e^{ipx} \psi(x - q)\,. \tag{3.61}$$

A question naturally arises: which genuine coherent state properties are still valid?

(i) The states $|z\rangle_\psi$ are normalizable:

$$_\psi\langle z|z\rangle_\psi = \langle\psi|D(z)^\dagger D(z)|\psi\rangle = \|\psi\|^2\,. \tag{3.62}$$

(ii) They solve the identity. Indeed, consider the operator

$$A = \int_{\mathbb{C}} |z\rangle_{\psi\psi}\langle z| \frac{d^2z}{\pi} = \int_{\mathbb{C}} D(z)|\psi\rangle\langle\psi|D(-z) \frac{d^2z}{\pi}\,. \tag{3.63}$$

This operator commutes with all operators $e^{is} D(z')$ of the unitary irreducible representation of the Weyl–Heisenberg group (exercise!). Hence, by applying the Schur lemma,[2] A = constant I_d, with consequently the same reproducing properties. The computation of the constant $\equiv c_\psi$ is straightforward:

$$c_\psi \, \|\psi\|^2 = \langle\psi|A|\psi\rangle = \int_{\mathbb{C}} |\langle\psi|D(z)|\psi\rangle|^2 \, \frac{d^2 z}{\pi} \, .$$

(iii) These coherent states enjoy the same covariance properties with respect to the action of W:

$$D(z')|z\rangle_\psi = e^{i z \wedge z'} |z + z'\rangle_\psi \, . \tag{3.64}$$

But then what have we lost?

(iv) We have lost an important property:

$$a|z\rangle_\psi \neq z|z\rangle_\psi \, , \tag{3.65}$$

since the equality only holds true for $|\psi\rangle \propto |0\rangle$.

Nevertheless, we get a serious improvement with regard to the freedom in the choice of ψ ! This is essentially the main interest in these states from the point of view of signal analysis. The so-called Gabor (or "windowed Fourier") transform of signal analysis [22–24] precisely rests upon the resolution of the identity

$$I_d = \int_{\mathbb{C}} |z\rangle_{\psi\psi}\langle z| \, \frac{d^2 z}{c_\psi \pi} \, . \tag{3.66}$$

This identity allows one to implement a Hilbertian analysis of any state $|\phi\rangle$ from the point of view of the continuous frame of coherent states $|z\rangle_\psi$:

$$|\phi\rangle = \int_{\mathbb{C}} {}_\psi\langle z|\phi\rangle \, |z\rangle_\psi \, \frac{d^2 z}{c_\psi \pi} \, . \tag{3.67}$$

The projection ${}_\psi\langle z|\phi\rangle$ of the state $|\phi\rangle$ onto the state

$$\psi_z(x) = e^{-\frac{i}{2} q p} e^{i p x} \psi(x - q) \, , \tag{3.68}$$

which is the "window" or "Gaboret" or even "wavelet", $|\psi\rangle$, translated *and* modulated, is called the *Gabor transform* or the *windowed Fourier transform* or the *time–frequency representation* of the "signal" ϕ. This transform reads as

$${}_\psi\langle z|\phi\rangle = \int_{-\infty}^{+\infty} e^{\frac{i}{2} q p} e^{-i p x} \overline{\psi(x - q)} \phi(x) \, dx \stackrel{\text{def}}{=} \mathcal{G}_\phi(q, p) \, . \tag{3.69}$$

Note that, in practice, one ignores the phase factor $e^{\frac{i}{2} q p}$ in the definition of the window. We give in Figure 3.3 an example of such a "time–frequency" representation of a signal.

2) Schur lemma: If T on vector space E and T' on E' are irreducible representations of a group G and $L : E \mapsto E'$ is a linear map such that $T'(g) L = L T(g)$ for all $g \in G$, then $L = 0$ or L is invertible. Furthermore, if $E = E'$ is a vector space over complex numbers, then L is a scalar.

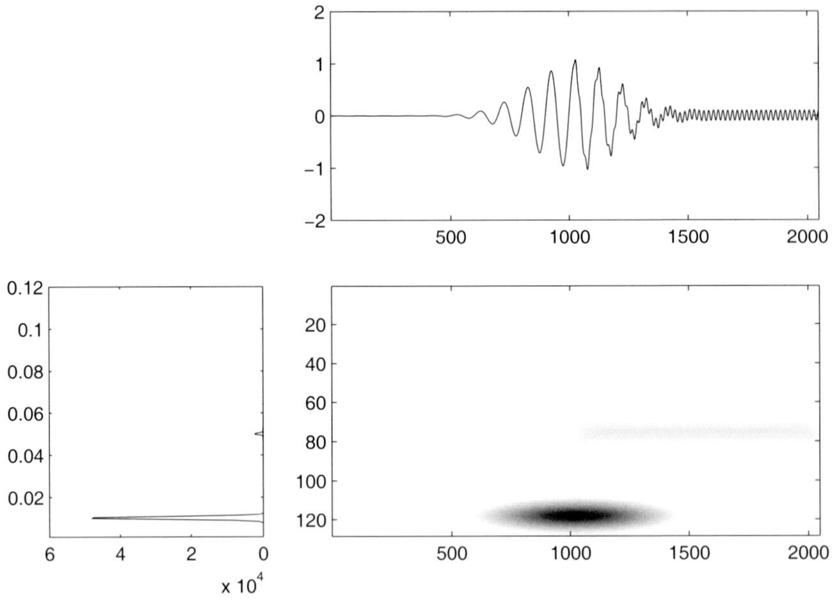

Fig. 3.3 Windowed Fourier (or Gabor, or time–frequency) transform of a signal (top). On the left, the Fourier transform, with two visible frequency peaks corresponding to the discernible regimes in the signal. On the right, the time–frequency representation of the signal. The two regimes clearly manifest themselves with their respective durations and frequencies (by courtesy of Pierre Vandergheynst, EPFL).

3.5 Quantum Distributions and Coherent States

In quantum mechanics, the existence, due to some classical statistical estimate, of a probability distribution on the states accessible to a system is described by a statistical operator called its *density matrix*. This operator is viewed as a "mixed state" superposition of pure state projectors $|\psi\rangle\langle\psi|$ with $\|\psi\| = 1$.

$$\text{“mixed” state: } \varrho = \sum_{\psi} p_{\psi} \underbrace{|\psi\rangle\langle\psi|}_{\text{“pure” state with norm 1}}, \tag{3.70}$$

where $0 \leq p_{\psi} \leq 1$, $\sum_{\psi} p_{\psi} = 1$. In the case of an orthonormal basis $|\psi_i\rangle$, $\varrho = \sum_i p_i |\psi_i\rangle\langle\psi_i|$, p_i is the probability (estimated from a classical point of view) for the system to be in state $|\psi_i\rangle$. So, a quantum mixed state is therefore a convex linear superposition of pure states (projectors).

Let $\{|n\rangle, n \in \mathbb{N}\}$ be an orthonormal basis (e.g., Fock number states) of the Hilbert space of states of the system. The mean value of a quantum observable $\mathcal{O}$ in a unit-norm state ψ is $\langle \mathcal{O} \rangle_{\psi} = \sum_n \langle n|\psi\rangle\langle\psi|\mathcal{O}|n\rangle$. Hence, the mean value of $\mathcal{O}$ in the state ϱ is

$$\langle \mathcal{O} \rangle_\varrho = \sum_\psi p_\psi \langle \mathcal{O} \rangle_\psi = \sum_n \langle n | \varrho \mathcal{O} | n \rangle = \mathrm{tr}(\mathcal{O}\varrho) \,. \tag{3.71}$$

In this section we give a short account of different types of phase space representations of the density matrix of a system, mainly pertaining to quantum optics [25, 26].

3.5.1 The Density Matrix and the Representation "$\mathcal{R}$"

In the Fock representation, the density matrix is just defined by its matrix elements $\varrho_{m,n}$:

$$\varrho = \sum_{m,n} \underbrace{\varrho_{m,n}}_{\langle m|\varrho|n\rangle} |m\rangle\langle n| \,.$$

In the "full" coherent state representation, the density matrix is determined by the (generally complex-valued) function named distribution "$\mathcal{R}$" defined as

$$\mathcal{R}(\overline{\alpha}, \beta) \stackrel{\text{def}}{=} \underset{s}{\langle \alpha|}\varrho\underset{s}{|\beta\rangle} = \langle \alpha|\varrho|\beta\rangle e^{\frac{1}{2}\left(|\alpha|^2+|\beta|^2\right)} \,. \tag{3.72}$$

This distribution appears in the Fock–Bargmann representation of ϱ:

$$\begin{aligned} \varrho &= \iint_{\mathbb{C}^2} \mathcal{R}(\overline{\alpha}, \beta)\, e^{-\frac{1}{2}\left(|\alpha|^2+|\beta|^2\right)} \, |\alpha\rangle\langle\beta| \, \frac{d^2\alpha}{\pi} \frac{d^2\beta}{\pi} \\ &= \iint_{\mathbb{C}^2} \mathcal{R}(\overline{\alpha}, \beta) \underset{s}{|\alpha\rangle}\underset{s}{\langle\beta|} \, \mu_s(d\alpha)\, \mu_s(d\beta) \,. \end{aligned} \tag{3.73}$$

In the case of a pure number state, $\varrho = |n\rangle\langle n|$, this distribution reads as

$$\mathcal{R}(\overline{\alpha}, \beta) = \frac{(\overline{\alpha}\beta)^n}{n!} \,. \tag{3.74}$$

In the case of a coherent state $\varrho = |z\rangle\langle z|$,

$$\mathcal{R}(\overline{\alpha}, \beta) = e^{-|z|^2} e^{z\overline{\alpha} + \bar{z}\beta} \,. \tag{3.75}$$

3.5.2 The Density Matrix and the Representation "$\mathcal{Q}$"

The distribution "Q", or "lower symbol" of ϱ, or *Husimi function* of the state ϱ, is the set of expected values of ϱ in coherent state representation:

$$\mathcal{Q}(\alpha, \overline{\alpha}) \stackrel{\text{def}}{=} \langle \alpha|\varrho|\alpha\rangle \,. \tag{3.76}$$

It is a positive function bounded by 1:

$$0 \leq Q(\alpha,\overline{\alpha}) = \sum_{\psi} p_\psi \underbrace{|\langle\psi|\alpha\rangle|^2}_{\leq 1} \leq 1\,.$$

Moreover, it is a true probability density on the phase space since we have

$$\int_{\mathbb{C}} Q(\alpha,\overline{\alpha})\, \frac{d^2\alpha}{\pi} = 1\,. \tag{3.77}$$

The simple choice of the pure state $\varrho = |n\rangle\langle n|$ illustrates its probabilistic meaning in terms of a Poisson distribution:

$$Q(\alpha,\overline{\alpha}) = |\langle n|\alpha\rangle|^2 = e^{-|\alpha|^2}\frac{|\alpha|^2}{n!}\,. \tag{3.78}$$

The Husimi function has a remarkable property (we will come back to this point): an operator $\mathcal{O}_A(a,a^\dagger) = \sum_{m,n} d_{mn}\, a^m {a^\dagger}^n$, in its "antinormal order" expansion, that is, all creation operators are placed on the right in monomials appearing in the series expansion of $\mathcal{O}(a,a^\dagger)$, has the following expectation value in state ϱ:

$$\left\langle \mathcal{O}_A(a,a^\dagger)\right\rangle_\varrho = \int_{\mathbb{C}} \mathcal{O}_A(\alpha,\overline{\alpha})\, Q(\alpha,\overline{\alpha})\, \frac{d^2\alpha}{\pi}\,. \tag{3.79}$$

3.5.3
The Density Matrix and the Representation "$\mathcal{P}$"

The distribution "P", or "upper symbol" of ϱ, is defined by the components of ϱ in its diagonal representation in terms of coherent state projectors:

$$\varrho = \int_{\mathbb{C}} P(\alpha,\overline{\alpha})\, |\alpha\rangle\langle\alpha|\, \frac{d^2\alpha}{\pi}\,. \tag{3.80}$$

It is a bounded function but has indeterminate sign. It is a pseudodensity of probability on the phase space in the sense that

$$\int_{\mathbb{C}} P(\alpha,\overline{\alpha})\, \frac{d^2\alpha}{\pi} = 1\,. \tag{3.81}$$

There exists an inversion formula to (3.80)

$$P(\alpha,\overline{\alpha}) = e^{|\alpha|^2} \int_{\mathbb{C}} \langle -\beta|\varrho|\beta\rangle\, e^{|\beta|^2}\, e^{-2i\alpha\wedge\beta}\, \frac{d^2\beta}{\pi}\,. \tag{3.82}$$

This formula is derived from the so-called *symplectic* Fourier transform of general functions of the complex variable α and its conjugate:

$$\begin{aligned} \widehat{f}^s(\alpha,\overline{\alpha}) &= \frac{1}{2\pi}\int_{\mathbb{C}} e^{i\alpha\wedge\beta} f(\beta,\overline{\beta})\, d^2\beta\,, \\ f(\beta,\overline{\beta}) &= \frac{1}{2\pi}\int_{\mathbb{C}} e^{i\beta\wedge\alpha} \widehat{f}^s(\alpha,\overline{\alpha})\, d^2\alpha\,, \\ \text{with } & \frac{1}{4\pi^2}\int_{\mathbb{C}} e^{i\alpha\wedge\beta}\, d^2\beta = \delta^{(2)}(\alpha)\,. \end{aligned} \tag{3.83}$$

Let us determine the distribution "P" for a coherent state itself, $\varrho = |\zeta\rangle\langle\zeta|$. One finds, as expected and up to the factor π, a Dirac distribution on $\mathbb{C}$ centered at ζ:

$$P(\alpha,\overline{\alpha}) = e^{-|\alpha|^2-|\zeta|^2}\int_{\mathbb{C}} e^{2i(\alpha-\zeta)\wedge\beta}\,\frac{d^2\beta}{\pi} = \pi\delta^2(\alpha-\zeta)\,. \tag{3.84}$$

Another interesting example [25] concerns a thermal radiation field, emitted by a source in thermal equilibrium at temperature T:

$$\varrho = \frac{e^{-\frac{H}{k_B T}}}{\operatorname{tr} e^{-\frac{H}{k_B T}}} = \sum_n \frac{\langle n\rangle^n}{(1+\langle n\rangle)^{n+1}}|n\rangle\langle n|\,,$$

where $\langle n\rangle \equiv \operatorname{tr}(a^\dagger a\varrho)$. One then gets the Gaussian distribution

$$P(\alpha,\overline{\alpha}) = \frac{1}{\langle n\rangle}e^{-\frac{|\alpha|^2}{\langle n\rangle}}\,.$$

Similarly to the Husimi function Q, a remarkable property holds concerning mean values of operators: an operator $\mathcal{O}_N(a,a^\dagger) = \sum_{m,n} c_{mn}\, a^{\dagger n}\, a^m$, in its "normal order" expansion, that is, all creation operators are placed on the left in monomials appearing in the series expansion of $\mathcal{O}(a,a^\dagger)$, has as the expected value in the state ϱ

$$\left\langle \mathcal{O}_N(a,a^\dagger)\right\rangle_\varrho = \int_{\mathbb{C}} \mathcal{O}_N(\alpha,\overline{\alpha})\,P(\alpha,\overline{\alpha})\,\frac{d^2\alpha}{\pi}\,. \tag{3.85}$$

3.5.4
The Density Matrix and the Wigner(–Weyl–Ville) Distribution

The Wigner distribution $W_c(\alpha,\overline{\alpha})$ associated with a matrix density ϱ is the symplectic Fourier transform of the mean value of the displacement operator or function D in the state ϱ:

$$\begin{aligned} W_c(\alpha,\overline{\alpha}) &\stackrel{\text{def}}{=} \frac{1}{2\pi}\int_{\mathbb{C}} \operatorname{tr}(D(\beta)\varrho)\,e^{2i\alpha\wedge\beta}\,\frac{d^2\beta}{\pi} \\ &= \frac{1}{2\pi}\int_{\mathbb{C}} \operatorname{tr}(e^{\beta a^\dagger-\overline{\beta}a}\varrho)\,e^{-(\beta\overline{\alpha}-\overline{\beta}\alpha)}\,\frac{d^2\beta}{\pi}\,. \end{aligned} \tag{3.86}$$

Similarly to the distribution "P", it is a pseudodensity of probability on the phase space

$$\int_{\mathbb{C}} W_c\left(\frac{1}{\sqrt{2}}(q+ip), \frac{1}{\sqrt{2}}(q-ip)\right)\,dq\,dp = 1\,.$$

A yet more remarkable property than those encountered for Q and P holds for mean values of operators. An operator $\mathcal{O}(a,a^\dagger)$ has as the mean value in the state ϱ

$$\left\langle \mathcal{O}(a,a^\dagger)\right\rangle_\varrho = \operatorname{tr}(\mathcal{O}\varrho) = \int_{\mathbb{C}} \mathcal{O}_S(\alpha,\overline{\alpha})\,W_c(\alpha,\overline{\alpha})\,d^2\alpha\,, \tag{3.87}$$

where $\mathcal{O}_S(\alpha,\overline{\alpha})$ is obtained from $\mathcal{O}(a,a^\dagger)$ by just substituting $a \mapsto \alpha$, $a^\dagger \mapsto \overline{\alpha}$ *after* symmetrization of the monomials in a and $a^\dagger$ appearing in the series expansion of $\mathcal{O}(a, a^\dagger)$. For instance,

$$\mathcal{O}(a, a^\dagger) = a^\dagger a = \tfrac{1}{2}(a^\dagger a + a a^\dagger) - \tfrac{1}{2} \to \mathcal{O}_S(\alpha, \overline{\alpha}) = |\alpha|^2 - \tfrac{1}{2} .$$

Let us establish the link between (3.86) and a more traditional expression of the Wigner(–Weyl–Ville) distribution. Putting $\alpha = \frac{1}{\sqrt{2}}(q + i p)$ and $\beta = \frac{-i}{\sqrt{2}}(\sigma + i\tau)$ in the integral

$$W_c(\alpha, \overline{\alpha}) = \frac{1}{2\pi} \int_{\mathbb{C}} \mathrm{tr}(e^{\beta a^\dagger - \overline{\beta} a} \varrho)\, e^{-(\beta\overline{\alpha} - \overline{\beta}\alpha)} \frac{d^2\beta}{\pi} ,$$

one gets the Fourier transform (actually with permuted variables) in the phase space:

$$W_c(\alpha, \overline{\alpha}) \stackrel{\mathrm{def}}{=} W(q, p) = \frac{1}{2\pi} \int_{\mathbb{C}} \mathrm{tr}(e^{-i(\tau P + \sigma Q)} \varrho)\, e^{i(\tau p + \sigma q)} \frac{d\tau\, d\sigma}{2\pi}$$
$$= \frac{1}{2\pi} \int_{-\infty}^{+\infty} \left\langle q - \frac{y}{2} \right| \varrho \left| q + \frac{y}{2} \right\rangle e^{i y p}\, dy . \tag{3.88}$$

For example, the Wigner function for a pure state $\varrho = |\psi\rangle\langle\psi|$ reads as

$$W(q, p) = \frac{1}{2\pi} \int_{-\infty}^{+\infty} \overline{\psi\left(q + \frac{y}{2}\right)}\, \psi\left(q - \frac{y}{2}\right) e^{i y p}\, dy . \tag{3.89}$$

Within a signal processing framework, this expression is precisely the *Wigner–Ville transform* of the "signal" ψ. The real-valued function W, with indeterminate sign, is a phase space representation of the state ϱ. In the case of a pure state, marginal integrations restore a true quantum probabilistic content:

$$\int_{-\infty}^{+\infty} W(q, p)\, dp = |\psi(q)|^2 , \qquad \int_{-\infty}^{+\infty} W(q, p)\, dq = |\widehat{\psi}(p)|^2 .$$

3.6
The Feynman Path Integral and Coherent States

The path integral was introduced by Feynman in 1948 as an alternative formulation of (nonrelativistic) quantum mechanics [27]. Starting from the Schrödinger equation (in which and in the sequel we put $\hbar = 1$)

$$i \frac{\partial \Psi(x,t)}{\partial t} = -\frac{1}{2m} \frac{\partial^2 \Psi(x,t)}{\partial x^2} + V(x)\, \Psi(x,t) \tag{3.90}$$

for a particle of mass m moving in a potential $V(x)$, a solution can be written as an integral,

$$\Psi(x,t) = \int K(x, t; x', t')\, \Psi(x', t')\, dx' , \tag{3.91}$$

which represents the wave function $\Psi(x,t)$ at time t as a linear superposition over the wave function $\Psi(x',t')$ at the initial time t', $t' < t$. The integral kernel or *propagator* $K(x,t;x',t')$ can be formally expressed as an integral running over all continuous paths $x''(u)$, $t' \leq u \leq t''$, where $x''(t) = x$ and $x''(t') = x'$ are fixed end points for all paths.

$$K(x,t;x',t') = \mathcal{N} \int e^{i \int [(m/2)\,\dot{x}''^2(u) - V(x''(u))]\,du}\, \mathcal{D}x''\,. \tag{3.92}$$

Note that the integrand involves the classical Lagrangian for the system.

To give some meaning to this mathematically ill-defined object, Feynman adopted the following *lattice regularization* with spacing ε:

$$K(x,t;x',t') = \lim_{\varepsilon\to 0} (m/2\pi i\varepsilon)^{(N+1)/2} \int \cdots \int \times \exp\left\{ i \sum_{l=0}^{N} \left[(m/2\varepsilon)(x_{l+1} - x_l)^2 - \varepsilon\, V(x_l) \right] \right\} \prod_{l=1}^{N} dx_l\,, \tag{3.93}$$

where $x_{N+1} = x$, $x_0 = x'$, and $\varepsilon \equiv (t - t')/(N+1)$, $N \in \{1, 2, 3, \dots\}$. This procedure that yields well-defined integrals has to be validated by the existence of a "continuum limit as $\varepsilon \to 0$. Following the original Feynman approach, various authors, such as Feynman himself, Kac, Gel'fand, Yaglom, Cameron, and Itô, attempted to find a suitable *continuous-time regularization* procedure along with a subsequent limit to remove that regularization that ultimately should yield the correct propagator (see the illuminating review by Klauder [28], from which a large part of this section is borrowed).

Now, it appears that a phase space formulation of path integrals is more natural, as was also suggested by Feynman [29], and later successfully carried further by Daubechies and Klauder [2, 30]. Feynman (1951) proposed for the propagator the following integral on paths in the phase space

$$K(q,t;q',t') = \mathcal{M} \int \exp\left\{ i \int \left[p''\dot{q}'' - H(q'', p'') \right] du \right\} \mathcal{D}p''\,\mathcal{D}q''\,. \tag{3.94}$$

Here one integrates over all paths $q''(u)$, $t' \leq u \leq t$, with $q''(t) \equiv q$ and $q''(t') \equiv q'$ held fixed, as well as over all paths $p''(u)$, $t' \leq u \leq t$, without restriction.

A lattice space version of this expression is commonly given by

$$K(q,t;q',t') = \lim_{\varepsilon\to 0} \int \cdots \int \exp\left\{ i \sum_{l=0}^{N} \left[\tfrac{1}{2}\, p_{l+1/2}(q_{l+1} - q_l) - \varepsilon\, H(p_{l+1/2}, \tfrac{1}{2}(q_{l+1} + q_l)) \right] \right\} \prod_{l=0}^{N} d p_{l+1/2}/(2\pi)\, \Pi_{l=1}^{N}\, dq_l\,. \tag{3.95}$$

Like before $\varepsilon = (t - t')/(N+1)$. The integration is performed over all p and q variables except for $q_{N+1} \equiv q$ and $q_0 \equiv q'$. It is important to observe the presence of restrictions due to the canonical formalism of quantum mechanics: since q_l implies a sharp q value at time $t' + l\varepsilon$, the conjugate variable has been denoted by $p_{l+1/2}$ to emphasize that a sharp p value must occur at a different time, here at $t' + (l + 1/2)\varepsilon$, since it is not possible to have sharp p and q values at the same time. Note that there is one more p integration than q integration in this formulation. This discrepancy becomes clear when one imposes the composition law that requires that

$$K(q, t; q', t') = \int K(q, t; q'', t'')\, K(q'', t''; q', t')\, dq'' , \tag{3.96}$$

a relation that implies, just on dimensional grounds, that there must be one more p integration than q integration in the definition of each K expression.

The contribution of Daubechies and Klauder (1985) was to reexamine (3.94) by using a complete phase space formalism combined with coherent states through the Fock–Bargmann representation of wave functions and operators. They introduced coherent states, denoted here by $\underset{\mathrm{f}}{|q, p\rangle}$ and defined as

$$\underset{\mathrm{f}}{|q, p\rangle} \overset{\text{def}}{=} e^{-iqP}\, e^{i\,p\,Q}|0\rangle = e^{-i\,p\,q/2}\, e^{i(pQ - qP)}|0\rangle . \tag{3.97}$$

Since they are the standard ones $|z\rangle$, $z = \frac{q+i\,p}{\sqrt{2}}$ times the (important here!) phase factor $e^{-i\,p\,q/2}$, they have unit norm and they resolve the unity as well:

$$\underset{\mathrm{f}}{\langle q, p|q, p\rangle} = 1 , \qquad \int_{\mathbb{R}^2} \underset{\mathrm{f}}{|q, p\rangle}\underset{\mathrm{f}}{\langle q, p|}\, dq\, d\,p/2\pi = I_d . \tag{3.98}$$

Let us introduce, with the phase space variables (q, p), the (nonanalytical) symbol of the wave function like was defined in (3.21):

$$\vec{\Psi}(q, p, t) \overset{\text{def}}{=} \underset{\mathrm{f}}{\langle q, p|}\Psi(\cdot, t)\rangle . \tag{3.99}$$

At this point, we should make clear the probabilistic interpretation of this object compared with $\Psi(x, t)$. We know that $|\psi(x, t)|^2$ is the probability density of finding the particle at position x (at time t). On the other hand, the quantity $|\vec{\Psi}(q, p, t)|^2$ is the probability that the state $|\Psi\rangle$ can be found in the state $\underset{\mathrm{f}}{|q, p\rangle}$. What is important is to be aware that in coherent state representation we can specify both values of the variables simultaneously at the same time.

In this Fock–Bargmann representation, position and momentum operators are given by

$$Q_{\mathrm{f}} = q + i\,\frac{\partial}{\partial p} , \qquad P_{\mathrm{f}} = -i\,\frac{\partial}{\partial q} , \tag{3.100}$$

and thus the coherent state representation of the Schrödinger equation with classical Hamiltonian $H(q, p)$ is given by

$$i\,\partial\vec{\Psi}(q, p, t)/\partial t = \mathcal{H}(-i\,\partial/\partial q, q + i\,\partial/\partial p)\,\vec{\Psi}(q, p, t) . \tag{3.101}$$

The solution to this form of Schrödinger's equation can be expressed in the form

$$\vec{\Psi}(q, p, t) = \int K(q, p, t; q', p', t')\, \vec{\Psi}(q', p', t')\, dq'\, dp'/2\pi\,, \tag{3.102}$$

where $K(q, p, t; q', p', t')$ denotes the propagator in the coherent state representation.

The propagator for the coherent state representation of Schrödinger's equation can also be given a formal phase space path integral form, namely,

$$K(q, p, t; q', p', t') = \mathcal{M} \int \exp\{i \textstyle\int [\, p''\, \dot{q}'' - H(q'', p'')]\, dt\}\, \mathcal{D}q''\, \mathcal{D}p''\,. \tag{3.103}$$

Despite the fact that this expression looks the same as (3.94), the pinned values and the lattice space formulations are different. In the coherent state case, we have

$$\begin{aligned} &K(q, p, t; q', p', t') = \\ &\lim_{\varepsilon \to 0} \int \ldots \int \exp\left\{ i \sum_{l=0}^{N} \left[\tfrac{1}{2}(p_{l+1} + p_l)(q_{l+1} - q_l) \right.\right. \\ &\quad \left.\left. - \varepsilon H\left(\tfrac{1}{2}(p_{l+1} + p_l) + i\tfrac{1}{2}(q_{l+1} - q_l),\, \tfrac{1}{2}(q_{l+1} + q_l) - i\tfrac{1}{2}(p_{l+1} - p_l) \right) \right] \right\} \\ &\quad \times \exp\left\{ -(1/4) \sum_{l=0}^{N} [(p_{l+1} - p_l)^2 + (q_{l+1} - q_l)^2] \right\} \prod_{l=1}^{N} dq_l\, dp_l/(2\pi)\,. \end{aligned} \tag{3.104}$$

Observe that there are now the same number of p and q integrations in this expression. Such a conclusion is fully in accord with the combination law as expressed in the coherent state representation, namely,

$$\begin{aligned} &K(q, p, t; q', p', t') = \\ &\quad \int K(q, p, t; q'', p'', t'')\, K(q'', p'', t''; q', p', t')\, dq''\, dp''/2\pi\,. \end{aligned} \tag{3.105}$$

Daubechies and Klauder [30] gave a mathematical rigor to the expressions (3.103) and (3.104) by introducing a *Brownian type* regularization term. They eventually proved the existence of the following limit:

$$\begin{aligned} &\lim_{\nu \to \infty} \mathcal{M}_\nu \int \exp\left\{ i \int [\, p\, \dot{q}'' - H(q'', p'')]\, du \right\} \\ &\qquad \times \exp\left\{ -\frac{1}{2\nu} \int [\dot{p}''^2 + \dot{q}''^2]\, du \right\} \mathcal{D}q''\, \mathcal{D}p'' \\ &\quad = \lim_{\nu \to \infty} 2\pi\, e^{\nu T/2} \int e^{i \int [p''\, dq'' - H(q'', p'')\, du]}\, d\mu_W^\nu(q'', p'') \\ &\quad \equiv \langle q, p |\, e^{-i(t-t')\mathcal{H}}\, | q', p' \rangle \equiv K(q, p, t; q', p', t')\,, \end{aligned} \tag{3.106}$$

where the second line of (3.106) is a mathematically rigorous formulation of the heuristic and formal first line, and μ_W^ν denotes the measure on continuous path

$q''(u)$, $p''(u)$, $t' \leq u \leq t$, said to be a *pinned Wiener measure*. Such a choice of regularization also justifies the choice of coherent states (3.97) and imposes the condition that $H(q, p)$ be precisely the upper symbol of the quantum Hamiltonian

$$\mathcal{H} = \int H(q, p)\, \underset{\mathrm{f}}{|q, p\rangle}\underset{\mathrm{f}}{\langle q, p|}\, dq\, dp/(2\pi)\,. \tag{3.107}$$

A sufficient set of technical assumptions ensuring the validity of this representation is given by

(a) $\int H(q, p)^2\, e^{-\alpha(p^2+q^2)}\, dq\, dp < \infty$, for all $\alpha > 0$,

(b) $\int H(q, p)^4 e^{-\beta(p^2+q^2)}\, dq\, dp < \infty$, for some $\beta < 1/2\hbar$,

(c) The quantum Hamiltonian $\mathcal{H}$ is essentially self-adjoint on the span of finitely many number eigenstates.

As a matter of fact, Hamiltonians that are semibounded, symmetric (Hermitian) polynomials of Q_f and P_f are admissible.
One can conclude that a continuous-time, Brownian motion regularization of the phase space path integral can be rigorously established. It applies to a wide class of Hamiltonians. It can also be proved that the formulation is fully covariant under general canonical coordinate transformations.

Recently, dos Santos and Aguiar [31] constructed a representation of the coherent state path integral using the Weyl symbol of the Hamiltonian operator. Their coherent state propagator provides an explicit connection between the Wigner and the Husimi representations of the evolution operator. The dos Santos–Aguiar representation is different from the usual path integral forms suggested by Klauder and Skagerstam presented in this section. These different representations, although equivalent quantum mechanically, lead to different semiclassical limits.

4
Coherent States in Quantum Information: an Example of Experimental Manipulation

4.1
Quantum States for Information

Quantum information processing is about exploiting quantum mechanical features in all facets of information processing (data communication, computing). Excellent textbooks, monographs, and reviews exist which give all the material necessary to understand this fast-developping field [32–34].

The states act as information carriers, while the communication is processed through a sequence of quantum operations constituting the channel. The sender encodes information by preparing the channel into a well-defined quantum state ϱ belonging to an alphabet $\mathcal{A} = \{\varrho_0, \varrho_1, \ldots, \varrho_M\}$. The receiver, following any relevant signal propagation, performs a measurement on the transmission channel to ascertain which state was transmitted by the sender. Quantum information theory is mainly based on superposition-basis and entanglement measurements. This requires high-fidelity implementation to be effective in the laboratory. Unfortunately, quantum measurements are "invasive" in the sense that little or no refinement is achieved by further observation of an already measured system. Some of the difficulties in implementing communication in quantum information stems from the fragility of Schrödinger-cat-like superpositions. Even with transmission of orthogonal codewords, decoherence, energy dissipation, and other imperfections deteriorate orthogonality.

If the states in the sender's alphabet are not orthogonal, no measurement can distinguish between overlapping quantum states without some ambiguity [21, 35–38]. Then errors seem unavoidable: there exists a nonzero probability that the receiver will misinterpret the transmitted codeword.

However, this impossibility of discriminating between nonorthogonal quantum states presents an advantage for quantum key distribution [39]. Indeed, nonorthogonality prevents an eavesdropper from acquiring information without disturbing the state. Also, in some cases it has been shown by Fuchs [36] that the classical information capacity of a noisy channel is actually maximized by a nonorthogonal alphabet.

Mathematically, the question of distinguishing between nonorthogonal states [36, 37] is addressed by optimizing a state-determining measurement over all pos-

Coherent States in Quantum Physics. Jean-Pierre Gazeau

ISBN: 978-3-527-40709-5

itive-operator-valued measures (POVM) [38]. But arbitrary POVMs are not easy to manipulate!

In a recent work, Cook, Martin, and Geremia [40] demonstrated that real-time quantum feedback can be used in place of a quantum superposition of the type "Schrödinger cat state" to implement an optimal quantum measurement for discriminating between optical coherent states. This work gives us an excellent opportunity of presenting standard coherent states as they are produced and used in realistic conditions. As a preliminary to the description of the experiment, we will also give an account of the theoretical background needed.

4.2
Optical Coherent States in Quantum Information

The optical field produced by a laser provides a convenient quantum system for carrying information. Since optical coherent states $|\alpha\rangle$ are not orthogonal, one would attempt to minimize the overlapping $\langle\alpha'|\alpha\rangle = e^{i\Im(\bar{\alpha}'\alpha)}e^{-|\alpha-\alpha'|^2/2}$ by using large-amplitude regimes. However, one faces power limitations and the appearance of nonlinear effects. So one is more inclined to develop optimization methods for communication processes based on small-amplitude optical coherent states and photodetection. When one tries to distinguish between two nonorthogonal states through some receiver device, there exists a quantum error probability. The latter is bounded below by some minimum, named the quantum limit or Helstrom bound [38] in this context.

Three types of receivers were described by Geremia [42]. Kennedy [44] proposed in 1972 a receiver based on simple photon counting to distinguish between two different coherent states. However, the Kennedy receiver error probability lies above the quantum mechanics minimum, that is, the Helstrom bound. Then, Dolinar [41] proposed in 1973 a measurement scheme capable of achieving the quantum limit. Dolinar's receiver, while still based on photon counting, approximates an optimal POVM by superposing a local feedback signal on the channel. A serious experimental drawback was that real-time adjustment of the local signal following each photon was considered as quite impracticable. As a result, Sasaki and Hirota [45] later proposed an alternative receiver that applies an open-loop unitary transformation to the incoming coherent state signals to render them more distinguishable by simple photon counting.

Geremia [42] compared, theoretically and numerically, the relative performance of the Kennedy, Dolinar, and Sasaki–Hirota receivers under *realistic* experimental conditions, insisting on the following aspects:

(i) subunity quantum efficiency, where it is possible for the detector to miscount incoming photons,

(ii) nonzero dark counts, where the detector can register photons even in the absence of a signal,

(iii) nonzero dead-time, or finite detector recovery time after registering the arrival of a photon,
(iv) finite bandwidth of any signal processing necessary to implement the detector,
(v) fluctuations in the phase of the incoming optical signal.

4.3 Binary Coherent State Communication

4.3.1 Binary Logic with Two Coherent States

Let us consider an alphabet consisting of two pure coherent states,

$$\varrho_0 = |\Psi_0\rangle\langle\Psi_0| \,, \quad \varrho_1 = |\Psi_1\rangle\langle\Psi_1| \,,$$

corresponding to the logic states "0" and "1," respectively. Without loss of generality, $\Psi_0(t)$ can be chosen as the vacuum, $\Psi_0(t) = 0$, that is, $|\Psi_0\rangle = |0\rangle$, while

$$\Psi_1(t) = \psi_1(t) \exp\left[-i(\omega t + \varphi)\right] + \text{c.c.} \,, \tag{4.1}$$

where ω is the frequency of the optical carrier and φ is (ideally) a fixed phase.

The envelope function, $\psi_1(t)$, is normalized such that

$$\int_0^\tau |\psi_1(t)|^2 \, dt = \bar{n} \,, \tag{4.2}$$

where $\bar{n}$ is the mean number of photons arriving at the receiver during the measurement interval, $0 \leq t \leq \tau$. That is, $\hbar\omega|\psi_1(t)|^2$ is the instantaneous average power of the optical signal for logic "1."

By combining the incoming signal with an appropriate local oscillator, one can always transform the amplitude keying with the alphabet of two coherent states $\mathcal{A} = \{|0\rangle\langle 0|, |\alpha\rangle\langle\alpha|\}$, with $|\alpha\rangle = |\Psi_1\rangle$, to the phase-shift keyed alphabet,

$$\left\{ \left|-\frac{1}{2}\alpha\right\rangle \left\langle -\frac{1}{2}\alpha\right| , \left|\frac{1}{2}\alpha\right\rangle \left\langle \frac{1}{2}\alpha\right| \right\} ,$$

via the unitary displacement, $D\left(-\frac{1}{2}\alpha\right) = \exp(-\frac{1}{2}(\alpha a^\dagger - \bar{\alpha} a))$. Similarly, if $|\Psi_0\rangle \neq |0\rangle$, a simple displacement can be used to restore $|\Psi_0\rangle$ to the vacuum state.

4.3.2 Uncertainties on POVMs

In the case of nonorthogonal quantum states as codewords, the receiver attempts to ascertain which state was transmitted by performing a quantum measurement, say, Π, on the channel. The operator Π is described by an appropriate POVM represented by a complete (here countable) set of positive operators [37] resolving the identity,

$$\sum_n \Pi_i = I_d\,, \quad \Pi_i \geq 0\,, \tag{4.3}$$

where n indexes the possible measurement outcomes.

We already gave a simple example of a continuous POVM in Section 3.2.3. In the same vein, an example of a finite POVM in the Euclidean plane is given by the following cyclotomic polygonal resolution of the unity:

$$\frac{2}{n}\sum_{q=0}^{n-1} \Pi_{\frac{2\pi q}{n}} = I_d\,, \quad \Pi_\theta = |\theta\rangle\langle\theta| = \begin{pmatrix} \cos^2\theta & \cos\theta\sin\theta \\ \cos\theta\sin\theta & \sin^2\theta \end{pmatrix}.$$

For binary communication, for which the POVM resolution of the unity reads $\Pi_0 + \Pi_1 = I_d$, the measurement by the receiver amounts to a decision between two hypotheses: H_0, that the transmitted state is ϱ_0, selected when the measurement outcome corresponds to Π_0, and H_1, that the transmitted state is ϱ_1, selected when the measurement outcome corresponds to Π_1.

4.3.3
The Quantum Error Probability or Helstrom Bound

Now, possibilities of errors mean that there is some chance that the receiver will select the null hypothesis, H_0 (or H_1), when ϱ_1 (or ϱ_0) is actually present. Thus, we have in terms of conditional probabilities

$$p\,(H_0|\varrho_1) = \mathrm{tr}[\Pi_0\varrho_1] = \mathrm{tr}[(I_d - \Pi_1)\varrho_1]\,, \quad p\,(H_1|\varrho_0) = \mathrm{tr}[\Pi_1\varrho_0]\,. \tag{4.4}$$

The total receiver error probability, say, $p[\Pi_0, \Pi_1]$, is then given by

$$p[\Pi_0, \Pi_1] = \xi_0\, p\,(H_1|\varrho_0) + \xi_1\, p\,(H_0|\varrho_1)\,, \quad \xi_0 + \xi_1 = 1\,, \tag{4.5}$$

where $\xi_0 = p_0(\varrho_0)$ and $\xi_1 = p_0(\varrho_1)$ are the probabilities that the sender will transmit ϱ_0 and ϱ_1, respectively; they reflect the prior knowledge that enters into the hypothesis testing process implemented by the receiver, and, in many cases $\xi_0 = \xi_1 = 1/2$.

Minimizing the error in receiver measurement over all possible POVMs (Π_0, Π_1) leads to the so-called *quantum error probability* or *Helstrom bound*,

$$P_H \equiv \min_{\Pi_0,\Pi_1} p[\Pi_0, \Pi_1]\,, \tag{4.6}$$

P_H is the smallest physically allowable error probability, given the overlap between ϱ_0 and ϱ_1.

4.3.4 The Helstrom Bound in Binary Communication

The receiver error probability

$$p[\Pi_0, \Pi_1] = \xi_0 \operatorname{tr}[\Pi_1 \varrho_0] + \xi_1 \operatorname{tr}\left[(I_d - \Pi_1)\varrho_1\right] = \xi_1 + \operatorname{tr}\left[\Pi_1(\xi_0\varrho_0 - \xi_1\varrho_1)\right]$$

is minimized by optimizing $\min_{\Pi_1} \operatorname{tr}[\Pi_1 \Gamma]$, $\Gamma \stackrel{\text{def}}{=} \xi_0\varrho_0 - \xi_1\varrho_1$, over Π_1 subject to $0 \le \Pi_1 \le I_d$.

Let $\Gamma = \sum_n \lambda_n |\gamma_n\rangle\langle\gamma_n|$ be the spectral decomposition of the operator Γ. One can write $\operatorname{tr}[\Pi_1\Gamma] = \sum_n \lambda_n \langle\gamma_n|\Pi_1|\gamma_n\rangle$. Then the Helstrom bound can be expressed as $P_{\text{H}} = \xi_1 + \sum_{\lambda_n<0} \lambda_n$, which corresponds to the case in which Π_1 is the projector on all eigenstates $|\gamma_n\rangle$ with negative λ_n.

For pure states, where $\varrho_0 = |\Psi_0\rangle\langle\Psi_0|$ and $\varrho_1 = |\Psi_1\rangle\langle\Psi_1|$, Γ has two eigenvalues, of which only one is negative,

$$\lambda_- = \frac{1}{2}\left(1 - \sqrt{1 - 4\xi_0\xi_1|\langle\Psi_0|\Psi_1\rangle|^2}\right) - \xi_1 < 0\,. \tag{4.7}$$

Then the quantum error probability is [38]

$$P_{\text{H}} = \xi_1 + \lambda_- = \frac{1}{2}\left(1 - \sqrt{1 - 4\xi_0\xi_1|\langle\Psi_1|\Psi_0\rangle|^2}\right)\,. \tag{4.8}$$

To prove this formula, let us consider an orthonormal basis $\{|e_0\rangle \equiv |\Psi_0\rangle, |e_1\rangle\}$ and the corresponding decomposition of $|\Psi_1\rangle$: $|\Psi_1\rangle = \mu_0|e_0\rangle + \mu_1|e_1\rangle$ with $\mu_0 = \langle\Psi_0|\Psi_1\rangle$. Thus, the operator Γ reads as the 2×2 matrix:

$$\Gamma = \xi_0\varrho_0 - \xi_1\varrho_1 = \begin{pmatrix} \xi_0 - \xi_1|\mu_0|^2 & -\xi_1\mu_0\bar{\mu}_1 \\ -\xi_1\bar{\mu}_0\mu_1 & -\xi_1|\mu_1|^2 \end{pmatrix}\,.$$

By taking into account $\xi_0 + \xi_1 = 1$ (completeness of the probabilities) and $|\mu_0|^2 + |\mu_1|^2 = 1$ (normalization of $|\Psi_1\rangle$), we get the eigenvalues of Γ as equal to

$$\lambda_\pm = \frac{1}{2}\left(1 \pm \sqrt{1 - 4\xi_0\xi_1|\langle\Psi_0|\Psi_1\rangle|^2}\right) - \xi_1\,,$$

from which we derive (4.8).

4.3.5 Helstrom Bound for Coherent States

From the expansion of coherent states over the number states,

$$|\alpha\rangle = e^{-|\alpha|^2/2} \sum_{n=0}^{\infty} \frac{\alpha^n}{\sqrt{n!}}|n\rangle\,,$$

the overlap between $|\Psi_1\rangle = |\alpha\rangle$ and $|\Psi_0\rangle = |0\rangle$ is just given in terms of the expected number of photons or the average value or lower symbol of the number operator in the coherent state $|\alpha\rangle$: $\tilde{n}_\alpha \equiv \langle\alpha|N|\alpha\rangle$:

$$\langle\Psi_1|\Psi_0\rangle = \langle\alpha|0\rangle = e^{-|\alpha|^2/2} = e^{-\tilde{n}_\alpha/2} . \tag{4.9}$$

So, the Helstrom bound is given by

$$P_\mathrm{H} = \frac{1}{2}\left(1 - \sqrt{1 - 4\xi_0\xi_1 e^{-\tilde{n}_\alpha}}\right) . \tag{4.10}$$

4.3.6
Helstrom Bound with Imperfect Detection

It is further possible to evaluate the Helstrom bound for imperfect detection. Nonunit efficiency of a photodetector leads to a photon count which is related to the ideal (efficiency $\eta = 1$) premeasured photon distribution by a Bernoulli transformation [43]. Accordingly, the probability $p_n(\eta)$ of detecting n photons using a nonideal photodetector ($\eta < 1$) is given in terms of the probability $p_m(\eta = 1)$ (using an ideal one) by

$$p_n(\eta) = \sum_{m=n}^{\infty} \binom{m}{n} \eta^n (1-\eta)^{m-n} \; p_m(\eta = 1) . \tag{4.11}$$

Coherent states have the convenient property that subunity quantum efficiency is equivalent to an ideal detector masked by a beam splitter with transmission coefficient, $\eta \le 1$. Indeed, in the case of coherent states, we have the Poisson distribution $p_m(\eta = 1) = e^{-|\alpha|^2} |\alpha|^{2m}/m!$, and so changing m into $s = m - n$ in the summation (4.11) gives

$$p_n(\eta) = \frac{\eta^n |\alpha|^{2n}}{n!} e^{-\eta|\alpha|^2} , \tag{4.12}$$

which amounts to replacing α by $\sqrt{\eta}\alpha$ in the expression of coherent states.

Accordingly, the Helstrom bound becomes

$$P_\mathrm{H}(\eta) = \frac{1}{2}\left(1 - \sqrt{1 - 4\xi_0\xi_1 e^{-\tilde{n}_\alpha \eta}}\right) . \tag{4.13}$$

This result and (4.8) indicate that there is a finite quantum error probability for all choices of $|\Psi_1\rangle$, even when an optimal measurement is performed.

4.4
The Kennedy Receiver

4.4.1
The Principle

The Kennedy receiver is based on the following principle. A near-optimal receiver simply counts the number of photon arrivals registered by the detector between

$t = 0$ and T. The receiver decides in favor of H_0 when the number of clicks is zero, otherwise H_1 is chosen.

This hypothesis testing procedure corresponds to the measurement operators

$$\Pi_0 = |0\rangle\langle 0| \,, \quad \Pi_1 = \sum_{n=1}^{\infty} |n\rangle\langle n| \,. \tag{4.14}$$

The receiver always correctly selects H_0 when the channel is in ϱ_0, since no photon can be registered when the vacuum state is present (ignoring background light and detector dark counts for now). Therefore, $p(H_1|\varrho_0) = 0$.

On the other hand, the Poisson statistics of coherent state photon numbers allows for the possibility that zero photons will be recorded even when ϱ_1 is present. So

$$p(H_0|\varrho_1) \equiv \mathrm{tr}[\Pi_0 \varrho_1] = |\langle 0|\Psi_1\rangle|^2 \tag{4.15}$$

is nonzero owing to the finite overlap of all coherent states with the vacuum.

4.4.2
Kennedy Receiver Error

Now, an imperfect detector, although able to count photons, can misdiagnose ϱ_1 if it fails to generate clicks for photons that do arrive at the detector.

The probability for successfully choosing H_1 when ϱ_1 is present is given by

$$p_\eta(H_1|\varrho_1) = \sum_{n=1}^{\infty} \sum_{k=1}^{\infty} p(n,k) |\langle n|\alpha\rangle|^2 \,, \tag{4.16}$$

where the Bernoulli distribution,

$$p(n,k) = \frac{n!}{k!(n-k)!} \eta^k (1-\eta)^{n-k} \,, \tag{4.17}$$

gives the probability that a detector with quantum efficiency, η will register k clicks when the actual number of photons is n.

The resulting Kennedy receiver error,

$$P_K(\eta) = 1 - p_\eta(H_1|\varrho_1) = \xi_1 e^{-\bar{n}_\alpha \eta} \,, \tag{4.18}$$

asymptotically gets closer to the Helstrom bound for large signal amplitudes, but is larger for small photon numbers.

4.5 The Sasaki–Hirota Receiver

4.5.1 The Principle

Still in the simple photon counting implementation, a unitary transformation to the incoming signal states prior to detection could help to approach better the Helstrom bound. With Sasaki and Hirota [45] let us consider the rotation

$$U(\theta) = \exp\left[\theta(|\Psi_0'\rangle\langle\Psi_1'| - |\Psi_1'\rangle\langle\Psi_0'|)\right] , \tag{4.19}$$

generated by the transformed alphabet, $\mathcal{A}' = \{|\Psi_0'\rangle\langle\Psi_0'|, |\Psi_1'\rangle\langle\Psi_1'|\}$, where the states are obtained from Gram–Schmidt orthogonalization of those of $\mathcal{A}$:

$$|\Psi_0'\rangle = |\Psi_0\rangle, \quad |\Psi_1'\rangle = \frac{|\Psi_1\rangle - c_0|\Psi_0\rangle}{\sqrt{1-c_0^2}}, \quad c_0 = \langle\Psi_1|\Psi_0\rangle = e^{-|\alpha|^2/2} \equiv e^{-\bar{n}_\alpha/2} . \tag{4.20}$$

The angle θ must be optimized to achieve the Helstrom bound.

The action of $U(\theta)$ on the incoming signal states is given by

$$U(\theta)|\Psi_0\rangle = \left(\cos\theta + \frac{c_0 \sin\theta}{\sqrt{1-c_0^2}}\right)|\Psi_0\rangle - \frac{\sin\theta}{\sqrt{1-c_0^2}}|\Psi_1\rangle$$

$$U(\theta)|\Psi_1\rangle = \frac{\sin\theta}{\sqrt{1-c_0^2}}|\Psi_0\rangle + \frac{\cos\theta\sqrt{1-c_0^2} - c_0\sin\theta}{\sqrt{1-c_0^2}}|\Psi_1\rangle .$$

4.5.2 Sasaki–Hirota Receiver Error

Since $|\Psi_0'\rangle$ is the vacuum state, hypothesis testing can still be performed by simple photon counting. However, unlike the Kennedy receiver, it is possible to misdiagnose ϱ_0 since $U(\theta)|\Psi_0\rangle$ contains a nonzero contribution from $|\Psi_1\rangle$.

The probability for a false-positive detection by a photon counter with efficiency η is given by

$$p_\eta^\theta(\mathrm{H}_1|\varrho_0) = \sum_{n=1}^{\infty}\sum_{k=1}^{\infty} p(n,k)|\langle n|U(\theta)|\Psi_0\rangle|^2 = \frac{c_0^{2\eta} - 1}{c_0^2 - 1}\sin^2\theta , \tag{4.21}$$

with

$$\langle n|U(\theta)|\Psi_0\rangle = \left[\cos\theta + \frac{c_0\sin\theta}{\sqrt{1-c_0^2}}\right]\delta_{n,0} - \frac{c_0\alpha^n\sin\theta}{\sqrt{n!(1-c_0^2)}} , \tag{4.22}$$

where α is the (complex) amplitude of $|\Psi_1\rangle$.

Similarly, the probability for correct detection is given by

$$p_\eta^\theta(\mathrm{H}_1|\varrho_1) = \sum_{n=1}^{\infty}\sum_{k=1}^{\infty} p(n,k)|\langle n|U(\theta)|\Psi_1\rangle$$

$$= \frac{c_0^{2\eta}-1}{c_0^2-1}\left[c_0\sin\theta - \sqrt{1-c_0^2}\cos\theta\right]^2 , \quad (4.23)$$

with

$$\langle n|U[\theta]|\Psi_1\rangle = \left[c_0\cos\theta - \frac{c_0^2\alpha^n\sin\theta}{\sqrt{n!(1-c_0^2)}}\right] + \frac{\sin\theta}{\sqrt{1-c_0^2}}\delta_{n,0} . \quad (4.24)$$

The total Sasaki–Hirota receiver error is then given by the weighted sum

$$P_{\mathrm{SH}}(\eta,\theta) = \xi_0 p_\eta^\theta(\mathrm{H}_1|\hat{\varrho}_0) + \xi_1\left[1 - p_\eta^\theta(\mathrm{H}_1|\varrho_1)\right] \quad (4.25)$$

and can be minimized over all possible values of θ to give

$$\theta = -\tan^{-1}\sqrt{\frac{\sqrt{1-4\xi_0\xi_1c_0^2}-1+2\xi_1c_0^2}{\sqrt{1-4\xi_0\xi_1c_0^2}+1-2\xi_1c_0^2}} . \quad (4.26)$$

For perfect detection efficiency, $\eta = 1$, (4.25) is equivalent to the Helstrom bound; however, for $\eta < 1$, it is larger.

4.6 The Dolinar Receiver

4.6.1 The Principle

The Dolinar receiver utilizes an adaptive strategy to implement a feedback approximation to the Helstrom POVM [41]. It operates by combining the incoming signal, $\Psi(t)$, with a separate local signal,

$$\mathrm{U}(t) = u(t)\exp\left[-i(\omega t + \phi)\right] + \mathrm{c.c.} \quad (4.27)$$

Here $u(t)$ is the “displacement” or “feedback” amplitude.

The detector counts photons with total instantaneous mean rate

$$\Phi(t) = |\psi(t) + u(t)|^2 , \quad (4.28)$$

where $\psi(t) = 0$ (for logic “0”) when the channel is in the state ϱ_0, and $\psi(t) = \psi_1(t)$ (for logic “1”) when the channel is in ϱ_1.

4.6.2 Photon Counting Distributions

Given the alphabet $\mathcal{A} = (\varrho_0, \varrho_1)$, the feedback amplitude $u(t)$, a transmission coefficient η, and some subdivision $(t_0 \equiv 0, t_1, \ldots, t_n, t_{n+1} \equiv \tau)$ of the measurement time interval (or "counting interval") $[0, \tau]$, the conditional probability $w\left[t_k|\varrho_i, u(t)\right]$ that a photon will arrive at time t_k and that it will be the only click during the half-closed interval $(t_{k-1}, t_k]$ [7] is called the exponential waiting time distribution for optical coherent states. It is defined as

$$w_\eta\left[t_k|\varrho_i, u(t)\right] = \eta\,\Phi(t_k)\,\exp\left(-\eta\int_{t_{k-1}}^{t_k}\Phi(t')\,dt'\right). \tag{4.29}$$

The corresponding exclusive counting densities for the measurement interval are then given by

$$p_\eta\left[t_1, \ldots, t_n|\varrho_i, u(t)\right] = \prod_{k=1}^{n+1} w_\eta\left[t_k|\varrho_i, u(t)\right]. \tag{4.30}$$

They allow one to evaluate, using the Bayes rule, the conditional arrival time probabilities $p_\eta\left[\varrho_i|t_1, \ldots, t_n, u(t)\right] = p_\eta\left[t_1, \ldots, t_n|\varrho_i, u(t)\right]\, p_0(\varrho_i)$. The latter reflect the likelihood that n photon arrivals occur precisely at the times $t_1, \ldots, t_n$,[3] given that the channel is in the state ϱ_i, the feedback amplitude is $u(t)$, and the detector quantum efficiency is η.

4.6.3 Decision Criterion of the Dolinar Receiver

The receiver decides between hypotheses H_0 and H_1 by selecting the one that is more consistent with the record of photon arrival times observed by the detector given the choice of $u(t)$. H_1 is selected when the ratio of conditional arrival time probabilities,

$$\Lambda = \frac{p_\eta\left[\varrho_1|t_1, \ldots, t_n, u(t)\right]}{p_\eta\left[\varrho_0|t_1, \ldots, t_n, u(t)\right]}, \tag{4.31}$$

is greater than one; otherwise it is assumed that ϱ_0 was transmitted.

By employing the Bayes rule, one can reexpress Λ in terms of the photon counting distributions

$$\Lambda = \frac{p_\eta\left[t_1, \ldots, t_n|\varrho_1, u(t)\right]\, p_0(\hat{\varrho}_1)}{p_\eta\left[t_1, \ldots, t_n|\varrho_0, u(t)\right]\, p_0(\hat{\varrho}_0)} = \frac{\xi_1\, p_\eta\left[t_1, \ldots, t_n|\varrho_1, u(t)\right]}{\xi_0\, p_\eta\left[t_1, \ldots, t_n|\varrho_0, u(t)\right]}, \tag{4.32}$$

3) Even though the term "arrival time" is not appropriate from an experimental point of view. Time interval is more appropriate.

In terms of error probabilities, the likelihood ratio is given by

$$\Lambda = \frac{p_\eta\left[\mathrm{H}_1|\varrho_1, u(t)\right]}{p_\eta\left[\mathrm{H}_1\varrho_0, u(t)\right]} = \frac{1 - p_\eta\left[\mathrm{H}_0|\varrho_1, u(t)\right]}{p_\eta\left[\mathrm{H}_1|\varrho_0, u(t)\right]}, \quad \text{for } \Lambda > 1 \tag{4.33}$$

(i.e., the receiver definitely selects H_1), and

$$\Lambda = \frac{p_\eta\left[\mathrm{H}_0|\varrho_1, u(t)\right]}{p_\eta\left[\mathrm{H}_0|\varrho_0, u(t)\right]} = \frac{p_\eta\left[\mathrm{H}_0|\varrho_1, u(t)\right]}{1 - p_\eta\left[\mathrm{H}_1|\varrho_0, u(t)\right]}, \quad \text{for } \Lambda < 1 \tag{4.34}$$

(i.e., the receiver definitely selects H_0).

4.6.4 Optimal Control

The minimization over $u(t)$ of the Dolinar receiver error probability,

$$P_\mathrm{D}[u(t)] = \xi_0\, p_\eta\left[H_1|\varrho_0, u(t)\right] + \xi_1\, p_\eta\left[H_0|\varrho_1, u(t)\right], \tag{4.35}$$

can be accomplished by employing the technique of dynamical programming [47].

The optimal control policy, $u^*(t)$, is identified by solving the Hamilton–Jacobi–Bellman equation,

$$\min_{u(t)}\left[\frac{\partial}{\partial t}\mathcal{J}[u(t)] + \nabla_\mathbf{p}\mathcal{J}[u(t)]^\mathrm{T}\frac{\partial}{\partial t}\mathbf{p}(t)\right] = 0, \tag{4.36}$$

where the "control cost" $\mathcal{J}[u(t)] \equiv P_\mathrm{D}[u(t)] = \xi^\mathrm{T}\mathbf{p}$ in an effective state-space picture given by the conditional error probabilities,

$$\mathbf{p}(t) = \begin{pmatrix} p_\eta\left[H_1|\varrho_0, u(t)\right](t) \\ p_\eta\left[H_0|\varrho_1, u(t)\right](t) \end{pmatrix}. \tag{4.37}$$

The partial differential equation for $\mathcal{J}$ is based on the requirement that $\mathbf{p}(t)$ and $u(t)$ are smooth (continuous and differentiable) throughout the entire receiver operation. However, like all quantum point processes, our conditional knowledge of the system state evolves smoothly only *between* photon arrivals. Fortunately, the dynamical programming *optimality principle* allows us to optimize $u(t)$ in a piecewise manner [47]. Performing the piecewise minimization leads to the control policy

$$u_1^*(t) = -\psi_1(t)\left(1 + \frac{\mathcal{J}[u_1^*(t)]}{1 - 2\mathcal{J}[u_1^*(t)]}\right) \tag{4.38}$$

for $\Lambda > 1$ (see [42] for the proof), where $p_\eta[\mathrm{H}_0|\varrho_1, u_1^*(t)] = 0$ and

$$\mathcal{J}[u_1^*(t)] = \xi_1\, p_\eta[\mathrm{H}_1|\varrho_0, u_1^*(t)] = \frac{1}{2}\left(1 - \sqrt{1 - 4\xi_0\xi_1 e^{-\eta\bar{n}(t)}}\right).$$

Here, $\bar{n}(t) = \int_0^t |\psi_1(t')|^2\, dt'$ is the average number of photons expected to arrive at the detector by time t when the channel is in the state ϱ_1.

Conversely, the optimal control takes the form

$$u_0^*(t) = \psi_1(t)\left(\frac{\mathcal{J}[u_0^*(t)]}{1-2\mathcal{J}[u_0^*(t)]}\right) \tag{4.39}$$

for $\Lambda < 1$, where $p_\eta[\mathrm{H}_1|\varrho_0, u_0^*(t)] = 0$ and

$$\mathcal{J}[u_0^*(t)] = \xi_1\, p_\eta[\mathrm{H}_0|\varrho_1, u_0^*(t)] = \frac{1}{2}\left(1-\sqrt{1-4\xi_0\xi_1 e^{-\eta\bar{n}(t)}}\right) .$$

4.6.5
Dolinar Hypothesis Testing Procedure

The Hamilton–Jacobi–Bellman solution leads to a conceptually simple procedure for estimating the state of the channel. The receiver begins at $t = 0$ by favoring the hypothesis that is more likely based on the prior probabilities, $p_0(0) = \xi_0$ and $p_1(0) = \xi_1$. Note that if $\xi_0 = \xi_1$, then neither hypothesis is a priori favored and the Dolinar receiver is singular with $P_\mathrm{D} = \frac{1}{2}$. Assuming that $\xi_1 \geq \xi_0$ (for $\xi_0 > \xi_1$, the opposite reasoning applies), the Dolinar receiver always selects H_1 during the initial measurement segment. The probability of deciding on H_0 is exactly zero prior to the first photon arrival such that an error only occurs when the channel is actually in ϱ_0.

To see what happens when a photon does arrive at the detector, it is necessary to investigate the behavior of $\Lambda(t)$ at the boundary between two measurement segments. Substituting the optimal control policy, $u^*(t)$, which alternates between $u_1^*(t)$ and $u_0^*(t)$, into the photon counting distribution leads to

$$p(t_1,\ldots,t_n|\varrho_i) = \eta^n \prod_{k=0}^{n+1} \Phi_i[u_{k|2}(t_{k-1},t_k)]$$
$$\times \exp\left(-\eta \int_{t_{k-1}}^{t_k} \Phi_i\left[u_{k|2}(t'_{k-1},t'_k)\right]\, dt'\right) . \tag{4.40}$$

Here, the notation $k|2$ stands for $k \mod 2$. This expression can be used to show that the limit of $\Lambda(t)$ approaching a photon arrival time, t_k, from the left is the reciprocal of the limit approaching from the right:

$$\lim_{t\to t_k^-} \Lambda(t) = \left[\lim_{t\to t_k^+} \Lambda(t)\right]^{-1} . \tag{4.41}$$

That is, if $\Lambda > 1$ such that H_1 is favored during the measurement interval ending at t_k, the receiver immediately swaps its decision to favor H_0 when the photon arrives. Each photon arrival invalidates the current hypothesis and the receiver completely reverses its decision on every click. This result implies that H_1 is selected when the number of photons, n, is even (or zero) and that H_0 when the number of photons is odd.

Despite the discontinuities in the conditional probabilities, $p_\eta[\mathrm{H}_1|\varrho_0, u^*(t)]$ and $p_\eta[\mathrm{H}_0|\varrho_1, u^*(t)]$, at the measurement segment boundaries, the total Dolinar receiver error probability, $P_\mathrm{D}(\eta, t) = \frac{1}{2}\left(1 - \sqrt{1 - \xi_0\xi_1 e^{-\eta\bar{n}(t)}}\right)$, evolves smoothly since $\lim_{t\to t_k^-} \mathcal{J}[u^*(t)] = \lim_{t\to t_k^+} \mathcal{J}[u^*(t)]$ at the boundaries.

Recognizing that $\bar{n}(\tau) = \bar{n}_\alpha$ leads to the final Dolinar receiver error,

$$P_\mathrm{D}(\eta) = \frac{1}{2}\left(1 - \sqrt{1 - 4\xi_0\xi_1 c_0^{2\eta}}\right), \tag{4.42}$$

which is equal to the Helstrom bound for all values of the detector efficiency, $0 < \eta \leq 1$.

4.7 The Cook–Martin–Geremia Closed-Loop Experiment

4.7.1 A Theoretical Preliminary

Cook, Martin, and Geremia [40] demonstrated that shot noise can be surpassed and even the quantum limit can be approached by using the Dolinar real-time quantum feedback in place of the cat-state measurement. They exploited the finite duration of any real measurement, and quantum states $|0\rangle$ and $|\alpha\rangle$ are realized as optical wave packets with spatiotemporal extent.

Measurements on an optical pulse inherently persist for a time set by the pulse length τ. Photon counting generates a measurement record $\Xi_{[0,\tau]} \equiv (t_1, t_2, \ldots, t_n)$, $t_0 = 0$, $t_{n+1} = \tau$, consisting of the observed photon arrival times. The total number of photon arrivals in the counting interval $[0, \tau]$ is viewed as one aggregate "instantaneous" measurement of the number operator.

In the closed-loop measurement as sketched in Figure 4.1a, photon counting is combined with feedback-mediated optical displacements applied during the photon counting interval. The amplitude of the displacement $u(t)$, here denoted by u_t, applied at each time t during the measurement is conditional on the accumulated measurement record $\Xi_{[0,\tau]}$ and is based on an evolving Bayesian estimate of the incoming wave-packet state.

Discrimination is performed by selecting the state $|\psi\rangle \in \{|0\rangle, |\alpha\rangle\}$ that maximizes the conditional probability $P\left(\Xi_{[0,\tau]}|\psi, u_{[0,t]}\right)$ that the measurement record $\Xi_{[0,\tau]}$ would be observed given the state ψ and the history of applied displacements denoted by $u_{[0,t]}$.

The feedback controller determines which state is most consistent with the accumulating record $\Xi_{[0,\tau]}$ and chooses the feedback amplitude at each point in time to minimize the probability of error over the remainder of the measurement interval $(t, \tau]$.

The policy for determining the optimal displacement amplitude u_t^* is based on the optimal control theory presented in Section 4.6.4: it engineers the feedback

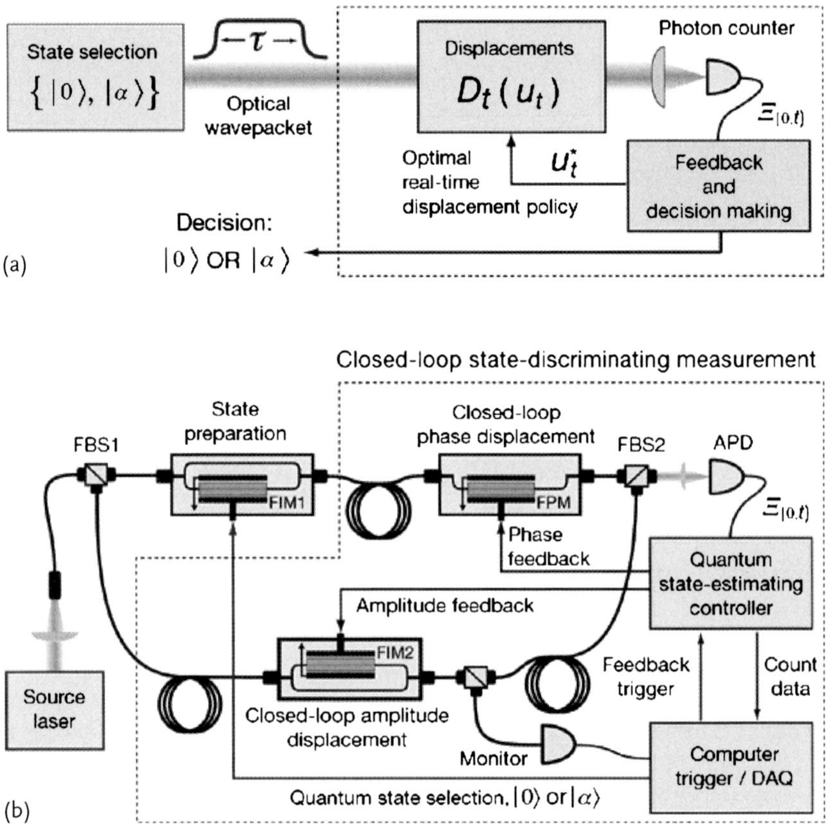

Fig. 4.1 A measurement that combines photon counting with feedback-mediated optical displacements to enact quantum-limited state discrimination between the coherent states $|0\rangle$ and $|\alpha\rangle$ is considered in (a). A diagram of the laboratory implementation of (a) is shown in (b). Source Cook *et al.* [40] (reprinted by permission from Macmillan Publishers Limited: [Nature] (Cook, R.L., Martin, P.J., and Geremia, J.M. 446, p. 774, 2007)).

such that the photon counter is least likely to observe additional clicks if it is correctly based on its best knowledge of the channel state at that time. In the present experiment, it is conveniently summarized as the minimization of the time-additive extension of total receiver error probabilities like (4.5), with $\xi_0 = \xi_1 = 1/2$ and where, at time t, the displacement history $u_{[0,t]}$ is also taken conditionally into account:

$$P_E[u_t] = \frac{1}{2}\int_0^\tau dt\, \left[P\left(\alpha|0, u_{[0,t]}\right) + P\left(0|\alpha, u_{[0,t]}\right)\right] . \tag{4.43}$$

The functional minimization of this expression leads to feedback policies (4.38) and (4.39) with $\psi_1(t) = \alpha$. They can be written as the unique formula

$$u_t^*\left(n_{[0,t]}\right) = \frac{\alpha}{2}\left(\frac{e^{i\pi(n_{[0,t]}+1)}}{\sqrt{1-e^{|\alpha|^2 t/\tau}}} - 1\right) \tag{4.44}$$

with the decision procedure that $|\alpha\rangle$ (or $|0\rangle$) is chosen when the number of photon counts $n_{[0,t]}$ in the measurement interval $[0, t]$ is even (or odd), that is, the ratio Λ is greater than 1 (or the ratio Λ is less than 1). We recall that this expression analytically achieves the fundamental quantum limit or Helstrom bound.

4.7.2
Closed-Loop Experiment: the Apparatus

As shown in Figure 4.1b the laboratory implementation of the closed-loop measurement consists in the following. Light from an external-cavity grating-stabilized diode laser[4] operating at 852nm is coupled into a polarization-maintaining fiber-optic Mach–Zehnder interferometer. A Mach–Zehnder interferometer is a device used to determine the phase shift caused by a small sample which is placed in the path of one of two collimated beams (thus having plane wave fronts) from a coherent light source. The input beamsplitter (FBS1) provides two optical fields with a well-defined relative phase: the upper arm of the interferometer acts as the target quantum system for state discrimination and the lower arm provides an auxiliary field used to perform closed-loop displacements at the second beamsplitter (FBS2). Photon counting on the outcoupled field is implemented using a gated silicon avalanche photodiode (APD). APDs are photodetectors that can be regarded as the semiconductor analog of photomultipliers. By applying a high reverse bias voltage (typically 100–200 V in silicon), APDs show an internal current gain effect (around 100) due to impact ionization (avalanche effect). The feedback controller is constructed from a combination of programmable waveform generators and high-speed digital signal processing electronics (feedback bandwidth 30 MHz). A digital counter records the number of photon counter clicks generated in each measurement interval $[0, \tau]$, during which time the feedback controller determines the feedback amplitude $u_t^*\left(n_{[0,t]}\right)$ in (4.44) via the accumulating count record $n_{[0,t)}$.

Coherent states for discrimination are realized as $\tau = 20\,\mu s$ optical pulses produced by a computer-controlled polarization-maintaining fiber-optic intensity modulator (FIM1) in the upper arm of the interferometer. The calibration parity between desired and observed values of α in Figure 4.2a (squares) highlights the ability to prepare arbitrary optical coherent states with amplitudes $0.1 \leq |\alpha| \leq 1$. Counting statistics for one such preparation with the mean photon number $\bar{n}_\alpha = \langle\alpha|N|\alpha\rangle = |\alpha|^2$ are shown in the inset. The circles in the inset are a Poisson fit to the counting data, for which $\chi^2 - 1$ has been computed below 1ppm, showing the quantum noise-limitation of the state preparation.

The phase of the prepared coherent states $\phi(\alpha)$ and the phase of the feedback displacements $\phi(u_t)$ are implemented by the modulator (FPM) in the upper arm of the interferometer. Without loss of generality, $\phi(\alpha) = 0$ was always chosen to simplify the interpretation of the displacements.

4) From Encyclopedia of Laser Physics and Technology, http://www.rp-photonics.com/: An external-cavity diode laser is a semiconductor laser based on a laser diode chip which typically has one end anti-reflection coated, and the laser resonator is completed with, e.g., a collimating lens and an external mirror.

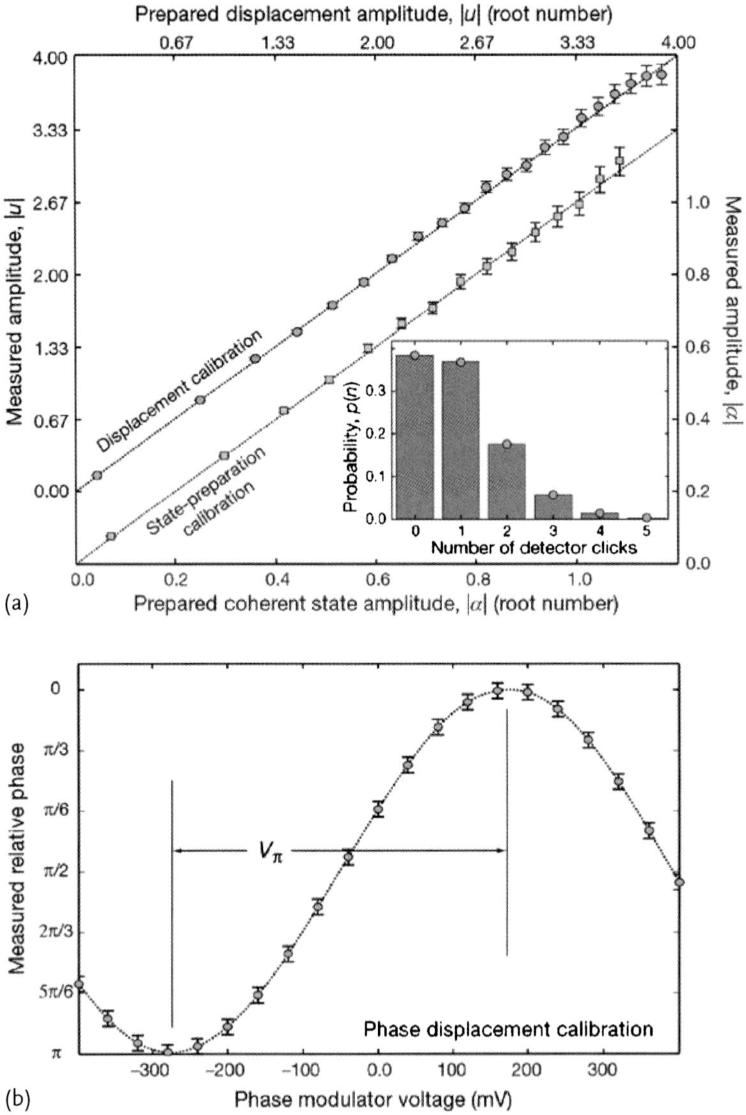

Fig. 4.2 (a) Quality of the preparation of the optical coherent states (circles) and the control of the displacements (squares). The inset shows a Poisson fit of the photon counting statistics for $\alpha \approx 1$, illustrating the quantum limitation of the state preparation. (b) Quality of the implementation of the controlled phase displacements (ϕ_{u_t}). The calibration in (b) illustrates the low control voltages (V_π) required to drive the apparatus, which allows for high-bandwidth application of the measurement feedback control. Each data point in (a) and (b) reflects a statistical ensemble of 100 000 replicate measurements, with error bars given by the estimated sample standard deviation. Source Cook *et al.* [40] (reprinted by permission from Macmillan Publishers Limited: [Nature] (Cook, R.L., Martin, P.J., and Geremia, J.M. 446, p. 774, 2007)).

Residual technical imperfections in the experiment result primarily from detector dark counts ($\bar{n}_d = 0.0078$), interferometer phase noise ($\delta\phi \approx 8$ mrad), and finite extinction of the modulators.

4.7.3
Closed-Loop Experiment: the Results

The measured probability of error versus mean photon number is shown in Figure 4.3a. The squares correspond to the case in which the feedback is disabled. The circles correspond to the closed-loop measurement.

4.7.3.1
Disabled Feedback

The measurement reduces to direct photon counting. The observed probability of error for discriminating between $|0\rangle$ and $|\alpha\rangle$ faithfully reproduces shot noise (line with squares) as a function of $\bar{n}_\alpha$. To get a precise view of the experimental challenge, data points were calculated using 100 000 optical pulses sampled randomly from $\{|0\rangle, |\alpha\rangle\}$ with equal probability. The label $|0\rangle$ has been used to signify the darkest field $\bar{n}_0 \approx 0.008$. The residual field appears to have a negligible effect, with a discrepancy of $\chi^2 - 1 = 1.13 \times 10^{-5}$ between the photon counting data in Figure 4.3a and the shot-noise error probability defined as $P_{\text{SN}} = e^{-\bar{n}_\alpha/2}$.

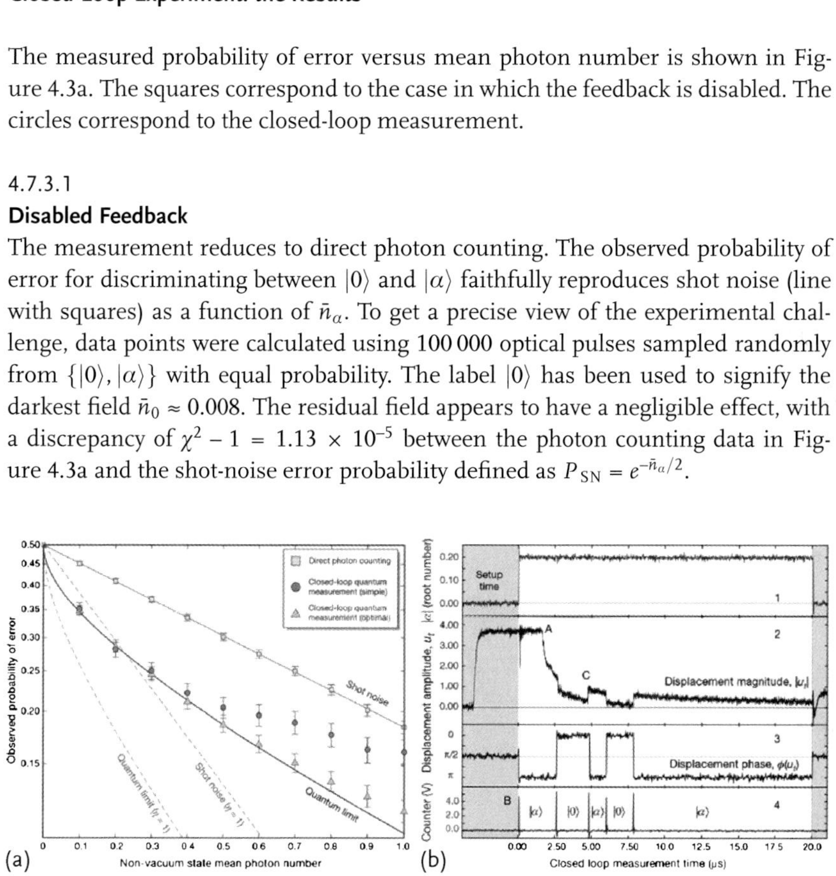

Fig. 4.3 (a) The measured probability of error versus mean photon number for both direct photon counting (squares) and the Cook–Martin–Geremia closed-loop measurement interpreted using a Bayesian estimator that assumes application of the optimal closed-loop control policy (circles) and that accounts for experimental imperfections (triangles). All data points were obtained from ensembles of 100 000 measurement trajectories, with error bars that reflect the sample standard deviation. The four traces in (b) (1–4) depict a single-shot closed-loop measurement trajectory. Attention should be paid to several technical issues: first, the finite dynamical range of the displacements (point A on the graph); second, the initial avalanche photodiode click is a timing signal, not a real detection event (B); and, third, the apparent rise time is that of the monitor photodiode not the feedback (C). Source Cook *et al.* [40] (reprinted by permission from Macmillan Publishers Limited: [Nature] (Cook, R.L., Martin, P.J., and Geremia, J.M. 446, p. 774, 2007)).

4.7.3.2
Closed-Loop Measurement

The premise behind the closed-loop measurement is to displace the field to the vacuum in each shot and decide which state is present on the basis of the displacement applied to cancel the field. As the controller gains increased confidence in its guess, it is better able to perform the correct nulling displacement. From (4.44), the displacement magnitude $|u_t^*|$ is inversely proportional to the time-dependent decision uncertainty $\sqrt{1 - e^{-\bar{n}_\alpha t/\tau}}$. Performing the optimization of (4.43) reveals that it is statistically optimal for the closed-loop measurement to reverse its state hypothesis with each detector click during the counting interval [41, 42].

Many aspects of this closed-loop measurement are evident from the single-shot trajectory in Figure 4.3b. At $t = 0$ there is no reason to prefer one state, $|0\rangle$ or $|\alpha\rangle$, over the other. But as more data become available, the controller refines its Bayesian estimate of the incoming optical state by updating the conditional probabilities $P(\psi|\Xi_{[0,t]}, u_{[0,t]})$. The sequence of hypothesis reversals in the example closed-loop trajectory is denoted along with the measurement record in Figure 4.3b, trace 4. As the measurement record accumulates, the controller eventually settles on its final (correct) decision, which in this case is $|\alpha\rangle$.

The data (circles) in Figure 4.3a demonstrate that the closed-loop state discrimination procedure (alternating guesses between $|0\rangle$ and $|\alpha\rangle$ with each photon arrival) surpasses the shot-noise error probability for amplitudes $|\alpha|$ less than about one. The fundamental quantum limit is essentially saturated over a nontrivial region of parameter space $\bar{n}_\alpha$.

The data (circles) in Figure 4.3a were determined assuming that the optimal feedback control policy in (4.44) was implemented perfectly by selecting the state $|\psi\rangle \in \{|0\rangle, |\alpha\rangle\}$ to maximize the conditional probability $P\left(\Xi_{[0,t]}|\psi, u^*_{[0,\tau]}\right)$. This approach is clearly suboptimal owing to the technical imperfections in the control displacements just described. Actually, the raw measurement data should be reinterpreted to account for deviations between the feedback actually performed in the experiment $u_{[0,\tau]}$ and the optimum policy $u^*_{[0,\tau]}$.

Owing to the nature of coherent states, the detection efficiency η (resulting from the combination of detector quantum efficiency η_d and optical efficiency η_e) factors out of a comparison between the shot-noise and quantum limits [42]. For comparison, the shot-noise error and quantum limits that would correspond to ideal detection ($\eta = 1$) have been plotted in Figure 4.3a. The intrinsic efficiency of the apparatus has been independently determined to be approximately $\eta \approx 0.35$.

The data (triangles) in Figure 4.3a reflect an analysis based on the true conditional probability $P\left(\Xi_{[0,t]}|\psi, u^*_{[0,\tau]}\right)$. It can be seen that the Cook–Martin–Geremia procedure nearly achieves the quantum limit (for the actual detection efficiency) over the full range of coherent states investigated. It can be observed that even with detection efficiency $\eta = 0.35$ the closed-loop measurement slightly outperforms the ideal shot-noise error that would be achieved in a technically lossless experiment $\eta = 1$ for photon numbers $\bar{n} < 0.2$.

4.8 Conclusion

Quantum feedback can be viewed as manipulating the outcome statistics of the number operator N. In the absence of feedback, the detailed measurement record consisting of photon arrival times $\Xi_{[0,\tau]} = (t_1, t_2, \ldots, t_n)$ provides no more information than the total number n: Poisson processes are stationary in time, but with feedback the significance of each click depends on when it occurs, even though the field is described by some coherent state at each point in time. The optimal feedback policy applies displacements in a manner that extracts as much information out of each photon arrival as possible. It is in this manner that shot noise can be surpassed to achieve the fundamental quantum limit over a nontrivial range of $|\alpha|$.

Furthermore, the Cook–Martin–Geremia procedure that has been described here appears as less demanding on the measurement resources needed to achieve optimal statistics than a direct implementation of a cat state: at no point in time has a superposition between optical coherent states been generated.

It should be added to these comments that very recently Wittman *et al.* [48] experimentally realized a new quantum measurement that detects binary optical coherent states with fewer errors than the homodyne and the Kennedy receiver for all amplitudes of the coherent states. Although the scheme is not capable of achieving the Helstrom bound, the implementation discriminates between binary coherent states with an error probability lower than the optimal Gaussian receiver, namely, the homodyne receiver. For more details on the theoretical background, see [49].

5
Coherent States: a General Construction

5.1
Introduction

We now depart from the standard situation and present a general method of construction of coherent states, starting from a few observations on the structure of these objects as superpositions of eigenstates of some self-adjoint operator, as was the harmonic oscillator Hamiltonian for the standard coherent states. It is the essence of quantum mechanics that this superposition has a probabilistic flavor. As a matter of fact, we notice that the probabilistic structure of the standard coherent states involves *two* probability distributions that underlie their construction. There are, in a sort of duality, a Poisson distribution ruling the probability of detecting n excitations when the quantum system is in a coherent state $|z\rangle$, and a gamma distribution on the set $\mathbb{C}$ of complex parameters, more exactly on the range $\mathbb{R}_+$ of the square of the radial variable. The generalization follows that duality scheme. Given a set X, equipped with a measure ν and the resulting Hilbert space $L^2(X,\nu)$ of square-integrable functions on X, we explain how the choice of an orthonormal system of functions in $L^2(X,\nu)$, precisely $\{\phi_j(x) \mid j \in \mathcal{J}\}$, $\int_X \overline{\phi_j(x)}\phi_{j'}(x)\,\nu(dx) = \delta_{jj'}$, carrying a probabilistic content, $\sum_{j\in\mathcal{J}} |\phi_j(x)|^2 = 1$, determines the family of coherent states $|x\rangle = \sum_j \overline{\phi_j(x)}|\phi_j\rangle$. The relation to the underlying existence of a reproducing kernel space will be briefly explained. This will be the guideline ruling the content of the subsequent chapters concerning each family of coherent states examined (in a generalized sense).

Coherent States in Quantum Physics. Jean-Pierre Gazeau

ISBN: 978-3-527-40709-5

5.2
A Bayesian Probabilistic Duality in Standard Coherent States

5.2.1
Poisson and Gamma Distributions

In the first chapters we reviewed the standard coherent states,

$$|z\rangle = \sum_{n=0}^{\infty} e^{-|z|^2/2} \frac{z^n}{\sqrt{n!}} |n\rangle \,, \tag{5.1}$$

in their principal physical and mathematical aspects. The key ingredients of these objects are

(i) The original set of parameters, namely, the complex plane $\mathbb{C}$, equipped with its Lebesgue measure d^2z/π. This set may or not be given a phase space status. The latter takes place within the framework of the classical motion of a particle on the real line. Otherwise, we could think about the set of time–frequency parameters in signal analysis.
(ii) Another set of parameters, namely, the natural numbers $\mathbb{N} \equiv \mathcal{J}$. Within the context of quantum physics, this set labels the possible issues of a certain experiment, such as counting the number of elementary excitations or "quanta" of a certain physical entity.

In relation with these two sets, the quantity

$$p_n(|z|^2) \stackrel{\text{def}}{=} |\langle n|z\rangle|^2 = e^{-|z|^2} \frac{|z|^{2n}}{n!} \tag{5.2}$$

leads to the following two interpretations.

(i) The discrete probability distribution $n \mapsto p_n(|z|^2)$ is a Poisson distribution on $\mathbb{N}$ with parameter $u \equiv |z|^2$ equal to the average number of occurrences. Clearly, this probability concerns experiments performed on the system within some experimental protocol, say, $\mathcal{E}$, and might be viewed as a stochastic model. Note that in the Poisson experiment, state preparation is determined only up to the unknown parameter $|z|^2$.
(ii) The continuous distribution $z \mapsto p_n(|z|^2)$ on $\mathbb{C}$ with measure d^2z/π is a gamma distribution when it is considered with respect to the $u = |z|^2$ variable, with the Lebesgue measure du on $\mathbb{R}_+$, and with n as a shape parameter.

The duality of interpretations that appears here is reminiscent of a similar duality observed in the theory of Bayesian statistical inference [50, 51]. In that context, the Poisson experiment would be performed and the Bayesian method used to obtain information about the behavior of the unknown parameter $|z|^2$ in the form of a conditional probability distribution on the parameter space $\mathbb{R}_+$ given an observed Poisson experimental value [52–55].

5.2.2
Bayesian Duality

We have already encountered Bayesian probabilities in the previous chapter. Let us try to become more familiar with this Bayesian context [55] (see Appendix A).

Suppose we have an experiment for which we postulate an experimental model in the form of a (one-parameter) family $n \mapsto P(n, u)$ of discrete probability distributions, where the unknown parameter u takes values in a measure space $(U, m(du))$. Suppose that the experiment has been performed, producing the result k. In Bayesian parlance, $P(k, u)$ as a function of u is called the "likelihood function" (see Appendix A) and $m(du)$ is called the "prior" measure on the parameter space U. Then we have a conditional probability density function f on the parameter space via an "inverse probability" formula where $f(u; k)du$ is proportional to $P(k, u)\, m(du)$. In Bayesian language, this final probability distribution on U is called the "posterior" probability distribution.

Thus, we have a duality of two probability distributions. We have the original discrete family indexed by parameter u, wherein, if the "true" value of u were known, it would serve as a predictive model for experimentally obtained data. Then we have the Bayesian posterior probability distribution on the (continuous) parameter space, which, if an experimental value were known, would serve as an "inferred" or "retrodictive" probability distribution for the unknown parameter.

As can be seen from the Poisson–gamma duality described above, the choice of measure space $(U, m(du))$ and coherent states (5.1) along with the expression (5.2) leads to a similar duality of the two probability distributions.

5.2.3
The Fock–Bargmann Option

Now, the exclusive character of the possible outcomes $n \in \mathbb{N}$ or $f(n)$ in the measurement of some quantum observable $f(N)$, N being the number operator, is encoded by the orthogonality between elements of the set of functions

$$e_n(z) = e^{-|z|^2/2} \frac{z^n}{\sqrt{n!}} \, .$$

These functions are complex square roots of the probability distributions in both senses above: $|e_n(z)|^2 = p_n(|z|^2)$.

They are in one-to-one correspondence with the Fock or number states $|n\rangle$. The closure of their linear span within the Hilbert space $L^2(\mathbb{C}, d^2z/\pi)$ is a sub-Hilbert space, say, $\mathcal{FB}_e$. The latter is reproducing, isomorphic to the Fock–Bergmann space introduced in Chapter 2, and also to the Fock space $\mathcal{H}$ generated by the number states.

5.2.4
A Scheme of Construction

In summary, what do we have?

(i) A set of parameters or data (in a classical sense), $\mathbb{C}$, for example, the set of initial conditions for the motion of a particle on the line, equipped by the Lebesgue measure d^2z/π.

(ii) The "large" Hilbertian arena, $L^2(\mathbb{C}, d^2z/\pi)$, which could be viewed as the space of images (with finite energy) in a signal analysis framework.

(iii) An orthonormal set of functions, $\{e_n(z) \in L^2(\mathbb{C}, d^2z/\pi),\ n \in \mathbb{N}\}$, which obeys the probabilistic identity

$$\sum_{n=0}^{\infty} |e_n(z)|^2 = 1\,. \tag{5.3}$$

(iv) A resulting family of states, the standard coherent states $|z\rangle$, in $\mathcal{FB}_e$ (or in $\mathcal{H}$), with $e_n(z)$ as the orthogonal projection on the basis element e_n (or $|n\rangle$).

(v) The identity (5.3) entails the normalization $\langle z|z\rangle = 1$, and the orthonormality of the $e_n(z)$'s entails the resolution of the identity in $\mathcal{FB}_e$ (or in $\mathcal{H}$).

In the next section, we will extend this scheme of construction to any measure set X.

5.3
General Setting: "Quantum" Processing of a Measure Space

In a first approach, one notices that quantum mechanics and signal analysis have many aspects in common. As a departure point of their respective formalism, one finds a *raw* set X of basic parameters, which we denote generically by $X = \{x \in X\}$. This set may be a classical phase space in the former case, like the complex plane for the particle motion on the line, whereas it may be a time–frequency plane (for Gabor analysis) or a time-scale half-plane (for wavelet analysis) in the latter one. Actually, it can be any set of data accessible to observation. For instance, it might be a temporal line or the circle or some interval. The minimal significant structure one requires so far is the existence of a measure $\mu(dx)$ on X. As a measure space, (X,μ), or simply X, could be given the name of an *observation* set, and the existence of a measure provides us with a statistical reading of the set of all measurable real- or complex-valued functions $f(x)$ on X: it allows us to compute, for instance, average values on subsets with bounded measure. Actually, both theories deal with quadratic mean values, and the natural framework of study is the Hilbert space $L^2(X,\mu)$ of all square-integrable functions $f(x)$ on the observation set X: $\int_X |f(x)|^2 \mu(dx) < \infty$. The function f is referred to as a *finite-energy* signal in signal analysis and might be referred to as a (pure) quantum state in quantum mechanics. However, it is precisely at this stage that the "quantum processing" of X differs from signal processing in at least three points:

(i) not all square-integrable functions are eligible as quantum states,
(ii) a quantum state is defined up to a nonzero factor,
(iii) among the functions $f(x)$, those that are eligible as quantum states and that are of unit norm, $\int_X |f(x)|^2 \mu(dx) = 1$, give rise to a probabilistic interpretation: the correspondence $X \supset \Delta \mapsto \int_\Delta |f(x)|^2 \mu(dx)$ is a probability measure, which is interpreted in terms of localization in the measurable set Δ and which allows one to determine mean values of quantum observables, which are (essentially) self-adjoint operators defined in a domain that is included in the set of quantum states.

The first point lies at the heart of the *quantization* problem (to which we devote the second part of the book): what is the more or less canonical procedure allowing us to select quantum states among simple signals? In other words, how should we select the true (projective) Hilbert space of quantum states, denoted by $\mathcal{K}$, that is, a closed subspace of $L^2(X,\mu)$, or equivalently the corresponding orthogonal projector $I_{\mathcal{K}}$?
This problem can be solved if one finds a map from X to the Hilbert space $\mathcal{K}$, $x \mapsto |x\rangle \in \mathcal{K}$, defining a family of states $\{|x\rangle\}_{x\in X}$ obeying the following two conditions:

- *normalization*

$$\langle x|x\rangle = 1\,, \tag{5.4}$$

- *resolution of the unity in* $\mathcal{K}$

$$\int_X |x\rangle\langle x|\, \nu(dx) = I_{\mathcal{K}}\,, \tag{5.5}$$

where $\nu(dx)$ is another measure on X, usually absolutely continuous with respect to $\mu(dx)$: this means that there exists a positive measurable function $h(x)$ such that $\nu(dx) = h(x)\mu(dx)$.

The explicit construction of such a set of vectors as well as its physical relevance are clearly crucial. It is remarkable that signal and quantum formalisms meet again on this level, since the family of states is called, in a wide sense, a *wavelet* family [56] or a *coherent state* family [11] according to the practitioner's field of interest. Two methods for constructing such families are generally in use. The first one rests upon group representation theory: a specific state or *probe*, say, $|x_0\rangle$, is transported along the orbit $\{|g \cdot x_0 \equiv x\rangle\,,\ g \in G\}$ by the action of a group G for which X is a homogeneous space. Irreducibility (Schur lemma) and unitarity conditions, combined with square integrability of the representation in some restricted sense, automatically lead to properties (5.4) and (5.5). Various examples of such group-theoretical constructions are given in [10, 11]. The second method has a wave-packet flavor in the sense that the state $|x\rangle$ is obtained from the superposition of elements in a fixed family of states $\{|\lambda\rangle\}_{\lambda\in\Lambda}$ that is total in $\mathcal{H}$:

$$|x\rangle = \int_{\Lambda} |\lambda\rangle\, \sigma(x, d\lambda)\,. \tag{5.6}$$

Here, the complex-valued x-dependent measure σ has its support Λ contained in the support of the spectral resolution $E(d\lambda)$ of a certain self-adjoint operator A, and the $|\lambda\rangle$'s are precisely eigenstates of A: $A|\lambda\rangle = \lambda|\lambda\rangle$. The choice of the operator A is ruled by the existence of the experimental device that allows us to measure all possible and exclusive issues $\lambda \in \mathrm{Sp}(A)$ of the physical quantity precisely encoded by A. The eigenstates can be understood in a distributional sense so as to put into the game of the construction portions belonging to the possible continuous part of the spectrum of A. Examples of such wave-packet constructions are given in [57–59], and here we will follow a similar procedure.

For pedagogical purposes, we now suppose that A is a self-adjoint operator in a Hermitian space (with finite dimension, say, $N + 1$) or a separable Hilbert space (with infinite dimension $N = \infty$), say, $\mathcal{H}$, of quantum states or of something else, it does not matter. Let us assume that the spectrum of A has only a discrete component, say, $\{a_n, 0 \le n \le N\}$. Normalized eigenstates of A are denoted by $|e_n\rangle$ and they form an orthonormal basis of $\mathcal{H}$. Next, suppose that the basis $\{|e_n\rangle\}_{0\le n\le N}$ is in one-to-one correspondence with an orthonormal set $\{\phi_n(x)\}_{0\le n\le N}$ of elements of $L^2(X, \mu)$. The generic $\mathcal{H}$ could be the Hilbert space $\mathcal{K}$, subspace of $L^2(X, \mu)$, but we keep our freedom in the choice of realization of $\mathcal{H}$. Furthermore, and this a decisive step in the wave-packet construction, we assume, in the case $N = \infty$, that

$$0 < \mathcal{N}(x) \equiv \sum_n |\phi_n(x)|^2 < \infty \quad \text{almost everywhere on } X\,. \tag{5.7}$$

Then, the states *in* $\mathcal{H}$,

$$|x\rangle \equiv \frac{1}{\sqrt{\mathcal{N}(x)}} \sum_n \overline{\phi_n(x)}|e_n\rangle\,, \tag{5.8}$$

satisfy both of our requirements (5.4) and (5.5). Indeed, the normalization is ensured because of the orthonormality of the set $\{|e_n\rangle\}$ and the presence of the normalization factor (5.7). The resolution of the unity $I_{\mathcal{H}}$ in $\mathcal{H}$ holds by virtue of the orthonormality of the set $\{\phi_n(x)\}$ if $\nu(dx)$ is related to $\mu(dx)$ by

$$\nu(dx) = \mathcal{N}(x)\mu(dx)\,. \tag{5.9}$$

Indeed, we have

$$\begin{aligned}\int_X |x\rangle\langle x|\, \nu(dx) &= \sum_{n,n'} |e_n\rangle\langle e_{n'}| \int_X \overline{\phi_n(x)}\, \phi_{n'}(x)\, \mu(dx)\\ &= \sum_n |e_n\rangle\langle e_n| = I_{\mathcal{H}}\,,\end{aligned}$$

where we have used the orthonormality of the ϕ_n's and the resolution of the unity in $\mathcal{H}$ obeyed by the orthonormal basis $|e_n\rangle$. Note that Hilbertian superposition (5.8) makes sense provided that the set X is equipped with a mild topological structure for which this map is continuous.

A direct and important consequence of the resolution of the unity is the existence of a *positive-operator-valued measure* [11, 35, 60, 61] on the measure space $(X, \mathcal{F})$ for a σ-algebra $\mathcal{F}$ of subsets of X,

$$\mathcal{F} \ni \Delta \mapsto \int_\Delta |x\rangle\langle x| \mathcal{N}(x)\, \mu(dx) \in \mathcal{L}(\mathcal{H})^+ . \tag{5.10}$$

The resolution of the unity in $\mathcal{H}$ can alternatively be understood in terms of the scalar product $\langle x|x'\rangle$ of two states of the family. Indeed, (5.5) implies that, with any vector $|\phi\rangle$ in $\mathcal{H}$, one can isometrically associate the function

$$\phi(x) \equiv \sqrt{\mathcal{N}(x)}\langle x\,|\phi\rangle \tag{5.11}$$

in $L^2(X, \mu)$. In particular,

$$\phi_n(x) = \sqrt{\mathcal{N}(x)}\langle x|e_n\rangle , \tag{5.12}$$

and this justifies our hypothesis that the basis $\{|e_n\rangle\}$ is in one-to-one correspondence with the orthonormal set $\{\phi_n(x)\}$. Hence, $\mathcal{H}$ is isometric to the Hilbert subspace $\mathcal{K}$ of $L^2(X, \mu)$, closure of the linear span of the $\phi_n(x)$'s defined by (5.12).

Now, by direct application of the resolution of the unity, the function $\phi(x)$ in (5.11), as an element of $\mathcal{K}$, obeys

$$\phi(x) = \int_X \sqrt{\mathcal{N}(x)\mathcal{N}(x')}\langle x|x'\rangle \phi(x')\, \mu(dx') . \tag{5.13}$$

Hence, $\mathcal{K}$ is a reproducing Hilbert space with kernel

$$\mathcal{K}(x, x') = \sqrt{\mathcal{N}(x)\mathcal{N}(x')}\langle x\,|x'\rangle , \tag{5.14}$$

and the latter assumes finite diagonal values (almost everywhere), $\mathcal{K}(x, x) = \mathcal{N}(x)$, by construction.

A last point of this construction of the space of quantum states concerns its statistical aspects, already pointed out in Section 5.2. There is indeed an interplay between two probability distributions:

- For almost each x, a discrete distribution,

$$n \mapsto \frac{|\phi_n(x)|^2}{\mathcal{N}(x)} . \tag{5.15}$$

  Like for the standard coherent states, this probability could be considered as concerning experiments performed on the system within some experimental protocol, say, $\mathcal{E}$, to measure the spectral values of the "quantum observable" A.

- For each n, a "continuous" distribution on (X, μ),

$$X \ni x \mapsto |\phi_n(x)|^2 . \tag{5.16}$$

Here, we observe the previously encountered Bayesian duality. There are two interpretations. The resolution of the unity verified by the "coherent" states $|x\rangle$ introduces a preferred *prior measure* on the observation set X, which is the set of parameters of the discrete distribution, with this distribution itself playing the role of the *likelihood function*. The associated discretely indexed continuous distributions become the related *conditional posterior distribution*.

Hence, a probabilistic approach to experimental observations concerning A should serve as a guideline in choosing the set of the $\phi_n(x)$'s.

Coming back to the standard coherent states, one briefly states the way in which their construction fits perfectly the above procedure:

- the observation set X is the classical phase space $\mathbb{R}^2 \simeq \mathbb{C} = \{x \equiv z = \frac{1}{\sqrt{2}}(q + i p)\}$ of a particle with one degree of freedom, more exactly the set of initial conditions (position and velocity) for motions of the particle on the line,
- the measure on X is Lebesgue, $\mu(dx) = \frac{1}{\pi} d^2 z$,
- the Hilbert space $\mathcal{H}$ is the Fock space with orthonormal basis the eigenstates $|n\rangle$ of the number operator $A \equiv N$,
- the functions $\phi_n(x)$ are the normalized powers of the complex variable $\bar{z}$ weighted by a Gaussian factor

$$\phi_n(x) \equiv e^{-\frac{1}{2}|z|^2} \frac{\bar{z}^n}{\sqrt{n!}} ,$$

- The Hilbert subspace $\mathcal{K}$ is the Fock–Bargmann space of all square-integrable functions that are of the form $\phi(z) = e^{-\frac{|z|^2}{2}} g(\bar{z})$, where $g(z)$ is analytically entire,
- the coherent states read

$$|z\rangle = e^{-\frac{|z|^2}{2}} \sum_n \frac{z^n}{\sqrt{n!}} |n\rangle .$$

5.4
Coherent States for the Motion of a Particle on the Circle

We now illustrate the general construction given in the previous section by considering an elementary although nonstandard example of coherent states. These states were proposed independently by De Bièvre–González [62], Kowalski–Rembieliński–Papaloucas [63–66], and González–Del Olmo [67] through approaches that substantially differ from the method illustrated in this chapter.

The observation set X is the phase space of a particle moving on the circle. It is the Cartesian product of the set of angular positions with the set of all possible velocities. It is naturally described by the cylinder

$$S^1 \times \mathbb{R} = \{x \equiv (\beta, J) \mid 0 \le \beta < 2\pi, \; J \in \mathbb{R}\}. \tag{5.17}$$

The velocity or momentum or "angular momentum" J (depending on the choice of units for the physical constants involved in the description of the motion) and the angle β are canonically conjugate variables and $dJ\,d\beta$ is the invariant measure on the phase space. So, we choose $\mu(dx) = \frac{1}{2\pi}\,dJ\,d\beta$ as a measure on X.

The functions $\phi_n(x)$, for $n \in \mathbb{Z}$, are suitably weighted Fourier exponentials:

$$\phi_n(x) = \left(\frac{\varepsilon}{\pi}\right)^{1/4} e^{-\frac{\varepsilon}{2}(J-n)^2}\, e^{in\beta}, \quad n \in \mathbb{Z}, \tag{5.18}$$

where $\varepsilon > 0$ can be arbitrarily small. This parameter could be viewed as the analog of the inverse of the square of the Planck constant, since J, as conjugate to an angle, could be considered as an angular momentum or an action. Actually ε represents a regularization. Notice that the continuous distribution $x \mapsto |\phi_n(x)|^2$ is the normal law centered at n (for the momentum variable J).

The normalization factor

$$\mathcal{N}(x) \equiv \mathcal{N}(J) = \sqrt{\frac{\varepsilon}{\pi}} \sum_{n\in\mathbb{Z}} e^{-\varepsilon(J-n)^2} < \infty \tag{5.19}$$

is a periodic train of normalized Gaussians, and is proportional to an elliptic theta function. Note that the Poisson summation formula leads to the alternative form

$$\mathcal{N}(J) = \sum_{n\in\mathbb{Z}} e^{2\pi i n J}\, e^{-\frac{\pi^2}{\varepsilon}n^2}. \tag{5.20}$$

Hence, we derive its regular behavior at $\varepsilon = 0$: $\lim_{\varepsilon\to 0} \mathcal{N}(J) = 1$.

Choosing a separable Hilbert space $\mathcal{H}$ with orthonormal basis $\{|e_n\rangle,\; n \in \mathbb{Z}\}$, we now have all the ingredients to define the coherent states on the circle:

$$|J, \beta\rangle = \frac{1}{\sqrt{\mathcal{N}(J)}} \left(\frac{\varepsilon}{\pi}\right)^{1/4} \sum_{n\in\mathbb{Z}} e^{-\frac{\varepsilon}{2}(J-n)^2}\, e^{-in\beta} |e_n\rangle. \tag{5.21}$$

For instance, the states $|e_n\rangle$ can be considered as Fourier exponentials $e^{in\beta}$ forming the orthonormal basis of the Hilbert space $L^2(S^1)$. They are the *spatial modes* in this representation, and are eigenstates of the (angular momentum) operator $\mathcal{J} = -i\partial/\partial\beta$ with eigenvalues $n \in \mathbb{Z}$.

As already mentioned, they could also be defined as elements of the reproducing kernel Hilbert space $\mathcal{K}$, subspace of $L^2\left(\mathbb{R} \times S^1, \frac{1}{2\pi}\,dJ\,d\beta\right)$ with the set $\{\phi_n \equiv |\phi_n\rangle,\; n \in \mathbb{Z}\}$ as an orthonormal basis,

$$|J, \beta\rangle = \frac{1}{\sqrt{\mathcal{N}(J)}} \left(\frac{\varepsilon}{\pi}\right)^{1/4} \sum_{n\in\mathbb{Z}} e^{-\frac{\varepsilon}{2}(J-n)^2}\, e^{-in\beta} |\phi_n\rangle, \tag{5.22}$$

The coherent states (5.21) or (5.22) are, as expected, normalized and resolve the unity *in* the Hilbert space $\mathcal{H}$ (or $\mathcal{K} \subset L^2(X, \frac{1}{2\pi}\, d J\, d\beta)$):

$$\langle J, \beta | J, \beta \rangle = 1\,, \qquad \int_X \frac{d J\, d\beta}{2\pi} \, | J, \beta \rangle \langle J, \beta | = I_d\,. \tag{5.23}$$

The "quantum" processing of the observation set $\mathbb{R} \times S^1$ is hence achieved by selecting in the (modified) Hilbert space $L^2(S^1 \times \mathbb{R}, \sqrt{\frac{\varepsilon}{\pi}}\, \frac{1}{2\pi} e^{-\varepsilon J^2}\, d J\, d\beta)$ all Laurent series in the complex variable $z = e^{\varepsilon J - i\beta}$. The overlap of two coherent states can also be expressed in terms of an elliptic theta function:

$$\langle J, \beta | J', \beta' \rangle = \frac{e^{-\frac{\varepsilon (J-J')^2}{4}}}{\sqrt{\mathcal{N}(J)\mathcal{N}(J')}} \left(\frac{\varepsilon}{\pi}\right)^{1/2} \sum_{n \in \mathbb{Z}} e^{-\varepsilon \left(\frac{J+J'}{2} - n\right)^2} e^{i n (\beta' - \beta)}\,. \tag{5.24}$$

Again the Poisson summation formula leads to the other form:

$$\langle J, \beta | J', \beta' \rangle = \frac{e^{-\frac{\varepsilon (J-J')^2}{4}}}{\sqrt{\mathcal{N}(J)\mathcal{N}(J')}} e^{i\left(\frac{J+J'}{2}\right)(\beta' - \beta)} \sum_{n \in \mathbb{Z}} e^{-\frac{1}{4\varepsilon}(\beta' - \beta + 2\pi n)^2} e^{\pi i n (J + J')}\,. \tag{5.25}$$

It is easily proven that as $\varepsilon \to 0$ this expression goes to zero if $\beta \neq \beta'$ and goes to 1 if $\beta = \beta'$.

5.5
More Coherent States for the Motion of a Particle on the Circle

Our choice of the Gaussian distribution for the J variable in the construction of the coherent states for the motion on the circle was essentially determined by the existing literature on the subject [63, 68]. Now, it is interesting to note that we can replace this Gaussian distribution by *any* (possibly even) probability distribution $\mathbb{R} \ni J \mapsto \pi(J)$ such that $\mathcal{N}(J) = \sum_n \pi(J - n) < \infty$. Let us be more precise. For the measure on the cylinder X, we still choose $\mu(dx) = \frac{1}{2\pi}\, d J\, d\beta$. The functions $\phi_n(x)$, for $n \in \mathbb{Z}$, are now given by

$$\phi_n(x) = \frac{1}{\sqrt{\nu}} \sqrt{\pi(J - n)}\, e^{i n \beta}\,, \quad n \in \mathbb{Z}\,, \tag{5.26}$$

with

$$\int_{-\infty}^{+\infty} \pi(J)\, d J = \nu\,.$$

The corresponding family of coherent states on the circle reads as

$$| J, \beta \rangle = \frac{1}{\sqrt{\mathcal{N}(J)}} \frac{1}{\sqrt{\nu}} \sum_{n \in \mathbb{Z}} \sqrt{\pi(J - n)}\, e^{-i n \beta} | \phi_n \rangle\,. \tag{5.27}$$

They are normalized and resolve the unity.

6
The Spin Coherent States

6.1
Introduction

The *spin* or $SU(2)$ coherent states form the second most known family of coherent states. They were introduced in the early 1970s by Radcliffe [69], Gilmore[70, 71], and Perelomov [72]. They also bear the name *atomic* or *Bloch* coherent states. This diversity of appellations reflects the range of domains in quantum physics where these objects play some role. The way in which we will introduce these states follows the probabilistic and Hilbertian scheme explained in the previous chapter. The central object or observation set is now the two-dimensional unit sphere S^2, equipped with the usual rotationally invariant measure. As orthonormal systems necessary for the construction of coherent states, sets of special functions called *spin spherical harmonics* will be selected. These functions are closely related to the unitary irreducible representations of the rotation group $SO(3)$ and its covering $SU(2)$ and imply the rich set of mathematical properties described in the present chapter.

6.2
Preliminary Material

Within the classical mechanics framework, the free rotation of a rigid body is characterized by its angular momentum $\mathbf{J}^{\mathrm{Cl}}$, where the superscript "Cl" stands for "classical." If the norm $\|\mathbf{J}^{\mathrm{Cl}}\|$ or the rotational kinetic energy $\|\mathbf{J}^{\mathrm{Cl}}\|^2/2I$ is conserved, the (pseudo)-vector $\mathbf{J}^{\mathrm{Cl}}$ describes a sphere of radius $J^{\mathrm{Cl}} = \|\mathbf{J}^{\mathrm{Cl}}\|$.

The quantum version of this angular momentum or *classical spin* is well known. At given $j = 0, \frac{1}{2}, 1, \frac{3}{2}, \ldots$ let us consider the Hermitian ($\equiv$ finite-dimensional Hilbert) space

$$\mathcal{H}^j \stackrel{\text{def}}{=} \mathbb{C}^{2j+1}, \tag{6.1}$$

with the set of kets $\{|j, m\rangle,\ m = -j, -j+1, \ldots, j-1, j\}$, $\langle j, m'|j, m\rangle = \delta_{mm'}$, as an orthonormal basis.

Coherent States in Quantum Physics. Jean-Pierre Gazeau

ISBN: 978-3-527-40709-5

Writing $\mathbf{J} = (J_x, J_y, J_z)$ for the angular momentum operator, we have, with $\hbar = 1$, the commutation rules

$$[J_x, J_y] = i J_z\,, \quad [J_z, J_x] = i J_y\,, \quad [J_y, J_z] = i J_x\,. \tag{6.2}$$

The basis vectors $|j, m\rangle$ are selected as common eigenvectors of the commuting pair of operators $J_z\,, \mathbf{J}^2 \stackrel{\text{def}}{=} \mathbf{J}\cdot\mathbf{J}$:

$$\mathbf{J}^2\,|j,m\rangle = j(j+1)\,|j,m\rangle\,, \quad J_z\,|j,m\rangle = m\,|j,m\rangle\,. \tag{6.3}$$

The ladder operators $J_\pm \stackrel{\text{def}}{=} J_x \pm i J_y$ act on the basis considered as

$$J_+|j,m\rangle = \sqrt{(j-m)(j+m+1)}\,|j,m+1\rangle\,, \tag{6.4}$$

$$J_-|j,m\rangle = \sqrt{(j+m)(j-m+1)}\,|j,m-1\rangle\,, \tag{6.5}$$

$$J_+|j,j\rangle = 0\,, \quad J_-|j,-j\rangle = 0\,.$$

All elements of the basis of $\mathcal{H}^j$ are readily derived from those "extremal" states $|j,j\rangle$ and $|j,-j\rangle$.

$$|j,m\rangle = \sqrt{\frac{(j-m)!}{(j+m)!2j!}}\,(J_+)^{j+m}\,|j,-j\rangle = \sqrt{\frac{(j+m)!}{(j-m)!2j!}}\,(J_-)^{j-m}\,|j,j\rangle\,. \tag{6.6}$$

6.3
The Construction of Spin Coherent States

Following the guideline explained in the previous chapter, we start from the sphere as the observation set: $X = S^2 = \{x \stackrel{\text{def}}{=} \hat{r} \stackrel{\text{def}}{=} (\theta,\phi) \in S^2\}$ in spherical coordinates. This set is equipped with the rotationally invariant measure $\mu(dx) \stackrel{\text{def}}{=} d\hat{r} = \sin\theta d\theta d\phi$.[5] We then proceed to the selection of the following orthonormal set in $L^2(S^2, d\hat{r})$:

$$\phi_{j,m}(\hat{r}) = \frac{1}{\sqrt{4\pi}}\sqrt{\binom{2j}{j+m}}\left(\cos\frac{\theta}{2}\right)^{j+m}\left(\sin\frac{\theta}{2}\right)^{j-m} e^{-i(j-m)\phi}\,. \tag{6.7}$$

We immediately check by using a simple binomial expansion that

$$\mathcal{N}(\hat{r}) = \sum_{m=-j}^{j} |\phi_{j,m}(x)|^2 = \frac{1}{4\pi}\,. \tag{6.8}$$

5) In the original definitions of coherent states, the measure is normalized, $\mu(dx) \stackrel{\text{def}}{=} \frac{d\hat{r}}{4\pi} = \sin\theta d\theta d\phi/4\pi$, and so our definitions of spin coherent states differ by a factor $1/\sqrt{4\pi}$.

We naturally define the Hilbert space $\mathcal{H}$ as $\mathcal{H} = \mathcal{H}^j$. Then, the "spin" or "Bloch" or "atomic" coherent states $|\hat{r}\rangle \in \mathcal{H}$[6] are given by

$$|\hat{r}\rangle = |\theta, \phi\rangle = \sum_{m=-j}^{j} \sqrt{\binom{2j}{j+m}} \left(\cos\frac{\theta}{2}\right)^{j+m} \left(\sin\frac{\theta}{2}\right)^{j-m} e^{i(j-m)\phi} |jm\rangle . \tag{6.9}$$

By construction, they are normalized and solve the identity $I_d \equiv I_{\mathcal{H}}$

$$\langle \hat{r}|\hat{r}\rangle = 1 , \quad \int_{S^2} \frac{d\hat{r}}{4\pi} |\hat{r}\rangle\langle\hat{r}| = I_d . \tag{6.10}$$

There exists another useful parameter for labeling the elements of the family of spin coherent states. The number $\zeta = \tan\frac{\theta}{2} e^{i\phi}$ parameterizes the (Riemann) sphere through a stereographic projection onto $\mathbb{C}$ of unit vectors $(\theta/2, \phi)$ with respect to the south pole:

$$S^2 \ni \hat{r} = (x, y, z) \mapsto \zeta = \frac{x + iy}{1 + z} \in \mathbb{C} \Leftrightarrow \begin{cases} x = \dfrac{2\Re(\zeta)}{1 + |\zeta|^2} \\ y = \dfrac{2\Im(\zeta)}{1 + |\zeta|^2} \\ z = \dfrac{1 - |\zeta|^2}{1 + |\zeta|^2} . \end{cases} \tag{6.11}$$

In terms of it, spin coherent states read as

$$\mathbb{C} \ni \zeta \mapsto |\hat{r}\rangle \equiv |\zeta\rangle = \sum_{m=-j}^{j} \sqrt{\frac{(2j)!}{(j-m)!(j+m)!}} \frac{\zeta^{j-m}}{(1 + |\zeta|^2)^j} |j, m\rangle , \tag{6.12}$$

and they resolve the unity in $\mathcal{H}^j$ in the following way:

$$\frac{2j+1}{\pi} \int_{\mathbb{C}} \frac{d^2\zeta}{(1 + |\zeta|^2)^2} |\zeta\rangle\langle\zeta| = I_d . \tag{6.13}$$

One should notice here the similarity with the standard coherent states

$$\mathbb{C} \ni z \mapsto |z\rangle = \sum_{n=0}^{\infty} e^{-\frac{|z|^2}{2}} \frac{z^n}{\sqrt{n!}} |n\rangle , \tag{6.14}$$

which are easily obtained from the spin coherent state at the limit of high spins through a *contraction* process. The latter is carried out through a scaling of the complex variable ζ, namely, $z = \sqrt{N}\,\zeta$, with $N = 2j$ and $n = j - m$, $|j, m\rangle \equiv |n\rangle$, and the limit $N \to \infty$:

$$\left|\zeta = \frac{z}{\sqrt{N}}\right\rangle_{\text{spin}} \underset{N\to\infty}{\to} |z\rangle . \tag{6.15}$$

6) Gilmore, Radcliffe, Perelomov

6.4 The Binomial Probabilistic Content of Spin Coherent States

For a given polar angle θ, consider the Bernoulli process, that is, a sequence of $n = 2j$ independent trials, with two possible outcomes for each trial:

1. "+" (win) with probability $p = \cos^2 \dfrac{\theta}{2}$,
2. "–" (loss) with probability $1 - p = \sin^2 \dfrac{\theta}{2}$.

Thus, we are certain to always win at the north pole of the sphere and to always lose at the south pole! At the equator the chances are equal, $p = 1/2$. Then the probability of winning after $k = j + m$ trials is the discrete binomial distribution with parameter $p = \cos^2 \frac{\theta}{2}$:

$$k \to p_k^{(n)} = \binom{n}{k} p^k (1-p)^{n-k} = 4\pi |\phi_{j,m}(\theta, \phi)|^2 = |\langle j, m|\theta, \phi\rangle|^2 . \tag{6.16}$$

In duality, we have the continuous binomial distribution with parameters $k = j+m$ and $n = 2j$

$$p = \cos^2 \frac{\theta}{2} \to p_k^{(n)} = \binom{n}{k} p^k (1-p)^{n-k} = 4\pi |\phi_{j,m}(\theta, \phi)|^2 = |\langle j, m|\theta, \phi\rangle|^2 . \tag{6.17}$$

We recover here the existence of the Bayesian duality of interpretations. The resolution of the unity verified by the spin coherent states introduces a preferred *prior measure* on the parameter space of polar angles $\theta \in [0, \pi]$ of the discrete distribution, with this distribution itself playing the role of the likelihood function [55]. The associated discretely indexed continuous distributions become the related conditional posterior distributions. We will illustrate this probabilistic content in the next chapter with the example of a quantum spin in a magnetic field.

6.5 Spin Coherent States: Group-Theoretical Context

The way the operators $J_\pm$ and J_z act on the basis elements of the Hermitian space $\mathcal{H}^j$ is the infinitesimal version of the action of a unitary irreducible representation of the rotation group $SO(3)$ or of its covering $SU(2)$.

At this point, we recall that any proper rotation in space is determined by a unit vector $\hat{n}$ defining the rotation axis and a rotation angle $0 \le \omega < 2\pi$ about the axis.

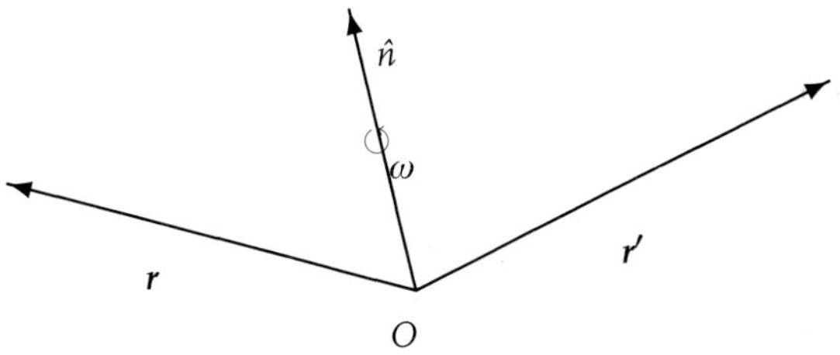

The action of such a rotation, $\mathcal{R}(\omega, \hat{n})$, on a vector $\vec{r}$ is given by

$$\vec{r'} \stackrel{\text{def}}{=} \mathcal{R}(\omega, \hat{n}) \cdot \vec{r} = \vec{r} \cdot \hat{n}\, \hat{n} + \cos\omega\, \hat{n} \times (\vec{r} \times \hat{n}) + \sin\omega\, (\hat{n} \times \vec{r}) , \tag{6.18}$$

$$\left(0, \vec{r'}\right) = \left(\cos\frac{\omega}{2}, \sin\frac{\omega}{2}\, \hat{n}\right) (0, \vec{r}) \left(\cos\frac{\omega}{2}, -\sin\frac{\omega}{2}\, \hat{n}\right) , \tag{6.19}$$

the latter being expressed in scalar–vector quaternionic form. Let us give here the minimal material necessary to understand the quaternionic formalism (Hamilton [73], 1843). We recall that the quaternion field as a multiplicative group is $\mathbb{H} \simeq \mathbb{R}_+ \times SU(2)$. The correspondence between the canonical basis of $\mathbb{H} \simeq \mathbb{R}^4$ $(1 \equiv e_0, e_1, e_2, e_3)$ and the Pauli matrices is $e_i \leftrightarrow (-1)^{i+1}\sigma_i$, with $i = 1, 2, 3$. Hence, the 2×2 matrix representation of these basis elements is the following:

$$\begin{aligned} &\begin{pmatrix} 1 & 0 \\ 0 & 1 \end{pmatrix} \leftrightarrow e_0 , \quad \begin{pmatrix} 0 & i \\ i & 0 \end{pmatrix} \leftrightarrow e_1 \equiv \hat{\imath} , \\ &\begin{pmatrix} 0 & -1 \\ 1 & 0 \end{pmatrix} \leftrightarrow e_2 \equiv \hat{\jmath} , \quad \begin{pmatrix} i & 0 \\ 0 & -i \end{pmatrix} \leftrightarrow e_3 \equiv \hat{k} . \end{aligned} \tag{6.20}$$

Any quaternion decomposes as $q = (q_0, \vec{q})$ (or $q^a e_a$, $a = 0, 1, 2, 3$) in scalar–vector notation (or in Euclidean metric notation). We also recall that the multiplication law explicitly reads in scalar–vector notation as $qq' = (q_0 q'_0 - \vec{q} \cdot \vec{q'}, q'_0 \vec{q} + q_0 \vec{q'} + \vec{q} \times \vec{q'})$. The (quaternionic) conjugate of $q = (q_0, \vec{q})$ is $\bar{q} = (q_0, -\vec{q})$, the squared norm is $\|q\|^2 = q\bar{q}$, and the inverse of a nonzero quaternion is $q^{-1} = \bar{q}/\|q\|^2$.

The rotations $\mathcal{R}(\omega, \hat{n})$, *direct* about their axis $\hat{n}$, are represented in $\mathcal{H}^j$ by the unitary operator:

$$\mathcal{R}(\omega, \hat{n}) \longrightarrow \exp\left(-i\omega\, \hat{n} \cdot \mathbf{J}\right) . \tag{6.21}$$

In particular, for a given unit vector

$$\hat{r} = (\sin\theta\, \cos\phi,\ \sin\theta\, \sin\phi,\ \sin\theta\, \cos\theta) \stackrel{\text{def}}{=} (\theta, \phi) , \quad 0 \le \theta \le \pi , \quad 0 \le \phi < 2\pi , \tag{6.22}$$

one considers the specific rotation $\mathcal{R}_{\hat{r}}$ that brings the unit vector pointing to the north pole, $\hat{k} = (0, 0, 1)$, to $\hat{r}$,

$$\hat{r} = \mathcal{R}_{\hat{r}}(\theta, \hat{u}_\phi) , \quad \hat{u}_\phi \stackrel{\text{def}}{=} (-\sin\phi, \cos\phi, 0) , \tag{6.23}$$

and the resulting unitary operator in $\mathcal{H}^j$:

$$\exp(-i\theta\,\hat{u}_\phi \cdot \mathbf{J}) \equiv \mathcal{D}^j(\mathcal{R}_{\hat{r}}) = e^{(\xi J_+ - \bar{\xi} J_-)}, \quad \xi \stackrel{\text{def}}{=} -\frac{\theta}{2}\, e^{-i\phi}. \tag{6.24}$$

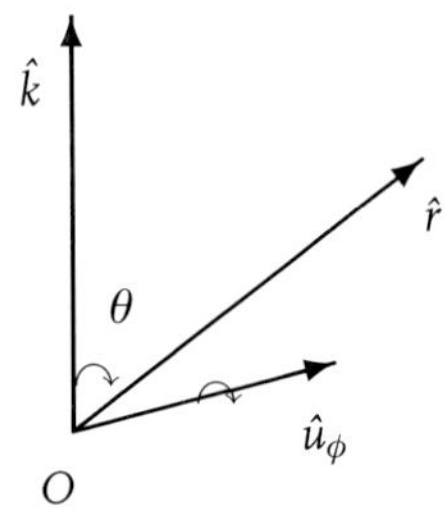

Spin coherent states then result from the "rotational" transport by $\mathcal{D}^j(\mathcal{R}_{\hat{r}})$ of the extremal state $|j, j\rangle$.

$$|x \equiv \hat{r}\rangle = \mathcal{D}^j(\mathcal{R}_{\hat{r}})\,|j, j\rangle. \tag{6.25}$$

As is indicated in (6.24), they are labeled as well by the complex numbers ξ in the closed disk centered at the origin and with radius $\pi/2$. The proof of (6.25) rests upon the application of the Baker–Campbell–Hausdorff or Zassenhaus formula,[7] which allows us to "disentangle" the exponential of a sum of operators:

$$e^{(\xi J_+ - \bar{\xi} J_-)} = e^{-\bar{\zeta} J_+}\, e^{\ln(1+|\zeta|^2)\, J_z}\, e^{\zeta J_-} = e^{\zeta J_-}\, e^{-\ln(1+|\zeta|^2)\, J_z}\, e^{-\bar{\zeta} J_+}, \tag{6.26}$$

where ζ is the alternative parameter introduced in (6.11):

$$\zeta = -\frac{\bar{\xi}}{|\xi|}\,\frac{\sin|\xi|}{\cos|\xi|} = \tan\frac{\theta}{2}\, e^{i\phi}.$$

Hence, $|\hat{r}\rangle$ is equal to

$$|\hat{r}\rangle = e^{\zeta J_-} \underbrace{e^{-\ln(1+|\zeta|^2)\, J_z} \underbrace{e^{-\bar{\zeta} J_+}\,|j, j\rangle}_{=|j,j\rangle}}_{=\left(\cos\frac{\theta}{2}\right)^{2j}|j,j\rangle} = \left(\cos\frac{\theta}{2}\right)^{2j} e^{\zeta J_-}\,|j, j\rangle$$
$$= \sum_{m=-j}^{j} \sqrt{\frac{(2j)!}{(j-m)!(j+m)!}} \left(\cos\frac{\theta}{2}\right)^{j+m} \left(\sin\frac{\theta}{2}\right)^{j-m} e^{i(j-m)\phi}\,|j, m\rangle. \tag{6.27}$$

Let us now describe the covariance property of the spin coherent states. Under a space rotation $\mathcal{R}(\omega, \hat{n})$ represented by a $(2j+1) \times (2j+1)$ unitary matrix acting

7) Let X and Y be two matrices and $[X, Y] = Z$ be their commutator. Then, $e^{(X+Y)} = e^X\, e^Y\, e^{-Z/2}\, e^{([X,Z]+2[Y,Z])/6} \ldots$

on $\mathcal{H}^j$, $\mathcal{R}(\omega, \hat{n}) \longrightarrow e^{-i\omega\,\hat{n}\cdot\mathrm{J}} \equiv \mathcal{D}^j(\mathcal{R}(\omega, \hat{n}))$, spin coherent states transform as

$$e^{-i\omega\,\hat{n}\cdot\mathrm{J}}\,|\hat{r}\rangle = e^{i\,j\,A(\hat{k},\hat{r},\hat{r}')}\,|\hat{r}'\rangle\,, \quad \hat{r}' = \mathcal{R}(\omega, \hat{n})\cdot\hat{r}\,, \tag{6.28}$$

where $A(\hat{k}, \hat{r}, \hat{r}')$ denotes the (symplectic) oriented area of the geodesic spherical triangle with vertices $\hat{k}, \hat{r}, \hat{r}'$.

This is a manifestation of the "quasi-classical" behavior of the spin coherent states, "quasi" since there appears a phase factor with topological interpretation in terms of $U(1)$ fiber, originating in the Hopf fibration of the three-dimensional sphere $S^3 \to S^2$. We recall here that the Lie group $SU(2)$ is, as a manifold, identical to S^3. Precisely, the rotation $\mathcal{R}_{\hat{r}}$ can be associated with the element $\xi = (\xi_0, \xi_1, \xi_2, \xi_3)$ of S^3, which, viewed as a unit quaternion, acts on the north pole of the 2-sphere S^2 through formula (6.24):

$$\xi\,(0, 0, 0, 1)\,\bar{\xi} = (0, x_1, x_2, x_3) = \hat{r}\,. \tag{6.29}$$

This gives rise to three quadratic relations,

$$\begin{aligned} x_1 &= 2(\xi_0\xi_2 + \xi_1\xi_3)\,, \\ x_2 &= 2(\xi_2\xi_3 - \xi_0\xi_1)\,, \\ x_3 &= 2(\xi_0^2 + \xi_3^2 - \xi_1^2 - \xi_2^2)\,, \end{aligned} \tag{6.30}$$

which exemplifies the so-called Hopf map $\xi \mapsto \hat{r}$. The Hopf inverse of a point of S^2 is a circle $\cong S^1 \subset S^3$. The Hopf inverse of a circle $\cong S^1 \subset S^2$ is a torus $\cong S^1 \times S^1 \subset S^3$, and so on.

The topological factor in (6.28) appears as well in the overlap of two spin coherent states

$$\langle\hat{r}|\hat{r}'\rangle = \left(\frac{1 + \hat{r}\cdot\hat{r}'}{2}\right)^j \left(e^{i\,A(\hat{k},\hat{r},\hat{r}')}\right)^j\,. \tag{6.31}$$

This overlap reads in terms of complex parameters ζ, ζ', as

$$\langle\zeta|\zeta'\rangle = \frac{(1 + \bar{\zeta}\zeta')^{2j}}{(1 + |\zeta|^2)^j(1 + |\zeta'|^2)^j}\,. \tag{6.32}$$

Note that the orthogonality holds for antipodal points: $\langle\hat{r}| - \hat{r}\rangle = 0$.

The spin coherent states saturate appropriate inequalities like standard coherent states do for Heisenberg inequalities. Those inequalities concern the triplet of operators J_x, J_y, J_z. From $[J_x, J_y] = i\,J_z$, one derives the inequalities for the product of variances calculated in an arbitrary state $|\psi\rangle \in \mathcal{H}^j$:

$$\Delta J_x\,\Delta J_y \geq \frac{1}{2}\,|\langle J_z\rangle|\,. \tag{6.33}$$

Precisely, the equality is achieved when $|\psi\rangle$ is chosen to be a spin coherent state $|\hat{r}\rangle\,, \hat{r} \in S^2$. The proof consists in checking first the equality in (6.33) for $|\hat{k}\rangle \equiv |j, j\rangle$ and next using the unitarity of (6.28) of the operator representing the rotation bringing $\hat{k}$ to an arbitrary $\hat{r}$.

6.6
Spin Coherent States: Fock–Bargmann Aspects

Like for the standard coherent states, the spin coherent states formalism allows a Fock–Bargmann realization (up to a nonanalytical factor) of the Hermitian space $\mathcal{H}^j$. The coherent state parameter used here is naturally the complex ζ parameterizing the Riemann sphere through (6.11)

$$\mathcal{H}^j \ni |\psi\rangle \mapsto \Psi(\zeta) \stackrel{\text{def}}{=} \frac{1}{\sqrt{4\pi}} \langle \bar{\zeta} | \psi \rangle \,. \tag{6.34}$$

From the expression of $|\zeta\rangle$, the function $\Psi(\zeta)$ assumes the form

$$\Psi(\zeta) = (1 + |\zeta|^2)^{-j} \, P(\zeta) \,, \tag{6.35}$$

where $P(\zeta)$ is an analytical polynomial of degree $\leq 2j$. The space of such functions $\Psi(\zeta)$ is the finite-dimensional Hilbert space, denoted by $\mathcal{K}^j$, of dimension $2j+1$, of all functions of the type (6.35) that are square-integrable with respect to the scalar product

$$\langle \Psi_1 | \Psi_2 \rangle \stackrel{\text{def}}{=} \frac{2j+1}{\pi} \int_{\mathbb{C}} \frac{d^2\zeta}{(1+|\zeta|^2)^2} \, \overline{\Psi_1(\zeta)} \, \Psi_2(\zeta) \,. \tag{6.36}$$

The map $\mathcal{H}^j \ni |\psi\rangle \mapsto \Psi(\zeta) \in \mathcal{K}^j$ is an isometry. It provides an "analytical" reading through a *reproducing* Hilbert space, of the quantum spin states in $\mathcal{H}^j = \mathbb{C}^{2j+1}$ (we could of course forget the quantum spin context and apply this isometry to any Hermitian space). Under this isometry, the orthonormal basis elements $|j, m\rangle$ in $\mathcal{H}^j$ are in one-to-one correspondence with the functions

$$u_{jm}(\zeta) = \sqrt{\frac{2j!}{(j-m)!(j+m)!}} \, \zeta^{j-m} \, (1 + |\zeta|^2)^{-j} \tag{6.37}$$

that form an orthonormal basis of $\mathcal{K}^j$.

Note that we could also consider the holomorphic purely polynomial realization of the Hermitian space $\mathcal{K}^j$. This finite-dimensional Fock–Bargmann space is denoted by $\mathcal{FB}^j$, consistent with the notation in Section 2.2.5, and is the space of all holomorphic polynomials $P(\zeta)$, of degree $\leq 2j$ equipped with the scalar product

$$\langle P_1 | P_2 \rangle \stackrel{\text{def}}{=} \frac{2j+1}{\pi} \int_{\mathbb{C}} \frac{d^2\zeta}{(1+|\zeta|^2)^{2j+1}} \, \overline{P_1(\zeta)} \, P_2(\zeta) \,. \tag{6.38}$$

6.7
Spin Coherent States: Spherical Harmonics Aspects

Another realization makes use of the well-known material of the functions $f(\hat{\nu})$, $\hat{\nu} \equiv (\theta, \phi)$, on the sphere, square-integrable with respect to the scalar product

$$\langle f_1 | f_2 \rangle \stackrel{\text{def}}{=} \int_{S^2} d\hat{\nu} \, \overline{f_1(\hat{\nu})} \, f_2(\hat{\nu}) \,, \tag{6.39}$$

and the orthonormal basis of spherical harmonics Y_{lm} at fixed integer $l \geq 0$. The isometry between the Fock–Bargmann realization $\mathcal{FB}^l$ and the Hermitian space $\mathcal{Y}^l$ generated by these Y_{lm}'s at fixed l is based on the following generating function for the spherical harmonics:

$$K(\bar{\zeta}, \hat{\nu} \equiv (\theta, \phi)) = \frac{1}{l!\, 2^l} \sqrt{\frac{(2l+1)\, 2l!}{4\pi}} \\ \times \left(-\sin\theta\, e^{i\phi} + 2\bar{\zeta} \cos\theta + \bar{\zeta}^2 \sin\theta\, e^{-i\phi}\right)^l (1 + |\zeta|^2)^{-l} . \quad (6.40)$$

We will see in the next sections the $SU(2)$ group representation origin of this expansion. The expression (6.40) *is* a spin coherent state in spherical representation. On the same footing, it is the kernel providing the link between the two spaces

$$K(\bar{\zeta}, \hat{\nu}) = \sum_{m=-l}^{l} u_{lm}(\bar{\zeta})\, Y_{lm}(\hat{\nu}) , \quad (6.41)$$

where the functions $u_{lm}(\zeta)$ are given in (6.37). Precisely, the isomorphism between the two Hermitian spaces is given by

$$\mathcal{Y}^l \ni f(\hat{\nu}) \mapsto \Psi(\zeta) = \int_{S^2} \overline{K(\bar{\zeta}, \hat{\nu})}\, f(\hat{\nu})\, d\hat{\nu} , \quad (6.42)$$

$$\mathcal{FB}^l \ni \Psi(\zeta) \mapsto f(\hat{\nu}) = \int_{\mathbb{C}} K(\bar{\zeta}, \hat{\nu})\, \Psi(\zeta)\, \frac{2l+1}{\pi} \frac{d^2\zeta}{(1 + |\zeta|^2)^2} . \quad (6.43)$$

6.8
Other Spin Coherent States from Spin Spherical Harmonics

A whole set of families of spin coherent states actually exists, as they were introduced by Perelomov within a group-theoretical context [10, 72]. Here, we will recover these Perelomov states by picking in our coherent state construction procedure the so-called spin spherical harmonics, of which the orthonormal system (6.7) represents a particular case. First, it is necessary to recall some technical facts about the group $SU(2)$ and its unitary irreducible representations. More details are given in Appendix C.

6.8.1
Matrix Elements of the *SU*(2) Unitary Irreducible Representations

Let $\xi = (\xi_0, \xi_1, \xi_2, \xi_3) = \xi^a e_a$, $\xi_0^2 + \xi_1^2 + \xi_2^2 + \xi_3^2 = 1$, be a vector pointing to the unit sphere S^3 in the Euclidean space $\mathbb{R}^4$ equipped with the orthonormal basis $\{e_a\,,\ a = 0, 1, 2, 3\}$ introduced in Section 6.5. With the correspondence established in (6.20), the group $SU(2)$ can be defined as the set of 2×2 matrices in one-to-one correspondence with such unit-norm vectors:

$$SU(2) \ni \xi = \begin{pmatrix} \xi_0 + i\xi_3 & -\xi_2 + i\xi_1 \\ \xi_2 + i\xi_1 & \xi_0 - i\xi_3 \end{pmatrix} , \quad (6.44)$$

where we use the same notation for a unit 4-vector as for the corresponding $SU(2)$

matrix. In bicomplex angular coordinates,

$$\xi_0 + i\xi_3 = \cos\omega e^{i\psi_1}\,, \quad \xi_1 + i\xi_2 = \sin\omega e^{i\psi_2}\,, \tag{6.45}$$

$$0 \leqslant \omega \leqslant \frac{\pi}{2}\,, \quad 0 \leqslant \psi_1, \psi_2 < 2\pi\,, \tag{6.46}$$

such $SU(2)$ matrices read as

$$SU(2) \ni \xi = \begin{pmatrix} \cos\omega e^{i\psi_1} & i\sin\omega e^{i\psi_2} \\ i\sin\omega e^{-i\psi_2} & \cos\omega e^{-i\psi_1} \end{pmatrix}, \tag{6.47}$$

in agreement with the notation of Talman [74].

Let us choose $j \in \mathbb{N}/2$ and $m \in \mathbb{Z}/2$ such that $-j \leq m \leq j$ and $j - m \in \mathbb{Z}$. With any vector $\mathbf{z} = \begin{pmatrix} z_1 \\ z_2 \end{pmatrix}$ in $\mathbb{C}^2$ let us associate the monomial function

$$e_j^m(\mathbf{z}) = \frac{z_1^{j+m}\, z_2^{j-m}}{\sqrt{(j+m)!(j-m)!}}\,. \tag{6.48}$$

At fixed j these monomials span a $(2j+1)$-dimensional vector space of polynomials $p_j(\mathbf{z})$. The group $SU(2)$ acts on this space through the following action:

$$p_j(\mathbf{z}) \overset{\xi \in SU(2)}{\longrightarrow} p_j(\xi^\dagger\, \mathbf{z})\,. \tag{6.49}$$

The corresponding matrix elements of this $(2j+1)$-dimensional unitary irreducible representation of $SU(2)$ are defined through the action (6.49) on the monomial basis elements:

$$\begin{aligned} e_j^{m_2}(\xi^\dagger\, \mathbf{z}) &= \\ &\frac{(\cos\omega e^{-i\psi_1} z_1 - i\sin\omega e^{i\psi_2} z_2)^{j+m_2} (-i\sin\omega e^{-i\psi_2} z_1 + \cos\omega e^{i\psi_1} z_2)^{j-m_2}}{\sqrt{(j+m_2)!(j-m_2)!}} \\ &= \sum_{m_1} D^j_{m_1 m_2}(\xi)\, e_j^{m_1}(\mathbf{z})\,. \end{aligned} \tag{6.50}$$

This expression provides a generating function for the matrix elements. The latter are given by [74]

$$\begin{aligned} D^j_{m_1 m_2}(\xi) &= (-1)^{m_1 - m_2} \left[(j+m_1)!(j-m_1)!(j+m_2)!(j-m_2)!\right]^{1/2} \\ &\times \sum_t \frac{(\xi_0 + i\xi_3)^{j-m_2-t}}{(j-m_2-t)!} \frac{(\xi_0 - i\xi_3)^{j+m_1-t}}{(j+m_1-t)!} \frac{(-\xi_2 + i\xi_1)^{t+m_2-m_1}}{(t+m_2-m_1)!} \frac{(\xi_2 + i\xi_1)^t}{t!}\,. \end{aligned} \tag{6.51}$$

With angular variables the matrix elements of the unitary irreducible representation of $SU(2)$ are given in terms of Jacobi polynomials [18] by

$$\begin{aligned} D^j_{m_1 m_2}(\xi) &= e^{-i m_1(\psi_1+\psi_2)} e^{-i m_2(\psi_1-\psi_2)}\, i^{m_2-m_1} \sqrt{\frac{(j-m_1)!(j+m_1)!}{(j-m_2)!(j+m_2)!}} \\ &\times \frac{1}{2^{m_1}} (1+\cos 2\omega)^{\frac{m_1+m_2}{2}} (1-\cos 2\omega)^{\frac{m_1-m_2}{2}} P_{j-m_1}^{(m_1-m_2, m_1+m_2)}(\cos 2\omega)\,, \end{aligned} \tag{6.52}$$

in agreement with Edmonds [75] (up to an irrelevant phase factor).

6.8.2
Orthogonality Relations

Let us equip the $SU(2)$ group with its invariant (*Haar*) measure:

$$\mu(d\xi) = \sin 2\omega \, d\omega \, d\psi_1 \, d\psi_2 \,, \tag{6.53}$$

in terms of the bicomplex angular parameterization. Note that the volume of $SU(2)$ with this choice of normalization is $8\pi^2$. The orthogonality relations satisfied by the matrix elements $D^j_{m_1 m_2}(\xi)$ read as

$$\int_{SU(2)} D^j_{m_1 m_2}(\xi) \, \overline{D^{j'}_{m'_1 m'_2}(\xi)} \, \mu(d\xi) = \frac{8\pi^2}{2j+1} \delta_{jj'} \delta_{m_1 m'_1} \delta_{m_2 m'_2} \,. \tag{6.54}$$

6.8.3
Spin Spherical Harmonics

The spin spherical harmonics, as functions on the 2-sphere S^2, are defined as follows:

$$\begin{aligned} {}_\sigma Y_{j\mu}(\hat{r}) &= \sqrt{\frac{2j+1}{4\pi}} \overline{D^j_{\mu\sigma}(\xi(\mathcal{R}_{\hat{r}}))} = (-1)^{\mu-\sigma} \sqrt{\frac{2j+1}{4\pi}} D^j_{-\mu-\sigma}(\xi(\mathcal{R}_{\hat{r}})) \\ &= \sqrt{\frac{2j+1}{4\pi}} D^j_{\sigma\mu}\left(\xi^\dagger(\mathcal{R}_{\hat{r}})\right) \,, \end{aligned} \tag{6.55}$$

where $\xi(\mathcal{R}_{\hat{r}})$ is a (nonunique) element of $SU(2)$ that corresponds to the space rotation $\mathcal{R}_{\hat{r}}$ introduced in (6.23) and that brings the unit vector $e_3 \equiv \hat{k}$ to the unit vector $\hat{r}$ with spherical coordinates (θ, ϕ):

We immediately infer from the definition (6.55) the following properties:

$$\overline{{}_\sigma Y_{j\mu}(\hat{r})} = (-1)^{\sigma-\mu} \, {}_{-\sigma} Y_{j-\mu}(\hat{r}) \,, \tag{6.56}$$

$$\sum_{\mu=-j}^{\mu=j} \left| {}_\sigma Y_{j\mu}(\hat{r}) \right|^2 = \frac{2j+1}{4\pi} \,. \tag{6.57}$$

Now, from the quaternionic description of 3-space rotations, we have the group homomorphism $\xi = \xi(\mathcal{R}) \in SU(2) \leftrightarrow \mathcal{R} \in SO(3) \cong SU(2)/\mathbb{Z}_2$:

$$\hat{r}' = \begin{pmatrix} i x'_3 & -x'_2 + i x'_1 \\ x'_2 + i x'_1 & -i x'_3 \end{pmatrix} = \mathcal{R} \cdot \hat{r} = \xi \begin{pmatrix} i x_3 & -x_2 + i x_1 \\ x_2 + i x_1 & -i x_3 \end{pmatrix} \xi^\dagger \,. \tag{6.58}$$

In the particular case of (6.55) the angular coordinates ω, ψ_1, ψ_2 of the $SU(2)$ element $\xi(\mathcal{R}_{\hat{r}})$ are constrained by

$$\cos 2\omega = \cos\theta, \quad \sin 2\omega = \sin\theta \,, \quad \text{so} \quad 2\omega = \theta \,, \tag{6.59}$$

$$e^{i(\psi_1+\psi_2)} = i e^{i\phi}, \quad \text{so} \quad \psi_1 + \psi_2 = \phi + \frac{\pi}{2} \,. \tag{6.60}$$

Here we should pay special attention to the range of values for the angle ϕ, depending on whether j and consequently σ and m are half-integer or not. If j is half-integer, then the angle ϕ should be defined $\mod (4\pi)$ whereas if j is integer, it should be defined $\mod (2\pi)$.

We still have one degree of freedom concerning the pair of angles ψ_1, ψ_2. We leave open the option concerning the σ-dependent phase factor by putting

$$i^{-\sigma} e^{i\sigma(\psi_1-\psi_2)} \stackrel{\text{def}}{=} e^{i\sigma\psi}\,, \tag{6.61}$$

where ψ is arbitrary. With this choice and considering (6.51) and (6.52) we get the expression for the spin spherical harmonics in terms of ϕ, $\theta/2$, and ψ, and of Jacobi polynomials, valid in the case in which $\mu \pm \sigma > -1$[8]:

$$\begin{aligned}{}_\sigma Y_{j\mu}(\hat{r}) &= (-1)^\mu e^{i\sigma\psi} \sqrt{\frac{2j+1}{4\pi}} \sqrt{\frac{(j-\mu)!(j+\mu)!}{(j-\sigma)!(j+\sigma)!}} \\ &\quad \times \frac{1}{2^\mu} (1+\cos\theta)^{\frac{\mu+\sigma}{2}} (1-\cos\theta)^{\frac{\mu-\sigma}{2}} P_{j-\mu}^{(\mu-\sigma,\mu+\sigma)}(\cos\theta)\, e^{i\mu\phi}\,. \end{aligned} \tag{6.62}$$

For other cases, it is necessary to use alternative expressions based on the relations [18]

$$P_n^{(-l,\beta)}(x) = \frac{\binom{n+\beta}{l}}{\binom{n}{l}} \left(\frac{x-1}{2}\right)^l P_{n-l}^{(l,\beta)}(x)\,, \quad P_0^{(\alpha,\beta)}(x) = 1\,. \tag{6.63}$$

Note that with $\sigma = 0$ we recover the expression for the normalized spherical harmonics (see Appendix C).

Finally, introducing the single complex variable $\zeta = z_2/z_1$, one should retain the following expression, directly emerging from (6.51), for the generating function of the spin spherical harmonics:

$$\begin{aligned}&\sqrt{\frac{2j+1}{4\pi}} \left(\cos\frac{\theta}{2} + \sin\frac{\theta}{2} e^{-i\phi}\zeta\right)^{j+\sigma} \left(-\sin\frac{\theta}{2} e^{i\phi} + \cos\frac{\theta}{2}\zeta\right)^{j-\sigma} \\ &\quad = \sum_{m=-j}^{j} \sqrt{\frac{(j+\sigma)!(j-\sigma)!}{(j+m)!(j-m)!}}\, \zeta^{j-m} (-1)^{-\sigma} e^{-i\sigma\phi}\, {}_\sigma Y_{jm}(\hat{r})\,. \end{aligned} \tag{6.64}$$

In particularizing the above equation to the case $\sigma = 0$, one recovers the kernel (6.40).

8) This expression is not exactly in agreement with the definitions of Newman and Penrose [77], Campbell [76] (note that there is a mistake in the expression given by Campbell, in which a $\cos\frac{\theta}{2}$ should read $\cot\frac{\theta}{2}$), and Hu and White [78]. Besides the presence of different phase factors, the disagreement is certainly due to a different relation between the polar angle θ and the Euler angle.

6.8.4
Spin Spherical Harmonics as an Orthonormal Basis

Specifying (6.54) to the spin spherical harmonics leads to the following orthogonality relations that are valid for integer j (and consequently integer σ):

$$\int_{S^2} {}_\sigma Y_{j\mu}(\hat{r}) \left({}_\sigma Y_{j'\nu}(\hat{r})\right)^* \mu(d\hat{r}) = \delta_{jj'}\delta_{\mu\nu}\,. \tag{6.65}$$

We recall that in the integer case, the range of values assumed by the angle ϕ is $0 \leqslant \phi < 2\pi$. Now, if we consider half-integer j (and consequently σ), the range of values assumed by the angle ϕ becomes $0 \leqslant \phi < 4\pi$. The integration above has to be carried out on the "doubled" sphere $\widetilde{S}^2$ and an extra normalization factor equal to $\frac{1}{\sqrt{2}}$ is needed in the expression of the spin spherical harmonics.

For a given integer σ the set $\left\{{}_\sigma Y_{j\mu},\ -\infty \leqslant \mu \leqslant \infty,\ j \geqslant \max(0, \sigma, m)\right\}$ forms an orthonormal basis of the Hilbert space $L^2(S^2)$. Indeed, at μ fixed so that $\mu \pm \sigma \geqslant 0$, the set

$$\left\{\sqrt{\frac{2j+1}{4\pi}}\sqrt{\frac{(j-\mu)!(j+\mu)!}{(j-\sigma)!(j+\sigma)!}}\,\frac{1}{2^\mu}\,(1+\cos\theta)^{\frac{\mu+\sigma}{2}}\,(1-\cos\theta)^{\frac{\mu-\sigma}{2}}\right.$$
$$\left.\times\, P_{j-\mu}^{(\mu-\sigma,\mu+\sigma)}(\cos\theta),\ j \geqslant \mu\right\}$$

is an orthonormal basis of the Hilbert space $L^2([-\pi,\pi], \sin\theta\, d\theta)$. The same holds for other ranges of values of μ by using alternative expressions such as (6.63) for Jacobi polynomials. Then it suffices to view $L^2(S^2)$ as the tensor product $L^2([-\pi,\pi], \sin\theta\, d\theta) \bigotimes L^2(S^1)$. Similar reasoning is valid for half-integer σ. Then, the Hilbert space to be considered is the space of "fermionic" functions on the doubled sphere $\widetilde{S}^2$, that is, such that $f(\theta, \phi + 2\pi) = -f(\theta, \phi)$.

6.8.5
The Important Case: $\sigma = j$

For $\sigma = j$, owing to the relations (6.63), the spin spherical harmonics reduce to their simplest expressions:

$${}_j Y_{j\mu}(\hat{r}) = (-1)^j e^{ij\psi}\sqrt{\frac{2j+1}{4\pi}}\sqrt{\binom{2j}{j+\mu}}\left(\cos\frac{\theta}{2}\right)^{j+\mu}\left(\sin\frac{\theta}{2}\right)^{j-\mu} e^{i\mu\phi}\,. \tag{6.66}$$

They are precisely the functions that appear, *up to* the factor $\sqrt{2j+1}$ and a phase factor also, in the construction of the spin coherent states described in Section 6.3.

6.8.6 Transformation Laws

Let $\mathcal{H}^{\sigma j}$ be the $(2j+1)$-dimensional Hermitian space, subspace of $L^2(S^2, \mu(d\hat{r}))$, spanned by the orthonormal set $\left\{ {}_\sigma Y_{j\mu}(\hat{r}), \ -j \leq m \leq j \right\}$ of spin spherical harmonics at fixed σ and j. Given a function $f(x)$ on the sphere S^2 belonging to $\mathcal{H}^{\sigma j}$ and a rotation $\mathcal{R} \in SO(3)$, we define the rotation operator $\mathcal{D}^{\sigma j}(\mathcal{R})$ for that representation by

$$\left(\mathcal{D}^{\sigma j}(\mathcal{R}) f\right)(x) = f(\mathcal{R}^{-1} \cdot x) = f({}^t\mathcal{R} \cdot x) . \tag{6.67}$$

We now consider the transformation law of the spin spherical harmonics under this representation of the rotation group. From the relation

$$\mathcal{R}\mathcal{R}_{{}^t\mathcal{R}\hat{r}} = \mathcal{R}_{\hat{r}} \tag{6.68}$$

for any $\mathcal{R} \in SO(3)$, and from the homomorphism $\xi(\mathcal{R}\mathcal{R}') = \xi(\mathcal{R})\xi(\mathcal{R}')$ between $SO(3)$ and $SU(2)$, we deduce from the definition (6.55) of the spin spherical harmonics the transformation law

$$\begin{aligned}
\left(\mathcal{D}^{\sigma j}(\mathcal{R}) {}_\sigma Y_{j\mu}\right)(\hat{r}) = {}_\sigma Y_{j\mu}({}^t\mathcal{R} \cdot \hat{r}) &= \sqrt{\frac{2j+1}{4\pi}} D^j_{\sigma\mu}\left(\xi^\dagger\left(\mathcal{R}_{{}^t\mathcal{R}\cdot\hat{r}}\right)\right) \\
&= \sqrt{\frac{2j+1}{4\pi}} D^j_{\sigma\mu}\left(\xi^\dagger\left({}^t\mathcal{R}\mathcal{R}_{\hat{r}}\right)\right) \\
&= \sqrt{\frac{2j+1}{4\pi}} D^j_{\sigma\mu}\left(\xi^\dagger\left(\mathcal{R}_{\hat{r}}\right)\xi\left(\mathcal{R}\right)\right) \\
&= \sqrt{\frac{2j+1}{4\pi}} \sum_\nu D^j_{\sigma\nu}\left(\xi^\dagger\left(\mathcal{R}_{\hat{r}}\right)\right) D^j_{\nu\mu}\left(\xi\left(\mathcal{R}\right)\right) \\
&= \sum_\nu {}_\sigma Y_{j\nu}(\hat{r}) D^j_{\nu\mu}\left(\xi\left(\mathcal{R}\right)\right) ,
\end{aligned} \tag{6.69}$$

as expected if we think of the special case ($\sigma = 0$) of the spherical harmonics [74].

6.8.7 Infinitesimal Transformation Laws

The generators of the representative of the three rotations $\mathcal{R}^{(a)}$, $a = 1, 2, 3$, around the three Cartesian axes, are the components of the angular momentum operator in the representation $\mathcal{D}^{\sigma j}$. When $\sigma = 0$, these generators are the usual angular momentum operators $J_a = -i\varepsilon_{abc} x^b \partial_c$ (short notation for $J_a^{(j)}$), which, in spherical coordinates, are given by

$$\begin{aligned}
J_3 &= -i\partial_\phi , \\
J_+ &= J_1 + iJ_2 = e^{i\phi}\left(\partial_\theta + i\cot\theta\partial_\phi\right) , \\
J_- &= J_1 - iJ_2 = -e^{-i\phi}\left(\partial_\theta - i\cot\theta\partial_\phi\right) .
\end{aligned} \tag{6.70}$$

In the general case $\sigma \neq 0$, we denote the generators by $\Lambda_a^{(\sigma j)}$. These "spin" angular momentum operators are given by

$$\Lambda_3^{\sigma j} = J_3 = -i\partial_\phi , \tag{6.71}$$

$$\Lambda_+^{\sigma j} = \Lambda_1^{\sigma j} + i\Lambda_2^{\sigma j} = J_+ + \sigma \csc\theta e^{i\phi} , \tag{6.72}$$

$$\Lambda_-^{\sigma j} = \Lambda_1^{\sigma j} - i\Lambda_2^{\sigma j} = J_- + \sigma \csc\theta e^{-i\phi} . \tag{6.73}$$

They obey the expected commutation rules,

$$[\Lambda_3^{\sigma j}, \Lambda_\pm^{\sigma j}] = \pm\Lambda_\pm^{\sigma j} , \quad [\Lambda_+^{\sigma j}, \Lambda_-^{\sigma j}] = 2\Lambda_3^{\sigma j} . \tag{6.74}$$

Their actions on the spin spherical harmonics are similar to the case $\sigma = 0$.

$$\Lambda_3^{\sigma j}\ {}_\sigma Y_{j\mu} = \mu\ {}_\sigma Y_{j\mu} \tag{6.75a}$$

$$\Lambda_+^{\sigma j}\ {}_\sigma Y_{j\mu} = \sqrt{(j-\mu)(j+\mu+1)}\ {}_\sigma Y_{j\mu+1} \tag{6.75b}$$

$$\Lambda_-^{\sigma j}\ {}_\sigma Y_{j\mu} = \sqrt{(j+\mu)(j-\mu+1)}\ {}_\sigma Y_{j\mu-1} . \tag{6.75c}$$

6.8.8 "Sigma-Spin" Coherent States

For a given pair (j, σ), we now define the family of coherent states in the $(2j+1)$-dimensional Hilbert space $\mathcal{H}^{\sigma j}$ by following our method of construction. We will call them "sigma-spin" coherent states because of the context, although they are, up to a constant, identical to the coherent states constructed by Perelomov [72] on purely group-theoretical arguments, as will be shown later.

Here, we just pick the orthonormal set $\{{}_\sigma Y^*_{j\mu}(\hat{r}), -j \leq \mu \leq j\}$, in one-to-one correspondence with an orthonormal basis $\{|\sigma j\mu\rangle, -j \leq \mu \leq j\}$ of $\mathcal{H}^{\sigma j}$, and consider the following superposition in $\mathcal{H}^{\sigma j}$:

$$|\hat{r}; \sigma\rangle = |\theta, \phi; \sigma\rangle = \frac{1}{\sqrt{\mathcal{N}(\hat{r})}} \sum_{\mu=-j}^{j} {}_\sigma Y_{j\mu}(\hat{r}) |\sigma j\mu\rangle ; \quad |\hat{r}\rangle \in \mathcal{H}_{\sigma j} , \tag{6.76}$$

with

$$\mathcal{N}(\hat{r}) = \sum_{\mu=-j}^{j} |{}_\sigma Y_{j\mu}(\hat{r})|^2 = \frac{2j+1}{4\pi} .$$

They are the sigma-spin coherent states. In particular, at the north pole of the sphere, they reduce to the state

$$|e_3; \sigma\rangle = |\sigma j\sigma\rangle . \tag{6.77}$$

For $\sigma = j$, they are equal to the spin coherent states, $|\hat{r}; 0\rangle \equiv |\hat{r}\rangle$. But, for a given j and two different $\sigma \neq \sigma'$, the corresponding families are distinct because they live in *different* Hermitian spaces of the same dimension $2j + 1$. This is due to the fact that the map between the two orthonormal sets is not unitary, since we should deal with expansions such as

$$ {}_\sigma Y_{j\mu} = \sum_{j'\mu'} \mathcal{M}_{j'\mu', j\mu}(\sigma', \sigma)\, {}_{\sigma'} Y_{j'\mu'} \,, \tag{6.78}$$

where

$$ \mathcal{M}_{j'\mu', j\mu}(\sigma', \sigma) = \int_{S^2} \overline{{}_{\sigma'} Y_{j'\mu'}(\hat{r})}\, {}_\sigma Y_{j\mu}(\hat{r})\, \mu(d\hat{r}) = [j' j \sigma' \sigma \mu]\, \delta_{\mu\mu'} \,, \tag{6.79}$$

the (nontrivial!) coefficient $[j' j \sigma' \sigma \mu]$ to be determined and forcing the sum to run on values of j' different from j.

The sigma-spin coherent states, by construction, are normalized and solve the identity in $\mathcal{H}^{\sigma j}$.

$$ \langle \hat{r}; \sigma | \hat{r}; \sigma \rangle = 1 \,, \quad \frac{2j+1}{4\pi} \int_{S^2} d\hat{r}\, |\hat{r}; \sigma\rangle\langle \hat{r}; \sigma| = I_d \,. \tag{6.80}$$

From the probabilistic point of view at the basis of the construction, the discrete distribution $\mu \mapsto \frac{4\pi}{2j+1} |{}_\sigma Y_{j\mu}(\hat{r})|^2$ has an interesting meaning in terms of the transition probability for a quantum spin interacting with a transient magnetic field. This point will be developed in the next chapter.

Their overlap $\langle \hat{r}; \sigma | \hat{r}'; \sigma \rangle$ is given, up to a phase, in terms of Jacobi polynomials and the dot product $\hat{r} \cdot \hat{r}'$:

$$ \langle \hat{r}; \sigma | \hat{r}'; \sigma \rangle = e^{-2i\sigma\psi} \left(\frac{1 + \hat{r} \cdot \hat{r}'}{2} \right)^{\sigma} P^{(0,2\sigma)}_{j-\sigma}(\hat{r} \cdot \hat{r}') \,, \tag{6.81}$$

with

$$ \tan \psi = -\frac{\sin(\phi - \phi')}{\cos(\phi - \phi') + \tan\frac{\theta}{2} \tan\frac{\theta'}{2}} \,.$$

The proof is based on the definition of the spin spherical harmonics (6.55) as particular cases of representation matrix elements of $SU(2)$, the group composition rule for the latter, and the expression (6.52)

$$ \begin{aligned} \langle \hat{r}; \sigma | \hat{r}'; \sigma \rangle &= \frac{4\pi}{2j+1} \sum_{\mu=-j}^{j} \overline{{}_\sigma Y_{j\mu}(\hat{r}')}\, {}_\sigma Y_{j\mu}(\hat{r}) \\ &= \sum_{\mu=-j}^{j} D^j_{\mu\sigma}(\xi(\mathcal{R}_{\hat{r}'}))\, D^j_{\sigma\mu}\left(\xi^\dagger(\mathcal{R}_{\hat{r}})\right) = D^j_{\sigma\sigma}(\xi(\mathcal{R}_{\hat{r}}\mathcal{R}_{\hat{r}'})) \,. \end{aligned}$$

It is then necessary to identify the angular parameters of the matrix elements of $\mathcal{R}_{\hat{r}}\mathcal{R}_{\hat{r}'} \in SU(2)$ by using (6.59) and (6.45):

$$ \mathcal{R}_{\hat{r}}\mathcal{R}_{\hat{r}'} \equiv \begin{pmatrix} \cos\omega\, e^{i\psi_1} & i\sin\omega\, e^{i\psi_2} \\ i\sin\omega\, e^{-i\psi_2} & \cos\omega\, e^{-i\psi_1} \end{pmatrix} .$$

We find that $\psi_1 = \psi$ and $\cos 2\omega = \hat{r} \cdot \hat{r}'$.

6.8.9
Covariance Properties of Sigma-Spin Coherent States

The definition of the rotation operator $\mathcal{D}^{\sigma j}(\mathcal{R})$ was given in (6.67). Starting from a coherent state $|\hat{r};\sigma\rangle$, let us consider the coherent state with rotated parameter $\mathcal{R}\cdot\hat{r}$. Owing to the transformation property (6.69), the invariance of $\mathcal{N}(\hat{r})$, and the unitarity of $\mathcal{D}^j$, we find

$$\begin{aligned}|\mathcal{R}\cdot\hat{r};\sigma\rangle &= \frac{1}{\sqrt{\mathcal{N}(\hat{r})}}\sum_{\mu=-j}^{j}\overline{{}_\sigma Y_{j\mu}({}^t\mathcal{R}\cdot\hat{r})}|\sigma j\mu\rangle \\ &= \frac{1}{\sqrt{\mathcal{N}(\hat{r})}}\sum_{\mu,\mu'=-j}^{j}\overline{{}_\sigma Y_{j\mu'}(\hat{r})}\,\overline{D^j_{\mu'\mu}\left(\xi\left(\mathcal{R}^{-1}\right)\right)}|\sigma j\mu\rangle \\ &= \frac{1}{\sqrt{\mathcal{N}(\hat{r})}}\sum_{\mu'=-j}^{j}\overline{{}_\sigma Y_{j\mu'}(\hat{r})}\sum_{\mu=-j}^{j}D^j_{\mu\mu'}\left(\xi\left(\mathcal{R}\right)\right)|\sigma j\mu\rangle \\ &= \mathcal{D}^{\sigma j}(\mathcal{R})|\hat{r};\sigma\rangle\,. \end{aligned} \tag{6.82}$$

Hence, we get the (standard) covariance property of the sigma-spin coherent state:

$$\mathcal{D}^{\sigma j}(\mathcal{R})|\mathcal{R}^{-1}\cdot\hat{r};\sigma\rangle = |\hat{r};\sigma\rangle\,. \tag{6.83}$$

From this relation and (6.77) we derive the $SU(2)$ theoretical content of the sigma-spin coherent states as the element of the orbit of the state $|\sigma j\sigma\rangle$ under the action of the representation $\mathcal{D}^{\sigma j}$ of $SO(3)$ (or rather $SU(2)$ in the case of half-integer j):

$$|\hat{r};\sigma\rangle = \mathcal{D}^{\sigma j}(\mathcal{R}_{\hat{r}})|\sigma j\sigma\rangle\,. \tag{6.84}$$

This equality generalizes (6.25) and can also be viewed as a definition of the sigma-spin coherent states. It is actually the Perelomov genuine definition.

7
Selected Pieces of Applications of Standard and Spin Coherent States

7.1
Introduction

We now proceed with the presentation of a first (small, but instructive) panorama of applications of the standard coherent state and spin coherent state to some problems encountered in physics, quantum physics, statistical physics, and so on. The selected models that are illustrated as examples in the present chapter were chosen because of their high pedagogical and illustrative content.

Coherent States and the Driven Oscillator The driven oscillator is a pedagogical model for presenting the S matrix as a unitary operator transforming an initial state into a final state. By introducing coherent states into the formalism as was done by Carruthers and Nieto [79], we show this operator to be nothing other than the displacement operator $D(z)$. Also, there will appear some interesting discrete probability distributions that generalize the Poisson distribution and for which there exists a limpid physical interpretation.

A Nice Application of Standard or Spin Coherent States in Statistical Physics: Superradiance This second part of this chapter is devoted to a nice example of the application of the coherent state formalism. The object pertains to atomic physics: two-level atoms in resonant interaction with a radiation field (Dicke model and superradiance). The content is based on two selected ancient papers, the first one by Wang and Hioe [80], and the second one, more mathematically oriented, by Hepp and Lieb [81].

Application to Quantum Magnetism Inspired by the Perelomov monograph [10], we explain how the spin coherent states can be used to solve exactly or approximately the Schrödinger equation for certain systems, such as a spin interacting with a variable magnetic field. Again, there will appear discrete probability distributions involving spin spherical harmonics and that generalize the binomial distribution.

Classical and Thermodynamical Limits Coherent states are useful in thermodynamics. Following a paper by Lieb [19], we establish a representation of the partition function for systems of quantum spins in terms of coherent states. After introducing the so-called Berezin–Lieb inequalities, we show how that coherent

Coherent States in Quantum Physics. Jean-Pierre Gazeau

ISBN: 978-3-527-40709-5

state representation makes crossed studies of classical and thermodynamical limits easier.

7.2
Coherent States and the Driven Oscillator

The Newton equation for the classical version of the driven oscillators is given by

$$m\ddot{x} + kx = F(t)\,, \tag{7.1}$$

where $F(t)$ represents the time-dependent driving force. Hence, we have to add to the free oscillator energy $\frac{1}{2m}p^2 + \frac{1}{2}kx^2$ the interaction potential energy $V_{\text{int}}(x,t) = -x\,F(t)$. The Hamiltonian of the quantum version of this model is just obtained by replacing x by $Q = l_c\left(a + a^\dagger\right)$, $l_c = 1/\sqrt{2\hbar m\omega}$ and p by $P = -i\,p_c\left(a - a^\dagger\right)$, $p_c = \sqrt{\hbar m\omega/2}$, while we still consider the driving force as classical:

$$H = \frac{1}{2m}P^2 + \frac{1}{2}kQ^2 - Q\,F(t) = \hbar\omega\left(aa^\dagger + \frac{1}{2}\right) - l_c(a + a^\dagger)\,F(t)\,. \tag{7.2}$$

In a first approach, let us suppose that there exists just a constant driving: $F(t) = F_0$. By shifting the lowering operator to

$$b = a - \frac{l_c F_0}{\hbar\omega}I_d\,, \quad [b, b^\dagger] = I_d\,, \tag{7.3}$$

one gets the Hamiltonian

$$H = \hbar\omega\left(bb^\dagger + \frac{1}{2}\right) - \frac{(l_c F_0)^2}{\hbar\omega}I_d\,. \tag{7.4}$$

The energy levels of the oscillator are, as expected, just translated by $\frac{(l_c F_0)^2}{\hbar\omega}$. The new ground state, say, $|0\rangle_b$, vanishes under the action of b:

$$b\,|0\rangle_b = 0 \Leftrightarrow a\,|0\rangle_b = \frac{l_c F_0}{\hbar\omega}|0\rangle_b\,. \tag{7.5}$$

This means that $|0\rangle_b$ is an element of the family of coherent states for the value of the parameter $\alpha = \frac{l_c F_0}{\hbar\omega}$:

$$|0\rangle_b = \left|\alpha = \frac{l_c F_0}{\hbar\omega}\right\rangle = e^{\frac{l_c F_0}{\hbar\omega}(a^\dagger - a)}\,|0\rangle\,. \tag{7.6}$$

Let us now examine the general situation for which we just suppose the asymptotic vanishing of the driving force,

$$F(\pm\infty) = 0\,, \tag{7.7}$$

a hypothesis that pertains to scattering theory. Otherwise said, the question is to determine the "S matrix" or unitary operator that transforms, from the Heisenberg viewpoint, the "ingoing" state $|\Psi_{\text{in}}\rangle$ into the "outgoing" state $|\Psi_{\text{out}}\rangle$:

$$|\Psi_{\text{in}}\rangle \mapsto |\Psi_{\text{out}}\rangle = S^{\dagger}\,|\Psi_{\text{in}}\rangle\,. \tag{7.8}$$

The method for finding S consists in first solving the classical version of this scattering process, and next translating the solution into the quantum language by using coherent states.

The treatment of the classical scattering described by

$$\ddot{x}(t) + \omega^2 x(t) = F(t)/m \tag{7.9}$$

consists in finding the corresponding Green function G or *elementary solution* within the framework of distribution theory and establishing the solution through a convolution with G. The Green function is a solution to

$$\left(\frac{d^2}{dt^2} + \omega^2\right) G(t) = \omega\,\delta(t)\,, \tag{7.10}$$

and is completely determined by stating initial conditions. Introducing the Heaviside function

$$\mathrm{H}(t) = \begin{cases} 0 & t < 0\,, \\ 1 & t > 0\,, \end{cases} \tag{7.11}$$

the *retarded* Green function $G_R(t) \stackrel{\text{def}}{=} \mathrm{H}(t)\sin\omega t$ vanishes for $t < 0$, whereas the *advanced* Green function $G_A(t) \stackrel{\text{def}}{=} -\mathrm{H}(-t)\sin\omega t$ vanishes for $t > 0$. The solution $x(t)$ of (7.9) is then expressed in terms of initial condition x_{in} or final condition x_{out} defined as

$$x(t) \sim \begin{cases} x_{\text{in}}(t) & t \to -\infty\,, \\ x_{\text{out}}(t) & t \to \infty\,, \end{cases} \tag{7.12}$$

both being solutions to the free oscillator equation

$$\ddot{x}_{\substack{\text{in}\\ \text{out}}} + \omega^2 x_{\substack{\text{in}\\ \text{out}}} = 0\,.$$

So, one obtains

$$x(t) = x_{\text{in}}(t) + \frac{1}{m\omega}\int_{-\infty}^{+\infty} G_R(t-t')\,F(t')\,dt' \tag{7.13}$$

$$= x_{\text{out}}(t) + \frac{1}{m\omega}\int_{-\infty}^{+\infty} G_A(t-t')\,F(t')\,dt'\,. \tag{7.14}$$

By eliminating $x(t)$ from the above equations, we obtain the classical counterpart of the S matrix:

$$x_{\text{out}}(t) = x_{\text{in}}(t) + \frac{1}{m\omega}\int_{-\infty}^{+\infty} G(t-t')\,F(t')\,dt'\,, \tag{7.15}$$

with

$$G(t) = G_A(t) - G_R(t) = \sin\omega t\,, \quad t \neq 0\,. \tag{7.16}$$

The scattering relation (7.15) is easily expressed in terms of the (actually inverse) Fourier transform of $F(t)$ defined here as

$$\hat{F}(\omega) \overset{\text{def}}{=} \int_{-\infty}^{+\infty} e^{i\omega t}\, F(t)\, dt\,.$$

$$\begin{aligned} x_{\text{out}}(t) &= x_{\text{in}}(t) - \left(\frac{1}{2im\omega}\int_{-\infty}^{+\infty} e^{i\omega t'}\, F(t')\, dt'\right) e^{-i\omega t} \\ &\quad + \left(\frac{1}{2im\omega}\int_{-\infty}^{+\infty} e^{-i\omega t'}\, F(t')\, dt'\right) e^{i\omega t} \\ &= x_{\text{in}}(t) + \frac{i\,\hat{F}(\omega)}{2m\omega} e^{-i\omega t} - \frac{\overline{i\,\hat{F}(\omega)}}{2m\omega} e^{i\omega t}\,. \end{aligned} \tag{7.17}$$

By writing the asymptotic free oscillator solutions as

$$x_{\text{in}}(t) = \frac{1}{2}\left(Ae^{-i\omega t} + \bar{A}e^{i\omega t}\right)\,, \quad x_{\text{out}}(t) = \frac{1}{2}\left(Be^{-i\omega t} + \bar{B}e^{i\omega t}\right)\,, \tag{7.18}$$

one gets from (7.17)

$$B = A + i\frac{\hat{F}(\omega)}{m\omega}\,. \tag{7.19}$$

Let us now transpose this into the quantum side through the correspondences

$$\tfrac{1}{2}A \mapsto l_c a\,, \quad \tfrac{1}{2}B \mapsto l_c b \tag{7.20}$$

so that the quantum counterpart of (7.18) reads as

$$Q_{\text{in}}(t) = l_c\left(a\,e^{-i\omega t} + a^\dagger\, e^{i\omega t}\right)\,, \quad Q_{\text{out}}(t) = l_c\left(b\,e^{-i\omega t} + b^\dagger\, e^{i\omega t}\right)\,, \tag{7.21}$$

with the relation between a and b being

$$a \overset{S}{\mapsto} b = a + i\frac{\hat{F}(\omega)}{\sqrt{2m\omega\hbar}} I_d \overset{\text{def}}{=} S^\dagger a S\,. \tag{7.22}$$

Hence, the effect of the interaction is to translate the ingoing normal mode a by an amount equal (up to a factor) to the Fourier component of $F(t)$ corresponding to the proper frequency of the free oscillator. The unitary operator that executes this task is precisely the displacement operator $D(z)$. Indeed, the following property holds true.

$$[a, D(\alpha)] = \alpha\, D(\alpha) \Leftrightarrow D^\dagger(\alpha) a D(\alpha) = a + \alpha\, I_d\,. \tag{7.23}$$

To prove (7.23), it is enough to check it on a coherent state, say, $|\beta\rangle$ with arbitrary β. We have

$$aD(\alpha)|\beta\rangle = e^{i\Im(\alpha\bar{\beta})} a|\alpha+\beta\rangle = (\alpha+\beta)e^{i\Im(\alpha\bar{\beta})}|\alpha+\beta\rangle = (\alpha+\beta)D(\alpha)|\beta\rangle\,,$$

whereas

$$D(\alpha)a|\beta\rangle = \beta D(\alpha)|\beta\rangle ,$$

and so $aD(\alpha) - D(\alpha)a = \alpha D(\alpha)$.

It follows that $b = S^\dagger a S$, where

$$S = D(\alpha) , \quad \text{with} \quad \alpha = i\frac{\hat{F}(\omega)}{\sqrt{2m\hbar\omega}} . \tag{7.24}$$

An alternative form of this S matrix is given by

$$\begin{aligned} S &= D(\alpha) = e^{(\alpha a^\dagger - \bar{\alpha} a)} \\ &= \exp\frac{il_c}{\hbar}\left(\hat{F}(\omega)a^\dagger + \overline{\hat{F}(\omega)a}\right) = \exp\frac{i}{\hbar}\int_{-\infty}^{+\infty} Q_{\text{in}}(t)\, F(t)\, dt . \end{aligned} \tag{7.25}$$

To understand better the physical significance of this S matrix, let us examine its matrix elements and the resulting transition probabilities when ingoing and outgoing states are energy eigenstates of the oscillator:

$$|\Psi_{\text{in}}\rangle \equiv |\Psi_{\text{in},m}\rangle , \quad |\Psi_{\text{out}}\rangle \equiv |\Psi_{\text{out},n}\rangle .$$

Let us expand the ingoing eigenstate in terms of the complete orthonormal set of outgoing eigenstates:

$$\begin{aligned} |\Psi_{\text{in},m}\rangle &= \sum_{n=0}^{+\infty} \langle\Psi_{\text{out},n}|\Psi_{\text{in},m}\rangle\, |\Psi_{\text{out},n}\rangle \\ &= \sum_{n=0}^{+\infty} \underbrace{\langle\Psi_{\text{out},n}|S|\Psi_{\text{out},m}\rangle}_{S_{nm}} |\Psi_{\text{out},n}\rangle . \end{aligned} \tag{7.26}$$

The matrix element S_{nm}, which is also equal to $\langle\Psi_{\text{in},n}|S|\Psi_{\text{in},m}\rangle$, is the matrix element of the displacement operator in the Fock basis:

$$S_{nm} = S_{nm}(\alpha) = \langle n|D(\alpha)|m\rangle , \quad \alpha = i\frac{\hat{F}(\omega)}{\sqrt{2m\hbar\omega}} . \tag{7.27}$$

The calculation of this matrix element can be carried out by using the resolution, by coherent states, of the identity and the subsequent reproducing property (3.17) of the kernel:

$$\begin{aligned} \langle n|D(\alpha)|m\rangle &= \iint_{\mathbb{C}^2} \frac{d^2z}{\pi}\frac{d^2z'}{\pi} \langle n|z\rangle\langle z|D(\alpha)|z'\rangle\langle z'|m\rangle \\ &= \frac{1}{\sqrt{n!m!}} \iint_{\mathbb{C}^2} \frac{d^2z}{\pi}\frac{d^2z'}{\pi}\, e^{-\frac{|z|^2}{2}}\, e^{-\frac{|z'|^2}{2}}\, z^n\, \bar{z}'^m\, e^{i\Im(\alpha\bar{z}')}\, \langle z|z' + \alpha\rangle \\ &= \frac{e^{-\frac{|\alpha|^2}{2}}}{\sqrt{n!m!}} \int_{\mathbb{C}} \frac{d^2z'}{\pi}\, e^{-|z'|^2}\, e^{-\bar{\alpha}z'}\, \bar{z}'^m\, (z' + \alpha)^n . \end{aligned}$$

After binomial and exponential expansions and integration, one ends up with the following expression:

$$S_{nm}(\alpha) = \sqrt{\frac{m!}{n!}}\, e^{-\frac{|\alpha|^2}{2}}\, \alpha^{n-m}\, L_m^{(n-m)}\left(|\alpha|^2\right) \quad \text{for} \quad m \le n\,,$$
$$= \sqrt{\frac{n!}{m!}}\, e^{-\frac{|\alpha|^2}{2}}\, (-\bar{\alpha})^{m-n}\, L_n^{(m-n)}\left(|\alpha|^2\right) \quad \text{for} \quad m > n\,, \tag{7.28}$$

where

$$L_n^{(\mu)}(x) = \sum_{k=0}^{n} (-1)^k \frac{\Gamma(n+\mu+1)}{\Gamma(\mu+k+1)(n-k)!} \frac{x^k}{k!} \tag{7.29}$$

is a generalized Laguerre polynomial.

Hence, the transition probability, say, in the case $m \to n \ge m$, is given by

$$P_{nm}(|\alpha|^2) = |S_{nm}(\alpha)|^2 = \frac{m!}{n!} e^{-|\alpha|^2} |\alpha|^{2(n-m)} \left(L_m^{(n-m)}\left(|\alpha|^2\right)\right)^2 . \tag{7.30}$$

This is a generalization of the discrete Poisson distribution in the sense that one recovers the latter for $m = 0$:

$$P_{n0}(|\alpha|^2) = \frac{|\alpha|^2}{n!} e^{-|\alpha|^2} .$$

One notices that, in the spirit of Chapter 5, the interpretation of the matrix elements (7.27) as "square roots" of a discrete probability distribution sheds interesting light on the following family of normalized coherent states that emerges from the unitary transport by $D(z)$ of the excited state $|m\rangle$:

$$|z; m\rangle \stackrel{\text{def}}{=} D(z)|m\rangle = \sum_{n=0}^{\infty} S_{nm}(z)\, |n\rangle\,, \tag{7.31}$$

with a normalization factor equal to 1 because of the unitarity of the matrix S.

As is well known, the Poisson distribution, as a function of its parameter, here $|\alpha|^2 = \frac{|\hat{F}(\omega)|^2}{2m\hbar\omega}$, reaches its maximum at $|\alpha|^2 = n$. Inverting this yields the most probable transition oscillator level $\nu = \frac{|\hat{F}(\omega)|^2}{2m\hbar\omega}$ for a given driving force $F(t)$. Otherwise said, the most probable energy transferred to the oscillator is $\Delta E = \nu\hbar\omega = \frac{|\hat{F}(\omega)|^2}{2m}$, and this coincides with the mean transfer of energy, namely,

$$\sum_{n=0}^{+\infty} P_{n0} n\hbar\omega = \nu\hbar\omega\,. \tag{7.32}$$

It should be noted that the most probable energy transferred to the oscillator would be the same whatever the initial state $|m\rangle$. Indeed, for all Ψ_{in}, and with $H_{\text{in}} = \hbar\omega(a^\dagger a + 1/2)$, $H_{\text{out}} = \hbar\omega(b^\dagger b + 1/2)$, $b = a + \alpha$,

$$\Delta E = \langle \Psi_{\text{in}} | H_{\text{out}} - H_{\text{in}} | \Psi_{\text{in}} \rangle$$
$$= \hbar\omega \left[|\alpha|^2 + i(\langle \Psi_{\text{in}} | a^\dagger | \Psi_{\text{in}} \rangle \alpha - \text{c.c.}) \right] , \tag{7.33}$$

and the factor of i vanishes for $|\Psi_{\text{in}}\rangle = |m\rangle$.

7.3
An Application of Standard or Spin Coherent States in Statistical Physics: Superradiance

7.3.1
The Dicke Model

In quantum optics, superradiance [85] is a phenomenon of collective emission of an ensemble of excited atoms or ions, first considered by Dicke [86–89]. It is similar to superfluorescence, but it starts with the coherent excitation of the ensemble, usually with an optical pulse. This coherence (i.e., a well-defined phase relationship between the excitation amplitudes of lower and upper electronic states) leads to a macroscopic dipole moment. The maximum intensity of the emitted light scales with the square of the number of atoms, because each atom contributes a certain amount to the emission amplitude, and the intensity is proportional to the square of the amplitude.

In the Dicke model [86] for the interaction between matter and radiation, one considers a system of N two-level atoms interacting with a radiation field. The atoms have fixed positions in a one-dimensional box of length L, and they are supposed to be far enough from each other so that their mutual interaction is negligible. Nevertheless, the assembly has to be thought of as a single quantum entity, because of its collective interaction with the radiation field. Hepp and Lieb [81] have obtained some exact results for the thermodynamical properties of this system in the limit $N \rightarrow +\infty,\ L \rightarrow +\infty,\ N/L = \text{const}$. In particular, they have shown the existence of a second-order phase transition, radiance $\rightarrow$ superradiance, at a certain critical temperature T_c, when the atom-field coupling is strong enough. This coupling is supposed to hold in the so-called dipolar approximation. On the other hand, the box of size L is supposed to be sufficiently small compared with the radiation wavelength so that all the atoms experience the same field. The latter is understood as collective with the hypothesis $L \ll \lambda_{\text{ray}}$. In the *rotating-wave approximation*, for which more details will be given below, the Hamiltonian of the system reads as (in units $\hbar = c = 1$)

$$H = H_0 + H_I\,, \tag{7.34}$$

where H_0 and H_I are the free and interaction Hamiltonians, respectively:

$$H_0 = H_{rad} + H_{at} = \sum_k \nu_k\, a_k^\dagger\, a_k + \frac{1}{2}\,\omega \sum_{j=1}^{N} \sigma_j^z\,, \tag{7.35}$$

$$H_I = \frac{1}{2\sqrt{L}} \left[\left(\sum_k \lambda'_k\, a_k \right) \left(\sum_{j=1}^{N} \sigma_j^+ \right) + \left(\sum_k \lambda'_k\, a_k^\dagger \right) \left(\sum_{j=1}^{N} \sigma_j^- \right) \right]. \tag{7.36}$$

Here, $a_k^\dagger$ and a_k are the creation and annihilation operators, respectively, for the kth radiation mode with frequency ν_k, ω is the two-level energy shift, and λ' is the atom-field coupling strength. In the dipolar approximation, the Hamiltonian H_I is supposed to be linear in $a_k^\dagger$ and a_k. Concerning the two-level atoms, they are described by Pauli matrices:

$$\sigma_j^{\pm} = \sigma_j^x \pm i\sigma_j^y ,$$
$$\sigma_j^x = \begin{pmatrix} 0 & 1 \\ 1 & 0 \end{pmatrix}_j , \quad \sigma_j^y = \begin{pmatrix} 0 & -i \\ i & 0 \end{pmatrix}_j , \quad \sigma_j^z = \begin{pmatrix} 1 & 0 \\ 0 & -1 \end{pmatrix}_j .$$

7.3.1.1
Rotating-Wave Approximation

The rotating-wave approximation consists in neglecting the terms $a_k\sigma_j^-$ (energy loss $\approx \hbar(\nu_k + \omega_j)$) and $a_k^\dagger\sigma_j^+$ (energy gain). Let us present here the elementary example of a two-level atom subjected to an oscillating electric field $E = \mathcal{E} \cos\nu t$. On a quasi-classical level [25], the time evolution of this system is described by the Schrödinger equation:

$$|\dot{\Psi}(t)\rangle = -i\,(H_0 + H_I)|\Psi(t)\rangle , \quad \text{with } |\Psi(t)\rangle = C_+(t)|+\rangle + C_-(t)|-\rangle , \tag{7.37}$$

where $H_0 = \frac{1}{2}\hbar\omega\sigma^z$, $H_I = -e\,Q\,\mathcal{E}\cos\nu t$, where Q is the position operator. This results in the differential system:

$$\begin{aligned} \dot{C}_+ &= -\tfrac{i}{2}\,\omega\, C_+ + I\,\Omega_R e^{-i\phi}\cos\nu t\; C_- , \\ \dot{C}_- &= \tfrac{i}{2}\,\omega\, C_- + I\,\Omega_R e^{i\phi}\cos\nu t\; C_+ , \end{aligned} \tag{7.38}$$

where ϕ is the argument of the complex $\langle -|Q|+\rangle$, $\langle -|Q|+\rangle = |\langle -|Q|+\rangle|e^{i\phi}$, and $\Omega_R \stackrel{\text{def}}{=} |\langle +|Q|-\rangle|\,\mathcal{E}/\hbar$ (Rabi frequency). Putting $c_\pm = C_\pm\, e^{\pm i\frac{\omega}{2}t}$, we get for the time evolution of the latter

$$\dot{c}_+ \simeq i\,\frac{\Omega_R}{2}\,e^{-i\phi}\,c_-\,e^{i(\omega-\nu)t} , \quad \dot{c}_- \simeq i\,\frac{\Omega_R}{2}\,e^{i\phi}\,c_+\,e^{-i(\omega-\nu)t} , \tag{7.39}$$

where terms proportional to $e^{\pm i(\omega+\nu)t}$ have been *neglected*.

7.3.1.2
Dicke Hamiltonian with a Single Mode

In the Dicke model dealing with a single mode of frequency ν only, the corresponding Hamiltonian reads as

$$H = a^\dagger a + \sum_{j=1}^{N}\left[\frac{1}{2}\,\varepsilon\,\sigma_j^z + \frac{\lambda}{2\sqrt{N}}\left(a\sigma_j^+ + a^\dagger\sigma_j^-\right)\right] , \tag{7.40}$$

where we have introduced the modified parameters

$$\varepsilon = \omega/\nu, \quad \lambda = \lambda'\sqrt{\varrho}/\nu, \quad \varrho = N/L,$$

which are better suited to thermodynamical considerations.

7.3.2 The Partition Function

The thermodynamical properties of the system are encoded by the partition function

$$Z(N,T) = \mathrm{Tr}\, e^{-\beta H}, \quad \beta = \frac{1}{k_B T}, \tag{7.41}$$

where k_B is the Boltzmann constant and T is the absolute temperature. It is precisely at this point in the explicit computation of the partition function that the standard coherent states $|z\rangle$ of the single-mode field fully play their simplifying role [80]. We have for a trace class operator A,

$$\mathrm{Tr}\, A = \sum_{n\geq 0} \langle n|A|n\rangle = \int \frac{d^2 z}{\pi} \langle z|A|z\rangle . \tag{7.42}$$

Thus, the partition function can be written as

$$Z(N,T) = \sum_{s_1=\pm 1} \cdots \sum_{s_N=\pm 1} \int \frac{d^2 z}{\pi} \langle s_1 \ldots s_N| \langle z|e^{-\beta H}|z\rangle |s_1 \ldots s_N\rangle , \tag{7.43}$$

where the sums are taken over all possible atomic sites (for simplicity, we have avoided a tensor-product notation). Therefore, it is necessary to estimate the partial matrix element

$$\langle z|e^{-\beta H}|z\rangle = \sum_{r=0}^{\infty} \frac{(-\beta)^r}{r!} \langle z|H^r|z\rangle , \tag{7.44}$$

bearing in mind the basic coherent state property $a|z\rangle = z|z\rangle$. To take into account the constancy of the ratio $\varrho = N/L$, let us rescale the mode operators as $b = \frac{a}{\sqrt{N}}, \quad b^\dagger = \frac{a^\dagger}{\sqrt{N}}$, so that

$$b b^\dagger = b^\dagger b + \frac{1}{N} . \tag{7.45}$$

Then the terms in the expansion of H^r, using normal ordering, are shown to be

$$\langle z|H^r|z\rangle = \left(\bar{z} z + \sum_{j=1}^{N} h_j \right)^r + O\left(\frac{1}{N}\right) , \tag{7.46}$$

where the h_j's are the individual atomic Hamiltonians

$$h_j = \frac{1}{2} \varepsilon \sigma_j^z + \frac{\lambda}{2\sqrt{N}} (z\sigma_j^+ + \bar{z}\sigma_j^-) .$$

Hence, we get the estimate

$$\langle z|e^{-\beta H}|z\rangle = e^{-\beta \bar{z} z} e^{-\beta \sum_{j=1}^{N} h_j} + O\left(\frac{1}{N}\right) , \tag{7.47}$$

and, for the partition function,

$$Z(N,T) = \int_{\mathbb{C}} \frac{d^2z}{\pi} e^{-\beta|z|^2} \sum_{s_1=\pm 1} \cdots \sum_{s_N=\pm 1} \left(\prod_{j=1}^{N} \langle s_j | e^{-\beta h_j} | s_j \rangle \right) + O\left(\frac{1}{N}\right)$$
$$= \int_{\mathbb{C}} \frac{d^2z}{\pi} e^{-\beta|z|^2} (\mathrm{Tr}\, e^{-\beta h})^N + O\left(\frac{1}{N}\right) . \tag{7.48}$$

The generic atomic Hamiltonian, $h = \frac{1}{2}\varepsilon\sigma^z + \frac{\lambda}{2\sqrt{N}}(z\sigma^+ + \bar{z}\sigma^-)$, has the eigenvalues

$$\pm\frac{1}{2}\varepsilon\,(1 + 4\lambda^2|z|^2/\varepsilon^2 N)^{1/2} .$$

Using this, we get, upon performing the angular integration and making use of standard asymptotic methods (Laplace or steepest descent [82]),

$$Z(N,T) = \int_{\mathbb{C}} \frac{d^2z}{\pi} e^{-\beta|z|^2} \left(2\cosh\frac{1}{2}\beta\varepsilon \left(1 + 4\lambda^2 \frac{|z|^2}{\varepsilon^2 N}\right)^{1/2} \right)^N + O\left(\frac{1}{N}\right)$$
$$\approx \text{const.}\,\sqrt{N} \max_{0\le |z|^2/N \le \infty} \exp N\varphi\left(\frac{|z|^2}{N}\right) , \quad \text{for large } N , \tag{7.49}$$

where $\varphi(\gamma) = -\beta\gamma + \ln\left(2\cosh\frac{1}{2}\beta\varepsilon\left(1 + 4\frac{\lambda^2}{\varepsilon^2}\gamma\right)^{1/2}\right)$.

7.3.3
The Critical Temperature

Let us now obtain the value of γ that maximizes $\varphi(\gamma)$, that is, for which

$$\varphi'(\gamma) = \beta\left(-1 + \frac{\lambda^2}{\varepsilon\eta}\tanh\frac{\beta\varepsilon\eta}{2}\right) = 0 ,$$

where we introduce the intermediate variable

$$\eta = \sqrt{\left(1 + \frac{4\lambda^2}{\varepsilon^2}\gamma\right)} \ge 1 .$$

It is necessary to examine the equation $\frac{\varepsilon}{\lambda^2}\eta = \tanh\frac{\beta\varepsilon}{2}\eta$. This is done through the graphical study shown in Figure 7.1.

Let us comment on the appearance of three different regimes.

7.3.3.1
Weak Coupling

If $\varepsilon/\lambda^2 > 1$, that is, $\lambda^2 < \varepsilon$ (weak coupling), the linear function $\varepsilon/\lambda^2\,\eta$ is already strictly above 1 at $\eta = 1$. Thus, $e^{\varphi(\gamma)}$ is monotone decreasing and its maximum

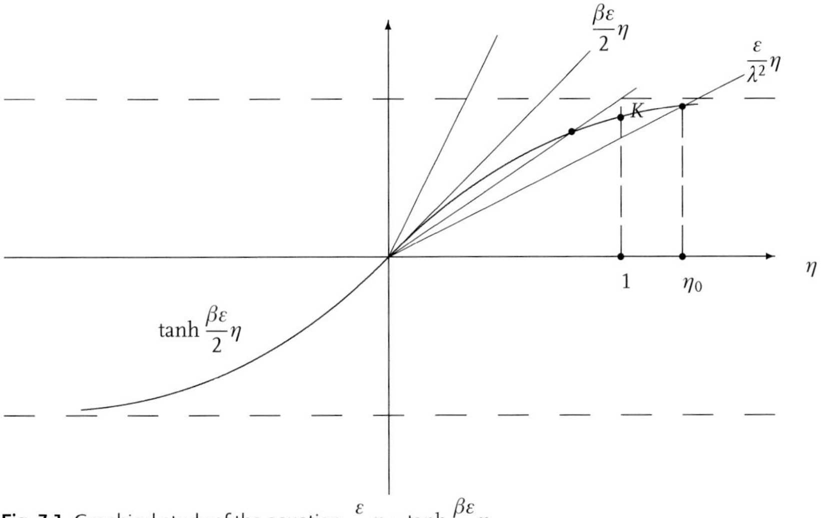

Fig. 7.1 Graphical study of the equation $\frac{\varepsilon}{\lambda^2}\eta = \tanh\frac{\beta\varepsilon}{2}\eta$.

holds at $y_0 = 0$. So

$$Z(N,T) = \text{const.}\,\sqrt{N}\,\left(2\cosh\tfrac{1}{2}\beta\varepsilon\right)^N\,, \tag{7.50}$$

and the free energy $f(T)$ per atom is given by

$$\begin{aligned}-\beta\, f(T) &\overset{\text{def}}{=} \lim_{N\to\infty}\frac{1}{N}\ \ln Z(N,T)\\ &= \lim_{N\to\infty}\frac{1}{N}\left(N\ \ln(2\cosh\tfrac{1}{2}\beta\varepsilon) + \ln\frac{\text{const.}}{\sqrt{N}}\right) = \ln(2\cosh\tfrac{1}{2}\beta\varepsilon)\,.\end{aligned} \tag{7.51}$$

This value would also be obtained with a free Hamiltonian

$$H = a^\dagger\, a + \sum_{j=1}^{N}\frac{1}{2}\,\varepsilon\,\sigma_j^z\,.$$

7.3.3.2
Intermediate Coupling

If $\lambda^2 > \varepsilon$ (intermediate or strong coupling), the solution of $\varphi'(y) = 0$ now depends on the value of β. If the linear function $\varepsilon/\lambda^2\ \eta$ is above the point K, this means that

$$\varepsilon/\lambda^2 > \tanh\frac{\beta\varepsilon}{2}\,,$$

that is, $\beta < \beta_c$, where β_c is determined by the equation $\varepsilon/\lambda^2 = \tanh\frac{\beta\varepsilon}{2}$. So, for $\beta \le \beta_c$, the free energy per atom is still given by

$$-\beta\, f(T) = \ln(2\cosh\tfrac{1}{2}\beta\varepsilon)\,.$$

7.3.3.3
Strong Coupling

If $\beta > \beta_c$, there exists a solution $1 < \eta_0 < \infty$ such that

$$2\sigma \stackrel{\text{def}}{=} \frac{\varepsilon}{\lambda^2}\eta_0 = \tanh\frac{\beta\varepsilon}{2}\eta_0$$

and so for y,

$$y = y_0 = \lambda^2\sigma^2 - \frac{\varepsilon^2}{4\lambda^2}\,.$$

Hence, in the case $\beta > \beta_c$ the free energy per atom is now different and is given by

$$-\beta f(T) = \ln(2\cosh\tfrac{1}{2}\beta\varepsilon) - \beta\lambda^2\sigma^2 + \beta\frac{\varepsilon^2}{4\lambda^2}\,, \tag{7.52}$$

with $2\sigma = \tanh\beta\lambda^2\sigma \neq 0$.

7.3.3.4
Summary

In summary, there does or does not exist a critical temperature. We have obtained the value of y that maximizes $\varphi(y)$, according to whether the coupling λ is strong ($\lambda > \sqrt{\varepsilon}$) or weak ($\lambda < \sqrt{\varepsilon}$). In the former case, one obtains a critical temperature T_c given by

$$\varepsilon/\lambda^2 = \tanh\frac{\varepsilon}{2k_B T_c}\,, \tag{7.53}$$

for which the system jumps from a "normal" state at $T > T_c$ to a superradiative state at $T < T_c$, whereas there is no such phase transition for weak coupling. As with every phase transition, a physical quantity presents a discontinuity at $T > T_c$, namely, in the present case, the specific heat.

7.3.4
Average Number of Photons per Atom

The physical difference between weak, intermediate, and strong regimes is better understood through the following expressions, obtained by using similar coherent state methods, for the average number of photons per atom:

$$\left\langle \left(\frac{a^\dagger a}{N}\right)^r \right\rangle = \delta_{r0}\,, \tag{7.54}$$

for $\lambda^2 > \varepsilon$ and all β, or for $\lambda^2 > 0$, $\beta < \beta_c$, while

$$\left\langle \left(\frac{a^\dagger a}{N}\right)^r \right\rangle = (\lambda^2\sigma^2 - \varepsilon^2/4\lambda)^r \tag{7.55}$$

for $\lambda^2 > \varepsilon$ and $\beta > \beta_c$, where σ is such that

$$2\sigma = \tanh \beta\lambda^2\sigma \neq 0 . \tag{7.56}$$

These expressions clarify the terminology. In the “normal” radiant state, where

$$\left\langle \left(\frac{a^\dagger a}{N} \right)^r \right\rangle = \delta_{r0}$$

holds, the number of photons emitted goes to zero as $N \to \infty$.

This is not so in the *superradiant* regime,

$$\left\langle \left(\frac{a^\dagger a}{N} \right)^r \right\rangle = (\lambda^2\sigma^2 - \varepsilon^2/4\lambda)^r , \tag{7.57}$$

where an infinite number of photons are emitted, as a consequence of the coherence of the maser light, a truly collective effect: as the number of photons rises in a kind of chain reaction, Dicke described this phenomenon as an “optical bomb.”

7.3.5
Comments

First of all, a generalization to multimode fields is straightforward, the same method based on the use of the standard coherent state again being useful for carrying out explicit computations.

Second, more general results have been given by Hepp and Lieb [81]. These authors have made mathematically rigorous the thermodynamical limit $N \to \infty$ carried out previously by Wang and Hioe. They have generalized the results to multilevel atoms and to the case of an infinite number of modes in finding the appropriate estimate. They have also extended the results to Hamiltonians (in the case of finite multimode) that have translation degrees of freedom for atoms and without the rotating-wave approximation. Finally, they have made use of the spin coherent states for multilevel atoms and standard coherent states for the multimode electromagnetic field.

7.4
Application of Spin Coherent States to Quantum Magnetism

In this third example of applications of coherent states, we consider a quantum spin in a variable magnetic field [10]. Let a particle of spin j (or simply a “spin”), with magnetic moment μ, be subjected to a variable magnetic field $\mathbf{H}(t)$. The time evolution of states is ruled by the Schrödinger equation:

$$i\frac{d}{dt}|\Psi(t)\rangle = -\boldsymbol{\mathcal{M}} \cdot \mathbf{J}\,|\Psi(t)\rangle , \quad \boldsymbol{\mathcal{M}} \overset{\text{def}}{=} \frac{\mu}{j}\mathbf{H} . \tag{7.58}$$

With $\mathcal{M}_\parallel \overset{\text{def}}{=} \frac{i}{2}(\mathcal{M}_x - i\mathcal{M}_y)$, one writes

$$i\frac{d}{dt}|\Psi(t)\rangle = i\left(\mathcal{M}_\parallel J_+ - \overline{\mathcal{M}_\parallel} J_- + i\mathcal{M}_z J_z\right)|\Psi(t)\rangle . \tag{7.59}$$

Furthermore, let us assume that the field $\mathbf{H}(t)$ tends to a definite limit in $t = \infty$ so that asymptotic states $\left|\Psi_{\substack{\text{in}\\ \text{out}}}\right\rangle$ exist.

We now introduce in this model the spin coherent states in complex parameterization, of which we recall the expression

$$|\zeta\rangle = \sum_{m=-j}^{j}\sqrt{\binom{2j}{j+m}}\frac{\zeta^{j-m}}{(1+|\zeta|^2)^j}|j,m\rangle . \tag{7.60}$$

They will be precisely used to build a solution of the Schrödinger solution. Let us put as an ansatz

$$|\Psi(t)\rangle = e^{-i\phi(t)}|\zeta(t)\rangle , \tag{7.61}$$

where the functions $\phi(t)$ and $|\zeta(t)\rangle$ remain to be determined.

Spin coherent states are *not* eigenstates of any lowering operator. We instead have

$$J_-|\zeta\rangle = \zeta^{-1}(j - J_z)|\zeta\rangle , \quad J_+|\zeta\rangle = \zeta(j + J_z)|\zeta\rangle . \tag{7.62}$$

On the other hand,

$$\frac{d}{dt}|\zeta(t)\rangle = \left[-\frac{j}{1+|\zeta|^2}\frac{d}{dt}(1+|\zeta|^2) + \frac{\dot{\zeta}}{\zeta}(j - J_z)\right]|\zeta(t)\rangle , \tag{7.63}$$

an identity that holds for any spin coherent state.

Now, behind (7.58) and the ansatz (7.61) lies a classical dynamical system obtained as follows. By identifying to zero the coefficients of the two operators involved, namely, the identity and J_z, one gets a differential system describing the time evolution of this system in the set of parameters ϕ and ζ:

$$\begin{aligned} i\dot{\phi} &= j\left(\overline{\mathcal{M}_\parallel}\,\bar{\zeta} - \mathcal{M}_\parallel\,\zeta - i\,\mathcal{M}_z\right) \\ \dot{\zeta} &= -\overline{\mathcal{M}_\parallel} - i\,\mathcal{M}_z\,\zeta - \mathcal{M}_\parallel\,\zeta^2 . \end{aligned} \tag{7.64}$$

Going back to the parameterization sphere S^2, the above equation is to be interpreted as the *coherent state quantization* of a precession motion for a magnetic moment $\boldsymbol{\mu} = (\mu/j)\,\mathbf{J}^{\text{Cl}}$ subjected to the field $\mathbf{H}$. Such a precession is described by the equation

$$\dot{\hat{n}} = \mu\mathbf{H}\times\hat{n} , \quad \text{where} \quad \hat{n} = \frac{\mathbf{J}^{\text{Cl}}}{j} . \tag{7.65}$$

At a first view of the system, if the interaction is such that $\mathcal{M}_\parallel(t) \to 0$ and $\mathcal{M}_z(t) \to$ const. in the limit $t \to \infty$, then the coherent state parameter localizes on a circle in the complex plane $|\zeta| \to$ const.. Indeed, $\dot{\zeta} \simeq -i\,\mathcal{M}_z\,\zeta$ and so $\zeta \simeq \zeta_0\, e^{-i\mathcal{M}_z t}$.

For the treatment of this "classical-like" dynamical system on the sphere, we follow a procedure similar to that presented in Section 7.2 for the driven oscillator. One easily derives the transition probability $P^{(j)}_{jm}$ for an extremal state as an initial state:

$$|\Psi_{\text{in}}\rangle = |j, j\rangle \longrightarrow |\Psi_{\text{out}}\rangle = |j, m\rangle \,, \tag{7.66}$$

$$P^{(j)}_{jm} = \binom{2j}{j+m}\left(\cos^2\frac{\theta}{2}\right)^{j+m}\left(\sin^2\frac{\theta}{2}\right)^{j-m} = \binom{2j}{j+m}\frac{|\xi^{j-m}|^2}{(1+|\xi|^2)^{2j}} \,. \tag{7.67}$$

We recognize here the binomial distribution discussed in Section 6.4 with the construction of coherent spin states. Actually we get more, since, for the general transition

$$|\Psi_{\text{in}}\rangle = |j, m'\rangle \longrightarrow |\Psi_{\text{out}}\rangle = |j, m\rangle \,,$$

the corresponding probability $P^{(j)}_{m'm}$ is precisely given by the squared modulus of a spin spherical harmonic:

$$P^{j}_{m'm} = \frac{4\pi}{2j+1}\left|{}_{m'}Y_{j,j-m}(\hat{r})\right|^2 = \left|D^{j}_{j-m,m'}\left(\xi\left(\mathcal{R}_{\hat{r}}\right)\right)\right|^2 \,. \tag{7.68}$$

Hence, we find in this elementary model a nice and deep physical interpretation of the discrete probability distribution appearing in the structure of general spin coherent states.

Let us end this section by presenting an elementary example of the above model. For a field of the form

$$\mathbf{H}(t) = A\,\hat{k} + B\,(\cos\omega t\,\hat{\imath} + \sin\omega t\,\hat{\jmath}) \,, \tag{7.69}$$

we obtain

$$\xi(t) = \frac{-\omega_\perp \,\sin\Omega t\; e^{i(\omega-\omega_\parallel)t}}{2\Omega\cos\Omega t + i(\omega-\omega_\parallel)\,\sin\Omega t} \tag{7.70}$$

with $\Omega = \frac{1}{2}\sqrt{(\omega-\omega_\parallel)^2+\omega_\perp^2}$, $\omega_\parallel = (\mu/j)\,A$, $\omega_\perp = (\mu/j)\,B$. If we choose $j = \frac{1}{2}$, we recover the well-known formula for the quantum precession "spin-flip":

$$P_{-\frac{1}{2}\frac{1}{2}}(t) = \frac{\omega_\perp^2\,\sin^2\Omega t}{(\omega-\omega_\parallel)^2+\omega_\perp^2} \,. \tag{7.71}$$

7.5 Application of Spin Coherent States to Classical and Thermodynamical Limits

We now consider a system of N quantum spins, all of them sharing the same spin value, say, j. The model was studied by Lieb [19] and revisited in [83, 84]. Lieb derived rigorous results concerning the intertwining between thermodynamical ($N \to \infty$) and classical ($\hbar \to 0$ or equivalently $j \to \infty$, whereas $\hbar j = \text{const.}$) limits. One can reasonably hope that in the $j \to \infty$ limit and after rescaling spin operators as

$$\mathbf{J} \to \frac{\mathbf{J}}{j}, \tag{7.72}$$

one gets the classical counterpart of the system, precisely objects living in that world where the quantum spin observables are replaced by classical vectors and integrals on the unit sphere S^2 are substituted for traces of operators. This has been proven for the Heisenberg model at fixed N. But what about the commutativity between the thermodynamical limit $N \to \infty$ and the classical limit $j \to \infty$?

7.5.1
Symbols and Traces

As a preliminary to the control of the commutativity of the two limits, Lieb has proven the following inequalities, derived also and independently by Berezin [90, 91]:

$$\underbrace{Z^{\text{Cl}}(j)}_{\text{classical-like partition fct.}} \leq \underbrace{Z^{\text{Qt}}(j)}_{\text{quantum partition fct.}} \leq Z^{\text{Cl}}(j+1)\,. \tag{7.73}$$

From them one can infer that, in the limit $j \to \infty$, the three quantities involved go to the same classical limit. Think of the analogy

$$\|\mathbf{J}^{\text{Cl}}\|^2 = j^2 \leq \underbrace{j(j+1) = (j+1/2)^2 - 1/4}_{\text{eigenvalue of } (\mathbf{J}^{\text{Qt}})^2} \leq (j+1)^2\,. \tag{7.74}$$

The proof makes use of the spin (or Bloch) coherent states:

$$|\hat{r}\rangle = |\theta, \phi\rangle = \sum_{m=-j}^{j} \sqrt{\begin{pmatrix} 2j \\ j+m \end{pmatrix}} \left(\cos\frac{\theta}{2}\right)^{j+m} \left(\sin\frac{\theta}{2}\right)^{j-m} e^{i(j-m)\phi} |jm\rangle\,. \tag{7.75}$$

With Berezin and Lieb, let us associate with an operator A acting in the Hermitian space $\mathcal{H}^j \cong \mathbb{C}^{2j+1}$ its two "symbols" with respect to the frame or "filter" provided by the overcomplete set of states $|\hat{r}\rangle$:

(L) Covariant or lower symbol or expected value in state $|\hat{r}\rangle$:

$$\check{A}(\hat{r}) = \langle \hat{r}|A|\hat{r}\rangle\,, \tag{7.76}$$

(U) Contravariant or upper symbol $\overset{\circ}{A}(\hat{r})$[9]:

$$A = \frac{2j+1}{4\pi} \int_{S^2} d\hat{r}\, \overset{\circ}{A}(\hat{r}) |\hat{r}\rangle\langle\hat{r}|\,. \tag{7.77}$$

9) We adopt here an alternative notation for upper symbols to avoid the appearance of too many hats!

Note here that $\overset{\circ}{A}(\hat{r})$ is not necessarily positive, even though A is a positive operator. If this integral of operators makes sense (in a weak sense), there is no reason to have uniqueness of the upper symbol $\overset{\circ}{A}(\hat{r})$. Nevertheless, it is always possible to choose it so that it is infinitely differentiable with respect to the angular coordinates θ, ϕ of $\hat{r}$ or with respect to the real and imaginary parts of the complex parameter $\zeta = \tan\frac{\theta}{2}\, e^{i\phi}$.

In the computation of operator traces, one may use alternatively lower and upper symbols. Indeed, since the kernel $K(\hat{r}', \hat{r}) \overset{\text{def}}{=} \langle \hat{r}' | \hat{r} \rangle$ is reproducing (owing to the resolution of the identity),

$$K(\hat{r}', \hat{r}) = \frac{2j+1}{4\pi} \int_{S^2} d\hat{r}\, K(\hat{r}', \hat{r})\, K(\hat{r}, \hat{r}'') , \tag{7.78}$$

the following result for the trace of an operator A (if, of course, the latter is trace class):

(i)

$$\operatorname{tr} A = \frac{2j+1}{4\pi} \int_{S^2} d\hat{r}\, \langle \hat{r} | A | \hat{r} \rangle = \frac{2j+1}{4\pi} \int_{S^2} d\hat{r}\, \check{A}(\hat{r}) , \tag{7.79}$$

on one hand, and

(ii)

$$\begin{aligned} \operatorname{tr} A &= \frac{2j+1}{4\pi} \int_{S^2} d\hat{r}\, \overset{\circ}{A}(\hat{r})\, \operatorname{tr} |\hat{r}\rangle\langle\hat{r}| = \frac{2j+1}{4\pi} \int_{S^2} d\hat{r}\, \overset{\circ}{A}(\hat{r})\, \underbrace{K(\hat{r}, \hat{r})}_{=1} \\ &= \frac{2j+1}{4\pi} \int_{S^2} d\hat{r}\, \overset{\circ}{A}(\hat{r}) \end{aligned} \tag{7.80}$$

on the other hand.

Let us give some examples of computation of lower $\check{A}(\hat{r})$ and upper $\overset{\circ}{A}(\hat{r})$ symbols for expressions involving spin operators, such as those appearing in Hamiltonians for spin systems. One easily proves that

$$\check{\mathbf{J}}(\hat{r}) = \langle \hat{r} | \mathbf{J} | \hat{r} \rangle = j\, \hat{r} , \tag{7.81}$$

whereas

$$\overset{\circ}{\mathbf{J}}(\hat{r}) = (j+1)\, \hat{r} . \tag{7.82}$$

For the dyad $\mathbf{JJ} \overset{\text{def}}{=} \{J_i\, J_j,\ i, j = x, y, z\}$,

$$\widecheck{\mathbf{JJ}}\,(\hat{r}) = j\,(j - \tfrac{1}{2})\, \hat{r}\, \hat{r} + \frac{j}{2} , \tag{7.83}$$

$$\overset{\circ}{\widehat{\mathbf{JJ}}}\,(\hat{r}) = (j+1)(j + \tfrac{3}{2})\, \hat{r}\, \hat{r} - \frac{j+1}{2} . \tag{7.84}$$

7.5.2
Berezin–Lieb Inequalities for the Partition Function

Let us first fix the Hilbertian framework for the quantum partition function. In considering the system of N quantum spins $\mathbf{J}_i$, $i = 1, 2, \ldots, N$, we suppose that the Hamiltonian of the system is polynomial in the $3N$ spin operators. We have

$$Z^{\mathrm{Qt}} = \alpha_N \operatorname{tr} e^{-\beta H}, \quad \alpha_N \overset{\text{def}}{=} \prod_{i=1}^{N} (2j_i + 1)^{-1}, \tag{7.85}$$

where the coefficient α_N is needed for normalization. The Hilbertian framework is precisely

$$\mathcal{H}_N = \bigotimes_{i=1}^{N} \mathcal{H}_i^j \cong \bigotimes_{i=1}^{N} \mathbb{C}^{2j+1}. \tag{7.86}$$

We next denote by $|\hat{r}_N\rangle$ the overcomplete set of normalized states in $\mathcal{H}_N$ built from the individual spin coherent states:

$$|\hat{r}_N\rangle = \bigotimes_{i=1}^{N} |\hat{r}_i\rangle, \quad \hat{r}_N \in \mathcal{S}_N \overset{\text{def}}{=} \underbrace{S^2 \times S^2 \times \cdots \times S^2}_{N}. \tag{7.87}$$

7.5.2.1
Classical Lower Bound for the Partition Function

First let us recall the *Peierls–Bogoliubov* inequality:

$$\langle \psi | e^{A} | \psi \rangle \geq e^{\langle \psi | A | \psi \rangle}, \tag{7.88}$$

which holds for any unit norm state ψ and any self-adjoint operator A in a finite-dimensional space. This inequality is derived from the spectral decomposition of A. The following inequality obeyed by the partition function results:

$$Z^{\mathrm{Qt}} = \frac{1}{(4\pi)^N} \int_{\mathcal{S}_N} d\hat{r}_N \, \langle \hat{r}_N | e^{-\beta H} | \hat{r}_N \rangle \geq \frac{1}{(4\pi)^N} \int_{\mathcal{S}_N} d\hat{r}_N \, e^{-\beta \check{H}(\hat{r}_N)}, \tag{7.89}$$

where $\check{H}(\hat{r}_N) \overset{\text{def}}{=} \langle \hat{r}_N | H | \hat{r}_N \rangle$. Now, if the quantum Hamiltonian H is linear in the components of each of the spin operators $\mathbf{J}_i$ (this case is said to be *normal*), then the lower symbol $\check{H}(\hat{r}_N)$ is readily obtained from the expression of H by replacing each $\mathbf{J}_i$ by $j_i \hat{r}_i$. So,

$$Z^{\mathrm{Qt}} \geq Z^{\mathrm{Cl}}(j_1, j_2, \ldots, j_N). \tag{7.90}$$

7.5.2.2
Classical Upper Bound for the Partition Function

To evaluate an upper bound, we consider the quantum Z^{Qt} as the limit of familiar approximations of the exponential

$$Z^{\mathrm{Qt}} = \operatorname{tr} e^{-\beta H} = \lim_{n \to \infty} Z(n)$$

where

$$Z(n) = \alpha_N \ \mathrm{tr} \left(1 - \frac{\beta H}{n}\right)^n \stackrel{\mathrm{def}}{=} \alpha_N \ \mathrm{tr}\left(F(n)\right)$$
$$= \alpha_N \int d\hat{r}_N^1 \int d\hat{r}_N^2 \cdots \int d\hat{r}_N^n \prod_{\mu=1}^{n} \overset{\circ}{F}_n\left(\hat{r}_N^\mu\right) L_j\left(\hat{r}_N^\mu, \hat{r}_N^{\mu+1}\right)$$

with $n + 1 \equiv 1$ in the last factor. The factor

$$\overset{\circ}{F}_n(\hat{r}_N^\mu) = \left(1 - \frac{\beta \overset{\circ}{H}(\hat{r}_N^\mu)}{n}\right)^n$$

is the upper symbol of

$$F(n) = \left(1 - \frac{\beta H}{n}\right)^n .$$

We now observe that

$$L_j(\hat{r}'_N, \hat{r}_N) \stackrel{\mathrm{def}}{=} \frac{1}{(4\pi)^N \alpha_N} \prod_{i=1}^{N} K(\hat{r}'_i, \hat{r}_i)$$

is reproducing since it inherits from the set of kernel factors $K(\hat{r}'_i, \hat{r}_i) = \langle \hat{r}'_i | \hat{r}_i \rangle$ their reproducing properties. We then write $Z(n) = \alpha_n \ \mathrm{tr}(F_n \mathcal{L}_j)^n$ in considering L_j as the kernel of a compact self-adjoint operator[10] L_j on $L^2(\mathcal{S}_N, d\hat{r}_N)$ and F_n as a multiplication operator. We next make use of the following set of inequalities (derived from Cauchy–Schwarz inequality and others, see, e.g., [92]) for two self-adjoint operators A and B:

$$\left|\mathrm{tr}(AB)^{2m}\right| \le \mathrm{tr}\left(A^2 \, B^2\right)^m \le \mathrm{tr}\, A^{2m} \, B^{2m} \,, \quad m = 2^l \,,\ l = 0, 1, 2, \ldots .$$

We eventually infer from them the upper bound:

$$Z^{\mathrm{Qt}} \le \frac{1}{(4\pi)^N} \int_{\mathcal{S}_N} d\hat{r}_N \, e^{-\beta \overset{\circ}{H}(\hat{r}_N)} .$$

In the normal case, it is enough to replace each spin operator $\mathbf{J}_i$ appearing in H by $(j_i + 1)\,\hat{r}_i$ to get the upper symbol $\overset{\circ}{H}(\hat{r}_N)$ of the Hamiltonian. We can conclude that

$$Z^{\mathrm{Qt}} \le Z^{\mathrm{Cl}}(j_1 + 1, j_2 + 1, \ldots, j_N + 1) \,, \tag{7.91}$$

and from (7.90) that the *Berezin–Lieb* inequalities hold true:

$$Z^{\mathrm{Cl}}(j_1, j_2, \ldots, j_N) \le Z^{\mathrm{Qt}} \le Z^{\mathrm{Cl}}(j_1 + 1, j_2 + 1, \ldots, j_N + 1) \,. \tag{7.92}$$

10) An operator A in a Hilbert space is said to be compact if it can be expanded as $A = \sum_n \lambda_n |f_n\rangle\langle g_n|$, where $\{|f_n\rangle\}$ and $\{|g_n\rangle\}$ are (not necessarily complete) orthonormal sets. The λ_n's form a sequence of positive numbers, called the singular values of the operator. The singular values can accumulate only at zero.

7.5.3 Application to the Heisenberg Model

Let us apply the inequalities (7.92) to the Heisenberg model in the elementary case of one spin value J only. We rescale the spin operators through $\mathbf{J}_i = \mathbf{S}_i/J$ to get for the N-spins Hamiltonian

$$H = \frac{1}{J^2}\sum_{i,j} \mathbf{S}_i \cdot \mathbf{S}_j \equiv H_N^{\mathrm{Qt}}(J)\,. \tag{7.93}$$

The quantum partition function reads as

$$Z_N^{\mathrm{Qt}}(J) = \frac{1}{(2J+1)^N}\, e^{-\beta H_N^{\mathrm{Qt}}(J)} \equiv e^{-\beta N f_N^{\mathrm{Qt}}(J)}\,, \tag{7.94}$$

where $f_N^{\mathrm{Qt}}(J)$ is the corresponding free energy per spin. Now consider the specific *Berezin–Lieb* inequalities:

$$\begin{aligned} Z^{\mathrm{Cl}}(J) &= \underbrace{\frac{1}{(4\pi)^N}\int_{S_N} d\hat{r}_N\, e^{-\beta \check{H}(\hat{r}_N)}}_{\equiv Z^{\mathrm{Cl}}\left(\frac{J^2}{J^2}\beta\right)} \le Z^{\mathrm{Qt}}(J) \le Z^{\mathrm{Cl}}(J+1) \\ &= \underbrace{\frac{1}{(4\pi)^N}\int_{S_N} d\hat{r}_N\, e^{-\beta \mathring{H}(\hat{r}_N)}}_{\equiv Z^{\mathrm{Cl}}\left(\frac{(J+1)^2}{J^2}\beta\right)}\,. \end{aligned} \tag{7.95}$$

The lower $Z^{\mathrm{Cl}}(\beta)$ and upper $Z^{\mathrm{Cl}}\left(\frac{(J+1)^2}{J^2}\beta\right)$ bounds are uniform with respect to the size N of the system. They are readily computed from

$$\begin{aligned} \check{H}(\hat{r}_N) &= \frac{1}{J^2}\sum_{i,j} J\hat{r}_i \cdot J\hat{r}_j = \sum_{i,j} \hat{r}_i \cdot \hat{r}_j \equiv H^{\mathrm{Cl}}\,, \\ \mathring{H}(\hat{r}_N) &= \frac{1}{J^2}\sum_{i,j} (J+1)\,\hat{r}_i \cdot (J+1)\,\hat{r}_j \approx_{J\ \mathrm{large}} H^{\mathrm{Cl}}\,. \end{aligned}$$

Hence, the combined classical and thermodynamical limits of the quantum spin system are just reached through a simple bound estimate for the free energy per spin:

$$\lim_{J\to\infty}\lim_{N\to\infty} f_N^{\mathrm{Qt}}(J) = f^{\mathrm{Cl}} = \lim_{N\to\infty} -\frac{1}{\beta}\ln Z_N^{\mathrm{Cl}}\,. \tag{7.96}$$

8
$SU(1,1)$ or $SL(2,\mathbb{R})$ Coherent States

8.1
Introduction

This chapter is devoted to the third most known family of coherent states, namely, the $SU(1,1)$ coherent states as they were established by Perelomov [10] in a group-theoretical approach. We adopt instead the construction set out in Chapter 4, choosing as an observation set the unit disk in the complex plane. Then we describe the main properties of these coherent states, that is, we list and comment on the sequence of properties as we did in Section 3.1, probabilistic aspects, link with $SU(1,1)$ representations, classical aspects, and so on. Finally, we make a short incursion in signal analysis by exploiting the fact that the unit disk has an unbounded representation that is the Poincaré half-plane of time-scale parameters. In this representation, the group $SU(1,1)$ is transformed into its real copy, namely, $SL(2,\mathbb{R})$. We then recover the continuous wavelet or time-scale transform of signals, which is precisely based on the subgroup of $SL(2,\mathbb{R})$ describing the affine transformations of the real line, $\mathbb{R} \ni t \mapsto b + a\,t$, $b, a \in \mathbb{R}$ with $a > 0$

Note that there exists another family of coherent states associated with $SU(1,1)$, namely, the Barut–Girardello coherent states. They will be considered in the next chapter through their appearance in the quantum motion problem of a particle in an infinite square well potential and also in the Pöschl–Teller potentials.

8.2
The Unit Disk as an Observation Set

Besides the complex plane, the infinite cylinder, and the sphere, there is the (open) unit disk $\mathcal{D} \stackrel{\text{def}}{=} \{z \in \mathbb{C}\,,\ |z| < 1\}$. There exist many situations in physics where the unit disk is involved as a fundamental model or at least is used as a pedagogical toy. For instance, it is a model of phase space for the motion of a material particle on a one-sheeted two-dimensional hyperboloid viewed as a $(1+1)$-dimensional space-time with negative constant curvature, namely, the two-dimensional *anti de Sitter* space-time [93–95]. In signal analysis, the *time-scale* half-plane, which repre-

Coherent States in Quantum Physics. Jean-Pierre Gazeau

ISBN: 978-3-527-40709-5

sents a nonbounded version of the unit disk, is the set of variables for (continuous) wavelet transform [11].

As a simple illustration of two-dimensional hyperbolic (*Lobatcheskian*) geometry, the unit disk has nice properties analogous to those of the sphere, except the fact that it is a noncompact bounded domain. It is commonly used as a model for the hyperbolic plane, by introducing a new metric on it, the Poincaré metric. The Poincaré metric is the metric tensor describing a two-dimensional surface of constant negative curvature. It reads in the present case (up to a constant factor) as

$$ds^2 = \frac{dz\, d\bar{z}}{(1-|z|^2)}\,. \tag{8.1}$$

The corresponding surface element is given by

$$\mu(d^2 z) = \frac{d(\Re z)\, d(\Im z)}{(1-|z|^2)^2} \equiv \frac{i}{2}\frac{dz \wedge d\bar{z}}{(1-|z|^2)^2}\,. \tag{8.2}$$

These quantities both emerge from a so-called *Kählerian potential* $\mathcal{K}_\mathcal{D}$:

$$\begin{aligned}
\mathcal{K}_\mathcal{D}(z,\bar{z}) &\stackrel{\text{def}}{=} \pi^{-1}(1-|z|^2)^{-2}\,,\\
ds^2 &= \frac{1}{2}\frac{\partial^2}{\partial z\,\partial \bar{z}}\ln \mathcal{K}_\mathcal{D}(z,\bar{z})\, dz\, d\bar{z}\,,\\
\mu(d^2 z) &= \frac{i}{4}\frac{\partial^2}{\partial z\,\partial \bar{z}}\ln \mathcal{K}_\mathcal{D}(z,\bar{z})\, dz \wedge d\bar{z}\,.
\end{aligned}$$

The unit disk equipped with such a potential has the structure of a two-dimensional *Kählerian manifold* [10, 96], an appellation shared by the complex plane, for which the potential is $\mathcal{K}_\mathbb{C} = \pi^{-1} e^{-z\bar{z}}$, the sphere S^2, or equivalently the projective complex line $\mathbb{CP}^1$, for which $\mathcal{K}_{S^2} = \pi^{-1}(1+|z|^2)^{-2}$, and the torus $\mathbb{C}/\mathbb{Z}^2 \simeq S^1 \times S^1$. Note that any Kählerian manifold is symplectic and so can be given a sense of phase space for some mechanical system. We will examine later the properties of these aspects in terms of symmetries of the unit disk.

Besides the unit disk, there are two other equivalent representations commonly used in two-dimensional hyperbolic geometry. One is the Poincaré half-plane, already mentioned, and defining a model of hyperbolic space on the upper half-plane. The disk $\mathcal{D}$ and the upper half-plane $\mathrm{P}_+ = \{Z \in \mathbb{C},\ \Im Z > 0\}$ are related by a conformal map, called Möbius transformation,

$$\mathrm{P}_+ \ni Z \mapsto z = e^{i\phi}\frac{Z - Z_0}{Z - \bar{Z}_0} \in \mathcal{D}\,, \tag{8.3}$$

ϕ and Z_0 being arbitrary. The canonical mapping is given by $Z_0 = i$ and $\phi = \pi/2$. It takes i to the center of the disk and the origin O to the bottom of the disk. The other representation is the punctured disk model, defined by $z = e^{i\pi Z}$, $\Im Z > 0$.

8.3 Coherent States

Let η be a real parameter such that $\eta > 1/2$ and let us equip the unit disk with a measure proportional to (8.2):

$$\mu_\eta(d^2z) \stackrel{\text{def}}{=} \frac{2\eta - 1}{\pi}\mu(d^2z) = \frac{2\eta - 1}{\pi}\frac{d^2z}{(1-|z|^2)^2}\,. \tag{8.4}$$

Consider now the Hilbert space $L^2_\eta = L^2(\mathcal{D}, \mu_\eta)$ of all functions $f(z, \bar{z})$ on $\mathcal{D}$ that are square-integrable with respect to μ_η. Within this "large" Hilbert space we select all functions of the form

$$\phi(z, \bar{z}) = (1 - |z|^2)^\eta g(\bar{z})\,, \tag{8.5}$$

where $g(z)$ is holomorphic on $\mathcal{D}$. The closure of the linear span of such functions is a Hilbert subspace of L^2_η denoted here by $\mathcal{K}_+$. An orthonormal basis of $\mathcal{K}_+$ is given by the countable set of functions

$$\phi_n(z, \bar{z}) \equiv \sqrt{\frac{(2\eta)_n}{n!}}(1 - |z|^2)^\eta \bar{z}^n \quad \text{with} \quad n \in \mathbb{N}, \tag{8.6}$$

where $(2\eta)_n = \frac{\Gamma(2\eta+n)}{\Gamma(2\eta)}$ is the Pochhammer symbol [18]. The proof is readily derived from the integral representation of the beta function,

$$B(x, y) = \frac{\Gamma(x)\Gamma(y)}{\Gamma(x+y)} = \int_0^1 t^{x-1}(1-t)^{y-1}\,dt\,.$$

Note that

$$\sum_{n=0}^{\infty} |\phi_n(z, \bar{z})|^2 = 1\,. \tag{8.7}$$

We are now in a position to define the coherent states resulting from this choice. They read as the following superpositions of vectors $|e_n\rangle$ forming an orthonormal basis in some separable Hilbert space $\mathcal{H}$:

$$|z;\eta\rangle \stackrel{\text{def}}{=} \sum_{n=0}^{\infty} \overline{\phi_n(z, \bar{z})}|e_n\rangle = (1-|z|^2)^\eta \sum_{n=0}^{\infty}\sqrt{\frac{(2\eta)_n}{n!}}z^n|e_n\rangle\,. \tag{8.8}$$

Again, by construction, these states are normalized and solve the identity $I_\mathcal{H}$ in $\mathcal{H}$:

$$\langle z;\eta|z;\eta\rangle = 1\,, \qquad \int_\mathcal{D} \mu_\eta(d^2z)\,|z;\eta\rangle\langle z;\eta| = I_\mathcal{H}\,. \tag{8.9}$$

Their mutual overlap offers an explicit representation of these states as elements of the Hilbert space $\mathcal{K}_+$. It is readily obtained from the binomial expansion:

$$\langle z';\eta|z;\eta\rangle = (1-|z|^2)^\eta\,(1-\bar{z}'\,z)^{-2\eta}\,(1-|z'|^2)^\eta\,. \tag{8.10}$$

It is also a reproducing kernel, for which the Hilbert space $\mathcal{K}_+$ is a Fock–Bargmann space, analogous to those encountered for the standard and spin coherent states.

8.4 Probabilistic Interpretation

As we could have guessed from the choice of the orthonormal set, we find beneath the structure of the states (8.8) a duality between two types of probability distributions. The first one is discrete and reads as

$$n \mapsto |\phi_n(z,\bar{z})|^2 = \frac{(2\eta)_n}{n!}(1-|z|^2)^{2\eta}|z|^{2n} \equiv P(2\eta, n; 1-|z|^2)\,. \tag{8.11}$$

It is, when 2η is an integer ≥ 1, a negative binomial distribution (see Appendix A). Recall that for a fixed integer $m \geq 1$, the negative binomial distribution is given by

$$P(m,n;\lambda) = \frac{\Gamma(m+n)}{\Gamma(n+1)\Gamma(m)}\,\lambda^m(1-\lambda)^n\,, \quad n = 0,1,2,\dots\,, \tag{8.12}$$

where the parameter λ lies in the interval $(0,1)$. The quantity $P(m,n,\lambda)$ can be thought of as being the probability that $m+n$ is the number of independent trials that are necessary to obtain the result of m successes (the $(m+n)$th trial being a success) when λ is the probability of success in a single trial. The term *negative binomial* stems from the fact that

$$(1-\lambda)^{-k} = \sum_{n=0}^{\infty} \frac{\Gamma(k+n)}{\Gamma(n+1)\Gamma(k)}\,\lambda^n\,,$$

from which it also follows that

$$\sum_{n=0}^{\infty} P(m,n;\lambda) = 1\,. \tag{8.13}$$

The second distribution is continuous in the variable $|z|^2$,

$$\begin{aligned} |z|^2 \mapsto& \frac{2\eta-1}{(1-|z|^2)^2}\,|\phi_n(z,\bar{z})|^2 \\ =& \frac{\Gamma(2\eta+n)}{\Gamma(2\eta-1)n!}(1-|z|^2)^{2\eta-2}|z|^{2n} \equiv \beta(1-|z|^2; 2\eta-1, n+1)\,. \end{aligned} \tag{8.14}$$

It is a beta distribution in the variable $\lambda = 1-|z|^2 \in [0,1]$, with parameters $2\eta - 1$ and $n+1 = 1,2,3,\dots,\infty$. We recall that the beta distribution in the variable $\lambda \in [0,1]$, with discrete parameters $m,n = 1,2,3,\dots$, is derived from the integral representation of the beta function and is given by

$$\beta(\lambda;m,n) = \frac{1}{B(m,n)}\lambda^{m-1}(1-\lambda)^{n-1}\,, \quad \int_0^1 \beta(\lambda;m,n)\,d\lambda = 1\,. \tag{8.15}$$

Thus, we observe that the prior measure on the parameter space $[0,1]$ is just $d\lambda$, whereas the associated Bayesian posteriors are the quantities $\frac{2\eta-1}{(1-|z|^2)^2}\,|\phi_n(z,\bar{z})|^2$ with $\lambda = 1-|z|^2$.

8.5 Poincaré Half-Plane for Time-Scale Analysis

As indicated in the introduction, we can use the Poincaré half-plane P_+ as a set of parameters for the coherent states described in this chapter. We choose the canonical Möbius transformation mapping the unit disk $\mathcal{D}$ onto P_+:

$$\mathcal{D} \ni z \mapsto Z = \frac{z+i}{iz+1} \in \mathrm{P}_+ \,, \tag{8.16}$$

$$\text{and conversely } z = \frac{Z-i}{1-iZ} \,. \tag{8.17}$$

Note that when extended to the boundaries, the bijection (8.16) is a Cayley transformation that maps in a stereographic way the unit circle S^1 onto the real line

$$S^1 \ni e^{i\theta} \mapsto t = \frac{e^{i\theta}+i}{ie^{i\theta}+1} \in \mathbb{R} \,, \; \theta \in [0, 2\pi) \,, \tag{8.18}$$

where $\theta = 0 \mapsto t = 1$, $\theta = \pi/2 \mapsto t = \infty$, $\theta = \pi \mapsto t = -1$ and $\theta = \frac{3\pi}{2} \mapsto 0$.

Let us introduce the (x, y), $y > 0$, variables as the real and imaginary parts of Z: $Z = x + iy$. In continuous wavelet analysis the real part x would have the meaning of a time variable, whereas y would stand for a scale. These coordinates are expressed in terms of the preimage z of Z as

$$x = \frac{2\mathfrak{R}(z)}{1+|z|^2 - 2\mathfrak{I}(z)} \,, \quad y = \frac{1-|z|^2}{1+|z|^2-2\mathfrak{I}(z)} \,. \tag{8.19}$$

The relation between respective Poincaré metrics is given by

$$ds^2 = (1-|z|^2)^{-1} dz\, d\bar{z} = \frac{dx^2+dy^2}{4y^2} \,. \tag{8.20}$$

The Lebesgue measures in the half-plane and in the open disk are related by

$$d^2Z = \frac{4}{(|z|^2+1-2\mathfrak{I}(z))^2} d^2z \,, \tag{8.21}$$

which, in terms of x and y, gives the relation between the respective Poincaré surface elements:

$$(1-|z|^2)^{-2}\, d^2z = \frac{dx\, dy}{4y^2} \,. \tag{8.22}$$

Inversely, going from the half-plane to the disk, we have

$$z = z(x, y) = \frac{2x}{x^2+(1+y)^2} + i\frac{x^2+y^2-1}{x^2+(1+y)^2} \,. \tag{8.23}$$

We also note also the useful formulas

$$|z|^2 = \frac{x^2+(y-1)^2}{x^2+(y+1)^2} \,, \quad 1-|z|^2 = \frac{4y}{x^2+(y+1)^2} \,. \tag{8.24}$$

In terms of parameters x and y, the coherent states (8.8) read as

$$|x, y; \eta\rangle \stackrel{\text{def}}{=} (4y)^\eta \sum_{n=0}^{\infty} \sqrt{\frac{(2\eta)_n}{n!}} \frac{(2x + i(y^2 + x^2 - 1))^n}{(x^2 + (1 + y)^2)^{\eta+n}} |e_n\rangle . \tag{8.25}$$

The resolution of the unity now reads as

$$\frac{2\eta - 1}{2\pi} \int_{\text{P}_+} |x, y; \eta\rangle\langle x, y; \eta| \, dx \, \frac{dy}{4y^2} = I_{\mathcal{H}} . \tag{8.26}$$

The expression (8.25) is a lot more involved than the one on the disk. However, it makes more transparent the time-scale content of such objects in view of utilization in analyzing a temporal signal s(t) belonging to the Hilbert space $L^2(\mathbb{R}, dt)$. Choosing the abstract $\mathcal{H}$ as the latter and some explicit orthonormal basis $\{|e_n\rangle, \ n \in \mathbb{N}\}$, we derive from the decomposition $|\text{s}\rangle = \sum_{n=0}^{\infty} s_n |e_n\rangle$ a *time-scale transform* of the signal as the following function of time and scale parameters (borrowing from wavelet analysis the usual notation $b \equiv x$, $a \equiv y$):

$$\begin{aligned} S_\eta(b, a) &\stackrel{\text{def}}{=} \langle b, a; \eta|\text{s}\rangle \\ &= (4a)^\eta \sum_{n=0}^{\infty} s_n \sqrt{\frac{(2\eta)_n}{n!}} \frac{(2b - i(a^2 + b^2 - 1))^n}{(b^2 + (1 + a)^2)^{\eta+n}} . \end{aligned} \tag{8.27}$$

However, whereas the *sensu stricto* continuous wavelet transform is based on the subgroup of $SL(2, \mathbb{R})$ describing the affine transformations of the real line, $\mathbb{R} \ni t \mapsto b + at$, $b, a \in \mathbb{R}$ with $a > 0$, as will be described in the last section of this chapter, the time-scale representation of signals based on the $SU(1, 1) \simeq SL(2, \mathbb{R})$ coherent states, as exemplified by (8.27), is of a different nature. As a matter of fact, it does not possess the affine covariant properties of the continuous wavelet transform. Furthermore, it depends on a prior Hilbertian decomposition of the signal versus a certain basis, and, at small scale $a \ll 1$ (habitually considered as the most interesting part of the analysis), it discriminates signals by their time evolution only, yielding a portrait of the signal in terms of the Fourier series:

$$S_\eta(b, a) \approx \left(\frac{4a}{b^2 + 1}\right)^\eta \sum_{n=0}^{\infty} s_n \sqrt{\frac{(2\eta)_n}{n!}} e^{in\theta_b} , \quad \text{with} \quad \theta_b = \arctan\left(\frac{2b}{b^2 - 1}\right) . \tag{8.28}$$

However, it offers the opportunity to play with the extra parameter $\eta > 1/2$.

8.6
Symmetries of the Disk and the Half-Plane

Like the sphere S^2 is invariant under space rotations forming the group $SO(3) \simeq SU(2)/\mathbb{Z}_2$, $\mathbb{Z}_2 = \{1, -1\}$, the unit disk $\mathcal{D}$ is invariant under transformations of the *homographic* or Möbius type:

$$\mathcal{D} \ni z \mapsto z' = (\alpha z + \beta)\,(\bar{\beta} z + \bar{\alpha})^{-1} \in \mathcal{D} , \tag{8.29}$$

with $\alpha, \beta \in \mathbb{C}$ and $|\alpha|^2 - |\beta|^2 \neq 0$. Since a common factor of α and β is unimportant in the transformation (8.29), one can associate with the latter the 2×2 complex matrix

$$\begin{pmatrix} \alpha & \beta \\ \bar{\beta} & \bar{\alpha} \end{pmatrix} \stackrel{\text{def}}{=} g\,, \quad \text{with} \quad \det g = |\alpha|^2 - |\beta|^2 = 1\,, \tag{8.30}$$

and we will write $z' = g \cdot z$. These matrices form the group $SU(1,1)$, the simplest example of a simple, noncompact Lie group (see Appendix B). It should be noted that $SU(1,1)$ leaves invariant the boundary $S^1 \simeq U(1)$ of $\mathcal{D}$ under the transformations (8.29).

The invariance of $\mathcal{D}$ under (8.29) is not only geometrical. It also holds for the Poincaré metric (8.1) and surface element (8.2), since both emerge from the invariant Kählerian potential $\mathcal{K}_{\mathcal{D}}$:

$$\begin{aligned} \mathcal{K}_{\mathcal{D}}(z, \bar{z}) &= \pi^{-1}(1 - |z|^2)^{-2} = \pi^{-1}(1 - |z'|^2)^{-2}\,, \\ ds^2 &= \frac{dz\, d\bar{z}}{(1 - |z|^2)} = \frac{dz'\, d\bar{z}'}{(1 - |z'|^2)}\,, \\ \mu(d^2 z) &= \frac{d(\Re z)\, d(\Im z)}{(1 - |z|^2)^2} = \frac{d(\Re z')\, d(\Im z')}{(1 - |z'|^2)^2}\,. \end{aligned}$$

This invariance is the essence of Lobatchevskian geometry [96].

Let us now turn our attention to the corresponding symmetries in the Poincaré half-plane. Let us write the canonical Möbius transformation (8.17), as

$$z = \frac{Z - i}{1 - iZ} \equiv \frac{1}{\sqrt{2}} \begin{pmatrix} 1 & -i \\ -i & 1 \end{pmatrix} \cdot Z \equiv \mathrm{m} \cdot Z\,, \quad Z = \mathrm{m}^{-1} \cdot z\,. \tag{8.31}$$

Therefore, the transformation $z' = g \cdot z$, where $g \in SU(1,1)$, becomes in the half-plane the transformation $Z' = s \cdot Z$, with

$$s = \mathrm{m}^{-1}\, g\, \mathrm{m} = \begin{pmatrix} \Re\alpha + \Im\beta & \Im\alpha + \Re\beta \\ -\Im\alpha + \Re\beta & \Re\alpha - \Im\beta \end{pmatrix} \equiv \begin{pmatrix} a & b \\ c & d \end{pmatrix}\,, \quad a, b, c, d \in \mathbb{R}\,. \tag{8.32}$$

Since $\det s = 1$, the set of such 2×2 real matrices form the group $SL(2, \mathbb{R})$, which leaves invariant the upper half-plane, and its Poincaré metric (8.20) and surface element (8.22) as well.

8.7 Group-Theoretical Content of the Coherent States

8.7.1 Cartan Factorization

In semisimple group theory there exists a well-known group factorization called the *Cartan decomposition* (see Appendix B) or "phase-space" decomposition in the present context when the unit disk, as a symplectic manifold, is given a phase-space meaning. The Cartan decomposition of $SU(1,1)$, denoted by $SU(1,1) = PH$, means that any $g \in SU(1,1)$ can be written as the product $g = p\,h$, with $p \in P$ and $h \in H$. It is defined by the *Cartan involution* $i_{\text{ph}} : g \mapsto (g^\dagger)^{-1}$ in the sense that P is made of all $p \in SU(1,1)$ such that $i_{\text{ph}}(p) = p^{-1}$, that is, $p = p^\dagger$ is Hermitian, while H has all its elements unchanged under i_{ph}, that is, $h^\dagger = h^{-1}$, which means that h is unitary. In consequence, $H \cong U(1)$ is the unitary subgroup of $SU(1,1)$. The decomposition reads explicitly

$$SU(1,1) \ni g = \begin{pmatrix} \alpha & \beta \\ \bar{\beta} & \bar{\alpha} \end{pmatrix} = p(z)\,h(\theta)\,, \tag{8.33}$$

with

$$p(z) = \begin{pmatrix} \delta & \delta z \\ \delta \bar{z} & \delta \end{pmatrix}, \quad z = \beta \bar{\alpha}^{-1}, \quad \delta = (1 - |z|^2)^{-1/2} \tag{8.34}$$

and

$$h(\theta) = \begin{pmatrix} e^{i\theta/2} & 0 \\ 0 & e^{-i\theta/2} \end{pmatrix}, \quad \theta = 2 \arg \alpha\,, \quad 0 \le \theta < 4\pi\,. \tag{8.35}$$

The *bundle section*[11] $\mathcal{D} \ni z \mapsto p(z) \in P$ gives the unit disk $\mathcal{D}$ a symmetric space realization identified with the coset space $SU(1,1)/H$. We remark that $p^2 = g g^\dagger$ and that $(p(z))^{-1} = p(-z)$. We can exploit the Cartan factorization by making $SU(1,1)$ act on $\mathcal{D}$ through a left action on the set of matrices $p(z)$. Explicitly,

$$g : p(z) \mapsto p(z') \quad \text{defined by} \quad g\,p(z) = p(z')\,h'\,. \tag{8.36}$$

It is then easily verified that z' is given by the Möbius action (8.29): $z' = g \cdot z$.

8.7.2
Discrete Series of *SU*(1, 1)

We now consider a class of unitary irreducible representations of $SU(1,1)$, precisely indexed by the parameter η appearing in the measure on the unit disk, and involved in the construction of the coherent states, to which this chapter is devoted. For a given $\eta > 1$, we introduce the Fock–Bargmann Hilbert space $\mathcal{FB}_\eta$ of all analytical functions $f(z)$ on $\mathcal{D}$ that are square-integrable with respect to the scalar product:

$$\langle f_1 | f_2 \rangle = \frac{2\eta - 1}{2\pi} \int_{\mathcal{D}} \overline{f_1(z)}\, f_2(z)\, (1 - |z|^2)^{2\eta - 2}\, d^2 z\,. \tag{8.37}$$

11) A nontrivial fiber bundle consists of four objects, (E, B, π, F), where E, B, and F are topological spaces and $\pi : E \mapsto B$ is a continuous surjection such that E is locally (but not globally!) homeomorphic to the Cartesian product $B \times F$. B is called the base space of the bundle, E the total space, and F the fiber. The map π is called the projection map (or bundle projection). A section (or cross section) is a continuous map, $s : B \mapsto E$, such that $\pi(s(x)) = x$ for all x in B.

Note that the elements of this space are just the conjugate of the elements of $\mathcal{K}_+$ cleared of their nonanalytical factor $(1-|z|^2)^{2\eta}$. The orthonormal basis given by (8.6) is now made of powers of z suitably normalized:

$$p_n(z) \equiv \sqrt{\frac{(2\eta)_n}{n!}}\, z^n \quad \text{with} \quad n \in \mathbb{N}\,. \tag{8.38}$$

We define, for $\eta = 1, 3/2, 2, 5/2, \ldots$ the unitary irreducible representation

$$g = \begin{pmatrix} \alpha & \beta \\ \bar{\beta} & \bar{\alpha} \end{pmatrix} \mapsto U^\eta(g)$$

of $SU(1,1)$ on $\mathcal{FB}_\eta$ by

$$\mathcal{FB}_\eta \ni f(z) \mapsto \left(U^\eta(g)\, f\right)(z) = (-\bar{\beta}\, z + \alpha)^{-2\eta}\, f\left(\frac{\bar{\alpha} z - \beta}{-\bar{\beta} z + \alpha}\right). \tag{8.39}$$

This countable set of representations constitutes the "almost complete" holomorphic discrete series of representations of $SU(1,1)$ [97–99]. It is "almost complete" because the lowest one, which corresponds to the value $\eta = 1/2$, requires a special treatment owing to the nonexistence of the inner product (8.37) in this case: there is no Fock–Bargmann realization in that case. We will come back to this important question in the last section. Had we considered the continuous set $\eta \in [1/2, +\infty)$, we would have been led to involving the universal covering of $SU(1,1)$ [100, 101].

The matrix elements of the operator $U^\eta(g)$ with respect to the orthonormal basis (8.38) are given (see, e.g., [102] or Appendix A in [103]) in terms of hypergeometrical polynomials by

$$\begin{aligned} U^\eta_{nn'}(g) &= \langle p_n | U^\eta(g) | p_{n'} \rangle = \left(\frac{n_>!\,\Gamma(2\eta + n_>)}{n_<!\,\Gamma(2\eta + n_<)}\right)^{1/2} \alpha^{-2\eta - n_>}\, \bar{\alpha}^{n_<} \\ &\quad \times \frac{(\gamma(\beta, \bar{\beta}))^{n_> - n_<}}{(n_> - n_<)!}\, {}_2F_1\left(-n_<\,,\ n_> + 2\eta\,;\ n_> - n_< + 1\,;\ \frac{|\beta|^2}{|\alpha|^2}\right), \end{aligned} \tag{8.40}$$

where

$$\gamma(\beta, \bar{\beta}) = \begin{cases} -\beta & n_> = n' \\ \bar{\beta} & n_> = n \end{cases}, \quad n_{\gtrless} = \begin{matrix}\max \\ \min\end{matrix}\,(n, n') \geq 0\,.$$

Owing to the relation $\frac{|\beta|^2}{|\alpha|^2} = 1 - \frac{1}{|\alpha|^2}$, this expression is alternatively given in terms of Jacobi polynomials as follows:

$$\begin{aligned} U^\eta_{nn'}(g) &= \left(\frac{n_<!\,\Gamma(2\eta + n_>)}{n_>!\,\Gamma(2\eta + n_<)}\right)^{1/2} \alpha^{-2\eta - n_>}\, \bar{\alpha}^{n_<} \\ &\quad \times \frac{(\gamma(\beta, \bar{\beta}))^{n_> - n_<}}{\sqrt{(n_> - n_<)!}}\, P_{n_<}^{(n_> - n_<\,,\, 2\eta - 1)}\left(\frac{1 - |\beta|^2}{1 + |\beta|^2}\right). \end{aligned} \tag{8.41}$$

8.7.3
Lie Algebra Aspects

Any element $g \in SU(1,1)$ can also be factorized, in a nonunique way, in terms of three one-parameter subgroup elements: $g = \pm h(\theta)\, s(u)\, l(v)$. Besides the sign $\pm$, the first factor was already encountered in the Cartan decomposition (8.35), whereas the others are of noncompact hyperbolic type and are given by

$$s(u) = \begin{pmatrix} \cosh u & \sinh u \\ \sinh u & \cosh u \end{pmatrix}, \quad l(v) = \begin{pmatrix} \cosh v & i\,\sinh v \\ -i\,\sinh v & \cosh v \end{pmatrix}, \quad u\,,\, v \in \mathbb{R}\,. \tag{8.42}$$

The first subgroup is isomorphic to $U(1)$, whereas the two others are isomorphic to $\mathbb{R}$. Their respective generators, $N_\mu, \mu = 0, 1, 2$, are defined by

$$h(\theta) = e^{\theta N_0}\,, \quad s(u) = e^{u N_1}\,, \quad h(\theta) = e^{v N_2}\,, \tag{8.43}$$

and are given in terms of the Pauli matrices by

$$N_0 = \frac{i}{2}\sigma_3\,, \quad N_1 = \frac{1}{2}\sigma_1\,, \quad N_2 = -\frac{1}{2}\sigma_2\,. \tag{8.44}$$

They form a basis of the Lie algebra $\mathfrak{su}(1,1)$ and obey the commutation relations

$$[N_0, N_1] = N_2\,, \quad [N_0, N_2] = -N_1\,, \quad [N_1, N_2] = -N_0\,. \tag{8.45}$$

Their respective self-adjoint representatives under the unitary irreducible representation (8.39), defined generically as $-i\,\partial/\partial t\; U^\eta(g(t))$, are the following differential operators on the Fock–Bargmann space $\mathcal{FB}_\eta$

$$i\,N_0 \mapsto K_0 = z\,\frac{d}{dz} + \eta\,, \tag{8.46a}$$

$$i\,N_1 \mapsto K_1 = -\frac{i}{2}(1 - z^2)\,\frac{d}{dz} + i\eta z\,, \tag{8.46b}$$

$$i\,N_2 \mapsto K_2 = \frac{1}{2}(1 + z^2)\,\frac{d}{dz} + \eta z\,, \tag{8.46c}$$

and obey the commutation rules

$$[K_0, K_1] = i\,K_2\,, \quad [K_0, K_2] = -i\,K_1\,, \quad [K_1, K_2] = -i\,K_0\,. \tag{8.47}$$

We may check that the elements of the orthonormal basis (8.38) are eigenvectors of the compact generator K_0 with equally spaced eigenvalues:

$$K_0\,|\,p_n\rangle = (\eta + n)\,|\,p_n\rangle\,. \tag{8.48}$$

The particular element $|\,p_0\rangle$ of the basis is a *lowest weight* or "vacuum" state for the representations U^η. Indeed, let us introduce the two operators with their commutation relation:

$$K_\pm = \mp i\,(K_1 \pm i K_2) = K_2 \mp i K_1\,, \quad [K_+, K_-] = -2K_0\,. \tag{8.49}$$

As differential operators, they read as $K_+ = z^2\, d/dz + 2\eta z$, $K_- = d/dz$. Adjoint of each other, they are raising and lowering operators, respectively,

$$\begin{aligned} K_+ \,|\, p_n\rangle &= \sqrt{(n+1)(2\eta+n)}\,|\, p_{n+1}\rangle\,, \\ K_- \,|\, p_n\rangle &= \sqrt{n\,(2\eta+n-1)}\,|\, p_{n-1}\rangle\,, \end{aligned} \tag{8.50}$$

and, as announced, $K_- \,|\, p_0\rangle = 0$. The states $|\, p_n\rangle$ are themselves obtained by successive ladder actions on the lowest state as follows:

$$|\, p_n\rangle = \sqrt{\frac{\Gamma(2\eta)}{\Gamma(2\eta+n)\,n!}}\,(K_+)^n\,|\, p_0\rangle\,. \tag{8.51}$$

One can also check directly from (8.46a) that the Casimir operator

$$C \stackrel{\text{def}}{=} K_1^2 + K_2^2 - K_0^2 = \frac{K_+K_- + K_-K_+}{2} - K_0^2 \tag{8.52}$$

is fixed at the value $C = -\eta(\eta-1)\, I_d$ on the space $\mathcal{FB}_\eta$ that carries the unitary irreducible representation U^η.

8.7.4
Coherent States as a Transported Vacuum

We are now in the position to explain the group-theoretical content of the coherent states (8.8), that is, the rationale behind the appellation $SU(1,1)$ coherent states, as they were introduced by Perelomov. To each element z in the unit disk corresponds the element $p(\bar{z})$ (note the conjugate variable) of $SU(1,1)$, defined in (8.34) from the Cartan decomposition. Let us now apply to the lowest state $|\, p_0\rangle$ the operators of the representation U^η restricted to the set P of such matrices, and expand the "transported" state in terms of the Fock–Bargmann basis:

$$U^\eta(p(\bar{z}))\,|\, p_0\rangle = \sum_{n=0}^{\infty} U^\eta_{n0}(p(\bar{z}))\,|\, p_n\rangle = (1-|z|^2)^\eta \sum_{n=0}^{\infty} \sqrt{\frac{(2\eta)_n}{n!}}\, z^n |\, p_n\rangle\,. \tag{8.53}$$

Thus, the coherent state defined in (8.8) is exactly this transported state:

$$U^\eta(p(\bar{z}))\,|\, p_0\rangle = |z;\eta\rangle\,. \tag{8.54}$$

In the same spirit, we could think of transporting any other element of the Fock–Bargmann basis under the action of the unit disk through the bundle section $z \mapsto p(\bar{z})$. We thus obtain a discretely indexed set of families of coherent states, defined by

$$|z;\eta;m\rangle \stackrel{\text{def}}{=} U^\eta(p(\bar{z}))\,|\, p_m\rangle = \sum_{n=0}^{\infty} U^\eta_{nm}(p(\bar{z}))\,|\, p_n\rangle\,, \tag{8.55}$$

with

$$\begin{aligned} U^\eta_{nm}(p(\bar{z})) = {} & \left(\frac{n_>!\,\Gamma(2\eta+n_>)}{n_<!\,\Gamma(2\eta+n_<)}\right)^{1/2} (1-|z|^2)^\eta \,\frac{|z|^{n_>-n_<}}{(n_>-n_<)!}\, e^{i(n-m)\phi} \\ & \times (\operatorname{sgn}(n-m))^{n-m}\, {}_2F_1\left(-n_<\,,\ n_> + 2\eta\,; n_> - n_< + 1\,;\, |z|^2\right)\,, \end{aligned} \tag{8.56}$$

with $z = |z|e^{i\phi}$. To fully justify the adjective "coherent," it is necessary to prove that these normalized states resolve the unity in $\mathcal{FB}_\eta$:

$$\int_{\mathcal{D}} \mu_\eta(d^2 z)\,|z;\eta;m\rangle\langle z;\eta;m| = I_{\mathcal{FB}_\eta}\,. \tag{8.57}$$

Indeed, from the representation property combined with (8.36), we have

$$U^\eta(g)\,U^\eta(p(\bar{z})) = U^\eta(g\;p(\bar{z})) = U^\eta(p(g\cdot\bar{z})\,U^\eta(h')\,,$$

which holds for any $g \in SU(1,1)$ and where $h' \in H$. Now, with

$$h' = \begin{pmatrix} e^{i\theta'/2} & 0 \\ 0 & e^{-i\theta'/2} \end{pmatrix}\,,\qquad U^\eta(h')\,|\,p_m\rangle = e^{-i(\eta+m)\theta'}\,|\,p_m\rangle\,.$$

So this phase factor disappears from

$$U^\eta(g)\int_{\mathcal{D}} \mu_\eta(dz\,d\bar{z})\,|z;\eta;m\rangle\langle z;\eta;m|\,U^\eta(g^{-1})$$
$$= \int_{\mathcal{D}} \mu_\eta(dz\,d\bar{z})\,|g\cdot z;\eta;m\rangle\langle g\cdot z;\eta;m|\,.$$

From the invariance of the measure and Schur's lemma, we deduce that the left-hand side of (8.57) is a multiple of the identity. It is straightforward to check that this factor is 1. Mutatis mutandis, we deduce from the resolution of the unity that the set of functions $\{U^\eta_{nm}(p(\bar{z}))\,,\; n \in \mathbb{N}\}$ is orthonormal with respect to the measure μ_η on the unit disk. The coherent states (8.55) are the counterparts of $SU(2)$ sigma-spin coherent states $|\hat{r};\sigma\rangle$ defined in (6.9) and Weyl–Heisenberg coherent states $|z;m\rangle$ defined in (7.31). It is possible to give the discrete probability distribution $n \mapsto |U^\eta_{nm}(p(\bar{z}))|^2$ with parameter $|z|^2$ or its counterpart on the Poincaré half-plane a physical meaning in terms of transition probability and S matrix, like we did for the two other types of coherent states in Chapter 7. Examples are provided by the Morse Hamiltonian (see [104] and references therein) and its supersymmetric aspects [105, 106], and also by the Schrödinger operator with a magnetic field on the Poincaré upper half-plane [107, 108].

There exists an alternative expression for the "$SU(1,1)$ displacement" operator $U^\eta(p(\bar{z}))$ used to build coherent states. It is, in the present context, the counterpart of the displacement operator $D(z) = \exp(za^\dagger - \bar{z}a)$ involved in the construction of the standard coherent states. With an element z of the unit disk $\mathcal{D}$ we associate the complex variable ζ with the same argument ϕ and modulus $|\zeta| = \tanh^{-1}|z|$, that is, $z = \tanh|\zeta|\,e^{i\phi}$. Then we have the relation

$$U^\eta(p(\bar{z})) = e^{\zeta K_+ - \bar{\zeta} K_-} \equiv D_\eta(\zeta)\,. \tag{8.58}$$

The demonstration exploits the 2×2 matrix representation of the elements of $SU(1,1)$ and its Lie algebra and the various factorizations on the group. Indeed, the result of the action of $D_\eta(\zeta)$ on the lowest state $|\,p_0\rangle$ will be made almost trivial if we are able to factorize it as

$$D_\eta(\zeta) = e^{A(z)K_+}\,e^{B(z)K_0}\,e^{C(z)K_-}\,,$$

since then, from $K_- \,|\, p_0\rangle = 0$, $K_0 \,|\, p_0\rangle = \eta \,|\, p_0\rangle$, and (8.51), we get

$$\begin{aligned} D_\eta(\zeta) \,|\, p_0\rangle &= e^{A(z)K_+}\, e^{B(z)K_0}\, e^{C(z)K_-} \,|\, p_0\rangle \\ &= e^{B(z)\eta} \sum_{n=0}^{\infty} \sqrt{\frac{(2\eta)_n}{n!}}\, (A(z))^n \,|\, p_n\rangle\,. \end{aligned} \tag{8.59}$$

Now, from the inverse of the correspondences (8.46a),

$$\zeta K_+ - \bar{\zeta} K_- \mapsto i\,(\zeta N_+ - \bar{\zeta} N_-) = \begin{pmatrix} 0 & \bar{\zeta} \\ \zeta & 0 \end{pmatrix},$$

we obtain by exponentiation

$$D_\eta(\zeta) \mapsto \begin{pmatrix} \cosh|\zeta| & e^{-i\phi} \sinh|\zeta| \\ e^{i\phi} \sinh|\zeta| & \cosh|\zeta| \end{pmatrix} \equiv \Delta(\zeta)\,.$$

This matrix is easily factorized as

$$\begin{aligned} \Delta(\zeta) &= e^{A(z)iN_+}\, e^{B(z)iN_0}\, e^{C(z)iN_-} \\ &= \begin{pmatrix} 1 & 0 \\ A(z) & 1 \end{pmatrix} \begin{pmatrix} e^{-B(z)/2} & 0 \\ 0 & e^{B(z)/2} \end{pmatrix} \begin{pmatrix} 1 & -C(z) \\ 0 & 1 \end{pmatrix} \\ &= \begin{pmatrix} e^{-B/2} & -Ce^{-B/2} \\ Ae^{-B/2} & e^{B/2} - Ae^{-B/2} \end{pmatrix}. \end{aligned}$$

A simple identification gives $A(z) = z$, $e^{-B(z)/2} = \cosh|\zeta| = (1 - |z|^2)^{-1/2}$, and $C(z) = -\bar{z}$. We thus recover in the action (8.59) of $D_\eta(z)$ the $SU(1,1)$ coherent states (8.8).

This kind of *disentangling* technique for factorizing elaborate representation operators is well known [9], and we will use it again in the chapters devoted to squeezed states and to fermionic coherent states.

8.8
A Few Words on Continuous Wavelet Analysis

The inner product (8.37) vanishes identically at the limit value $\eta = 1/2$. However, there exists a well-defined unitary irreducible representation of $SU(1,1)$ corresponding to this value. Its carrier space can be realized as the Hilbert space $L^2(S^1, d\theta/2\pi)$ of the exponential Fourier series, that is, the space of square-integrable complex-valued functions on the boundary S^1 of the unit disk $\mathcal{D}$. Indeed, the action (8.29) of

$$g = \begin{pmatrix} \alpha & \beta \\ \bar{\beta} & \bar{\alpha} \end{pmatrix} \in SU(1,1)$$

on $\mathcal{D}$ extends to the boundary as

$$g \cdot e^{i\theta} = (\alpha\, e^{i\theta} + \beta)\,(\bar{\beta}\, e^{i\theta} + \bar{\alpha})^{-1} \equiv e^{i\theta'} \in S^1\,, \tag{8.60}$$

and so leaves the latter invariant. It is then easy to check that the transformation $\mathcal{U}^{\eta=1/2}(g) \equiv \mathcal{U}(g)$ in $L^2(S^1, d\theta/2\pi)$, defined by

$$(\mathcal{U}(g)\, f)(e^{i\theta}) = (-\bar{\beta}\, e^{i\theta} + \alpha)^{-1}\, f(g^{-1} \cdot e^{i\theta}) \equiv f'(e^{i\theta}) \,, \tag{8.61}$$

is unitary with respect to the inner product for Fourier series,

$$\langle f_1 | f_2 \rangle = \frac{1}{2\pi} \int_0^{2\pi} \overline{f_1(e^{i\theta})}\, f_2(e^{i\theta})\, d\theta = \langle f_1' | f_2' \rangle \,. \tag{8.62}$$

This is due to the transformation of the measure:

$$\frac{d\theta'}{d\theta} = |\bar{\alpha} e^{i\theta} - \beta|^{-2} \,.$$

Let us now transport this material onto the real line, the boundary of the upper half-plane P_+, by restricting the transformations (8.31) and (8.32) to $S^1 \ni e^{i\theta} \mapsto x \in \mathbb{R}$:

$$x = \frac{e^{i\theta} + i}{1 + i e^{i\theta}} = \frac{\cos\theta}{1 - \sin\theta} \,, \qquad d\theta = \frac{2\, dx}{1 + x^2} \,. \tag{8.63}$$

This gives on the level of the inner product (8.62) the relation

$$\frac{1}{2\pi} \int_0^{2\pi} \overline{f_1(e^{i\theta})}\, f_2(e^{i\theta})\, d\theta = \frac{1}{\pi} \int_{-\infty}^{+\infty} \overline{f_1(\mathrm{m} \cdot x)}\, f_2(\mathrm{m} \cdot x)\, \frac{dx}{1 + x^2} \,, \tag{8.64}$$

where

$$\mathrm{m} = \frac{1}{\sqrt{2}} \begin{pmatrix} 1 & -i \\ -i & 1 \end{pmatrix}$$

is the Möbius–Cayley matrix introduced in (8.31). Hence, any square-integrable function on the circle yields a square-integrable function on the real line with respect to the Lebesgue measure along the map:

$$L^2(S^1, d\theta/2\pi) \ni f(e^{i\theta}) \mapsto F(x) \stackrel{\text{def}}{=} (1 - i x)\, f(\mathrm{m}.x) \in L^2(\mathbb{R}, dx) \,. \tag{8.65}$$

On the level of group representations, the unitary representation (8.61) of $SU(1,1)$ is transported into the following unitary representation of $SL(2,\mathbb{R})$ on $L^2(\mathbb{R}, dx)$:

$$(\mathcal{U}(s)\, F)(x) = (-c x + a)^{-1}\, f\left(s^{-1} \cdot x\right) \equiv F'(x) \,, \quad s = \begin{pmatrix} a & b \\ c & d \end{pmatrix} \,. \tag{8.66}$$

We now restrict the above representation to the subgroup Aff(ℝ) of *affine* transformations of the real line, which is defined as

$$\mathrm{Aff}(\mathbb{R}) \stackrel{\text{def}}{=} \left\{ (b, a) \equiv \begin{pmatrix} \mathrm{sgn}(|a|)\sqrt{a} & b \\ 0 & \frac{1}{\sqrt{|a|}} \end{pmatrix} \,, \ b \in \mathbb{R} \,,\ a \neq 0 \right\} \,, \tag{8.67}$$

with the action resulting from the Möbius transformation of the real line of the type (8.29) $(b,a)\cdot x = b + a\,x$ and the group law

$$(b,a)(b',a') = (b + ab', aa')\,. \tag{8.68}$$

Thus, Aff($\mathbb{R}$) is a semidirect product of the translation group $\mathbb{R}$ by the dilation group $\mathbb{R}_*$: Aff($\mathbb{R}$) = $\mathbb{R} \rtimes \mathbb{R}_*$. The unit element is $(0,1)$ and the inverse of (b,a) is $(-a^{-1}b, a^{-1})$. Continuous wavelet analysis rests on the essential result that [11, 109, 110]:

Continuous Wavelet Analysis rests on the essential result that [11, 109], up to unitary equivalence, Aff($\mathbb{R}$) has a unique UIR, acting in $L^2(\mathbb{R}, dx)$, namely

$$\left(U(b,a)f\right)(x) = |a|^{-1/2}\, f\left(\frac{x-b}{a}\right) \equiv f_{ba}(x) \quad (a \neq 0, b \in \mathbb{R})\,, \tag{8.69}$$

or, on Fourier transforms, $\hat{f}(\xi) = \frac{1}{\sqrt{2\pi}}\int_{-\infty}^{+\infty} e^{-i\xi x}\, f(x)\, dx$:

$$\left(\widehat{U}(b,a)\hat{f}\right)(\xi) = |a|^{1/2}\, \hat{f}(a\xi)e^{-ib\xi} \quad (a \neq 0, b \in \mathbb{R})\,. \tag{8.70}$$

The representation U is square integrable, i.e., for all *admissible* $\psi \in L^2(\mathbb{R}, dx)$, the function $(b,a) \mapsto \langle\psi|U(b,a)|\psi\rangle$ is square integrable on Aff($\mathbb{R}$) w.r.t. its left Haar measure $db\, da/a^2$. Now, a vector $\psi \in L^2(\mathbb{R}, dx)$ is said *admissible* if it satisfies the condition

$$c_\psi \equiv 2\pi \int_{-\infty}^{\infty} |\widehat{\psi}(\xi)|^2 \frac{d\xi}{|\xi|} < \infty\,. \tag{8.71}$$

In practice, the admissibility condition (8.71) (plus some regularity: $\psi \in L^1 \cap L^2$ suffices) is equivalent to a zero mean condition:

$$\psi \text{ admissible} \overset{(\Leftarrow)}{\Rightarrow} \widehat{\psi}(0) = 0 \quad \Leftrightarrow \quad \int_{-\infty}^{+\infty} \psi(x)\, dx = 0\,. \tag{8.72}$$

An admissible function will be called a *wavelet*. Thus a wavelet ψ is by necessity an *oscillating* function, real or complex-valued (see the examples below), and this is in fact the origin of the term "wavelet". Let us explain in what sense this admissibility condition is crucial in signal analysis. Let ψ be a wavelet and $\mathsf{s} \in L^2(\mathbb{R})$ a function viewed as a signal within the present context. Then the continuous wavelet transform (CWT) of s with respect to ψ is the function $\mathsf{S} \equiv T_\psi \mathsf{s}$ on the time-scale half-plane, which is given by the scalar product of s with the transformed wavelet ψ_{ba}:

$$\begin{aligned} \mathsf{S}(b,a) = \langle\psi_{ba}|\mathsf{s}\rangle &= \int_{-\infty}^{+\infty} \frac{\overline{\psi(a^{-1}(x-b))}}{\sqrt{|a|}}\, \mathsf{s}(x)\, dx \\ &= \sqrt{|a|}\int_{-\infty}^{+\infty} \overline{\widehat{\psi}(a\xi)}\, \hat{\mathsf{s}}(\xi)\, e^{ib\xi}\, d\xi\,. \end{aligned} \tag{8.73}$$

Then (8.71) ensures that the wavelet transform preserves the "energy" of the signal:

$$\|s\|^2 = \frac{1}{c_\psi} \int_{-\infty}^{\infty} db \int_0^{\infty} \frac{da}{a^2} |S(b,a)|^2 . \tag{8.74}$$

To a certain extent, this equation is equivalent to the resolution of the unity by the continuous family $\{\psi_{ba}\}$:

$$I_d = \frac{1}{c_\psi} \int_{-\infty}^{\infty} db \int_0^{\infty} \frac{da}{a^2} |\psi_{ba}\rangle \langle \psi_{ba}| . \tag{8.75}$$

In this regard, the states $|\psi_{ba}\rangle$ can be viewed as coherent states for the affine group.

In practice one often imposes on the analyzing wavelet ψ a number of additional properties, for instance, restrictions on the support of ψ and of $\widehat{\psi}$. Or ψ may be required to have a certain number $N \geq 1$ of *vanishing moments* (by the admissibility condition (8.72), the moment of order 0 must always vanish):

$$\int_{-\infty}^{\infty} x^n \psi(x)\, dx = 0, \quad n = 0, 1, \ldots, N . \tag{8.76}$$

This property improves its efficiency at detecting singularities in the signal. Indeed, the transform (8.73) is then blind to the smoothest part of the signal that is polynomial of degree up to N — and less interesting, in general. Only the sharper part remains, including all singularities (like jumps in the signal or one of its derivatives). For instance, if the first moment ($n = 1$) vanishes, the transform will erase any linear *trend* in the signal.

Let us give two well-known examples of wavelets, which are depicted in Figure 8.1.

The Mexican hat or Marr wavelet: This is simply the second derivative of a Gaussian:

$$\psi_H(x) = (1 - x^2)\, e^{-x^2/2}, \quad \widehat{\psi_H}(\xi) = \xi^2\, e^{-\xi^2/2} . \tag{8.77}$$

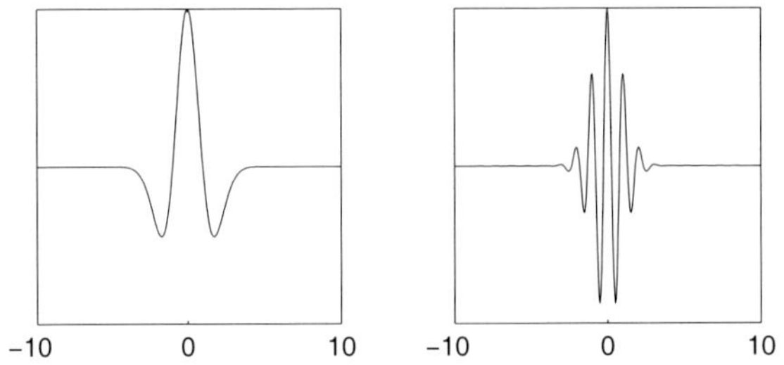

Fig. 8.1 Two usual wavelets: The Mexican hat or Marr wavelet (left); the real part of the Morlet wavelet (right), ($\xi_0 = 5.6$).

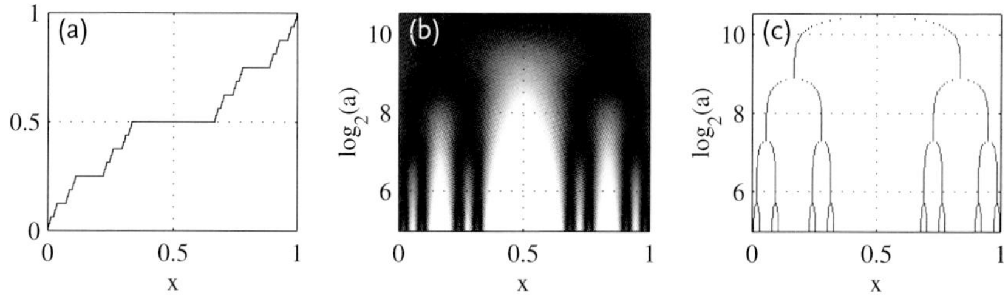

Fig. 8.2 Continous wavelet transform of a fractal function: (a) the devil's staircase; (b) its wavelet transform (with the first derivative of a Gaussian); (c) the corresponding skeleton, that is, the set of lines of local maxima (by courtesy of Pierre Vandergheynst, EPFL).

It is a real wavelet, with two vanishing moments ($n = 0, 1$). Similar wavelets, with more vanishing moments, are obtained by taking higher derivatives of the Gaussian:

$$\psi_H^{(m)}(x) = \left(\frac{1}{i}\frac{d}{dx}\right)^m e^{-x^2/2}, \quad \widehat{\psi_H^{(m)}}(\xi) = \xi^m \, e^{-\xi^2/2}. \tag{8.78}$$

The Morlet wavelet: This is just a modulated Gaussian:

$$\begin{aligned} \psi_M(x) &= \pi^{-1/4}\left(e^{i\xi_0 x} - e^{-\xi_0^2/2}\right) e^{-x^2/2} \\ \widehat{\psi_M}(\xi) &= \pi^{-1/4}\left[e^{-(\xi-\xi_0)^2/2} - e^{-\xi^2/2}\, e^{-\xi_0^2/2}\right]. \end{aligned} \tag{8.79}$$

In fact the first term alone does *not* satisfy the admissibility condition; hence the necessity for a correction. However, for ξ_0 large enough (typically $\xi_0 \geq 5.5$), this correction term is numerically negligible ($\leq 10^{-4}$). The Morlet wavelet is complex; hence, the corresponding transform $S(b, a)$ is also complex. This enables one to deal separately with the phase and the modulus of the transform, and the phase turns out to be a crucial ingredient in most algorithms used in applications such as feature detection.

We give in Figure 8.2 an example of such a "time-scale" representation of a signal, namely, the quite transient "devil's staircase", an increasing fractal function whose the derivative is zero everywhere with the exception of a Cantor set of Lebesgue measure zero.

9
Another Family of $SU(1,1)$ Coherent States for Quantum Systems

9.1
Introduction

This chapter is devoted to another family of $SU(1,1)$ coherent states that naturally emerges from the quantum motion in infinite-well and trigonometric Pöschl–Teller potentials. The latter are defined by [111]:

$$V(x) \equiv V_{\lambda,\kappa}(x) = \frac{1}{2} V_0 \left(\frac{\lambda(\lambda-1)}{\cos^2 \frac{x}{2a}} + \frac{\kappa(\kappa-1)}{\sin^2 \frac{x}{2a}} \right), \quad 0 \leqslant x \leqslant \pi a\,, \tag{9.1}$$

and are represented in Figure 9.1 for different values of the parameters (λ, κ).

The material is mainly borrowed from [112], and from more recent research on the subject. It is also a direct illustration of a construction of coherent states that was proposed by Klauder and the author in [58]. In the construction of these states, we take advantage of the simplicity of the solutions, which ultimately stems from the fact they share a common $SU(1,1)$ symmetry à la Barut–Girardello [113]. Indeed, the Pöschl–Teller potentials share with their infinite-well limit the nice property of being analytically integrable. The reason behind this can be understood within a group-theoretical context: these potentials possess an underlying dynamical algebra, namely, $\mathfrak{su}(1,1)$ and the discrete series representations of the latter. We know from the previous chapter that the discrete series unitary irreducible representations of the Lie algebra $\mathfrak{su}(1,1)$ are labeled by a parameter η, which takes its values in $\{\frac{1}{2}, 1, \frac{3}{2}, 2, \dots\}$ for the discrete series *sensu stricto*, and in $[\frac{1}{2}, +\infty)$ for the extension to the universal covering of the group $SU(1,1)$. The relation between the Pöschl–Teller parameters and η is given by $2\eta - 1 = \lambda + \kappa$, and the limit case $\lambda, \kappa \to 1^+$ corresponds to $\eta = \frac{3}{2}$.

9.2
Classical Motion in the Infinite-Well and Pöschl–Teller Potentials

Let us consider a particle trapped in an infinite square well, that is, confined in the interval $0 < x < \pi a$, and also in the Pöschl–Teller family (9.1) viewed as regularizations of such an infinite potential.

Coherent States in Quantum Physics. Jean-Pierre Gazeau

ISBN: 978-3-527-40709-5

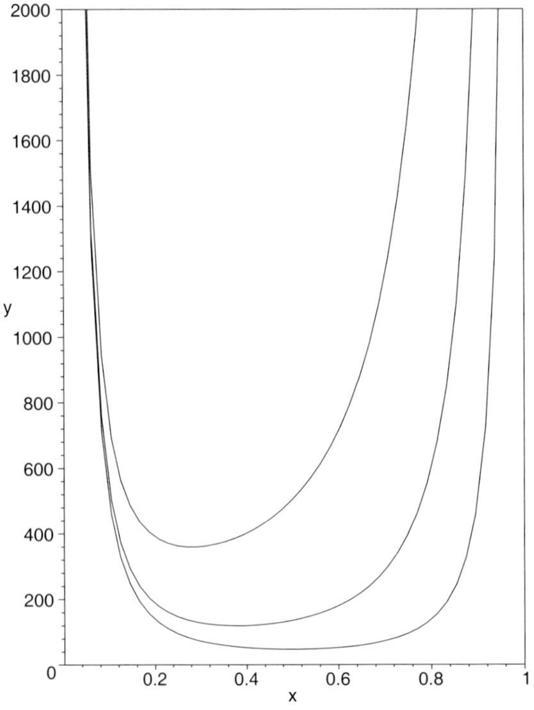

Fig. 9.1 The Pöschl–Teller potential $V(x) = \frac{1}{2}V_0\left[\lambda(\lambda-1)\cos^{-2}\frac{x}{2a} + \kappa(\kappa-1)\sin^{-2}\frac{x}{2a}\right]$, with $a = \pi^{-1}$ and for $(\lambda,\kappa) = (4,4), (4,8), (4,16)$ (from bottom to top). Source Antoine *et al.* [112] (reprinted with permission from [Antoine, J.-P., Gazeau, J.-P., Monceau, P., Klauder J.R., Penson K.A., Temporally stable coherent states for infinite well, J. Math. Phys., 42, p. 2349, 2001], American Institute of Physics).

9.2.1
Motion in the Infinite Well

Let us first review the classical behavior of a particle of mass m trapped in an infinite well of width πa, an elementary but not so trivial model. For a nonzero energy $E = \frac{1}{2}mv^2$, there corresponds a speed $v = \sqrt{\frac{2E}{m}}$ for a position $0 < x < \pi a$. There are perfect reflections at the boundaries of the well. So the motion is periodic with period (the "round-trip time") T equal to

$$T = \frac{2\pi a}{v} = 2\pi a\sqrt{\frac{m}{2E}}\,. \tag{9.2}$$

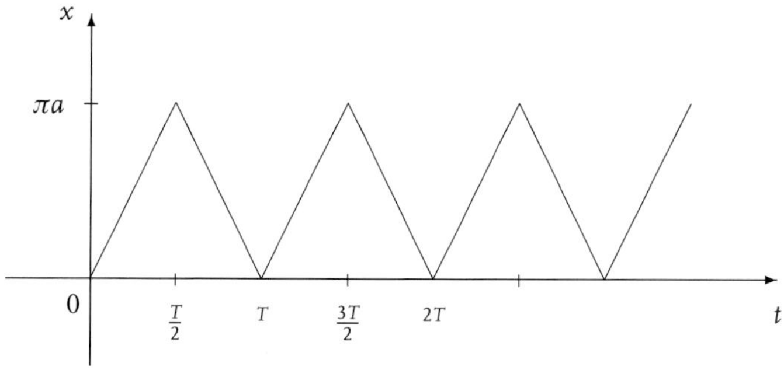

Fig. 9.2 The position $x(t)$ of the particle trapped in an infinite square well of width πa, as a function of time.

With the initial condition $x(0) = 0$, the time behavior of the position is then given by (see Figure 9.2)

$$\begin{aligned} 0 \leqslant t \leqslant \tfrac{1}{2}T: &\quad x = vt\,, \\ \tfrac{1}{2}T \leqslant t \leqslant T: &\quad x = 2\pi a - vt\,, \end{aligned} \tag{9.3}$$

and of course $x(t + nT) = x(t)$.

Consequently the velocity is a periodized Haar function:

$$v = v \sum_{n=0}^{+\infty} \left[\chi_{[nT,(n+\frac{1}{2})T]} - \chi_{[(n+\frac{1}{2})T,(n+1)T]}\right] \tag{9.4}$$

(here χ_B denotes the characteristic function of a set $B \in \mathbb{R}$), whereas the acceleration is the superposition of two Dirac combs on the half-line:

$$\gamma = \sum_{n=0}^{+\infty} \left[\delta_{nT} - \delta_{(n+\frac{1}{2})T}\right] .$$

The average position and average velocity of the particle are then

$$\overline{x} = \frac{1}{T}\int_0^T x(t)\,dt = \frac{\pi a}{2}\,, \quad \overline{v} = 0\,, \tag{9.5}$$

whereas the mean square dispersions are

$$\sqrt{\overline{x^2} - \overline{x}^2} = \frac{\pi a}{2\sqrt{3}}\,, \quad \sqrt{\overline{v^2} - \overline{v}^2} = \sqrt{\frac{2E}{m}}\,. \tag{9.6}$$

Figure 9.3 shows the phase trajectory of the system. This trajectory encircles a surface of area equal to the action variable $A = \frac{1}{2\pi}\oint p\,dq = mva$, where $q = x$ and $p = mv$ are canonically conjugate. Note the other expressions for A:

$$A = \frac{2\pi a^2 m}{T} = \frac{mv^2 T}{2\pi} = a\sqrt{2mE}\,. \tag{9.7}$$

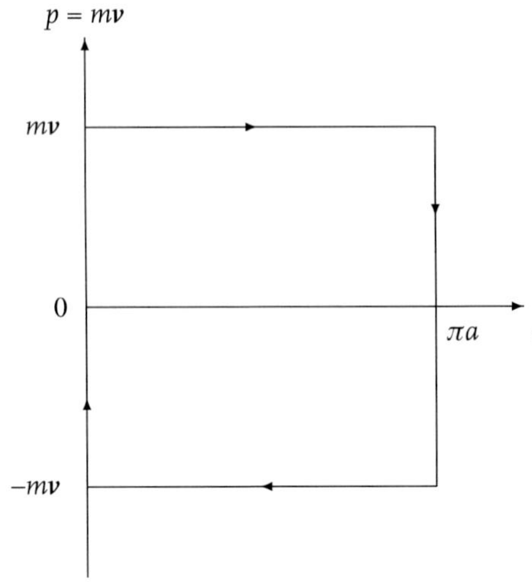

Fig. 9.3 Phase trajectory of the particle in an infinite square well.

9.2.2
Pöschl–Teller Potentials

The solution to the equations of motion with the potentials (9.1) is straightforward, in spite of the rather heavy expression of the latter. The turning points $x_\pm$ of the periodic motion at a given energy E are given by

$$x_\pm = a \arccos\left[\frac{\alpha-\beta}{2} \pm \sqrt{\Delta}\right], \tag{9.8}$$

where

$$\Delta = (1 - \tfrac{1}{2}(\sqrt{\alpha}+\sqrt{\beta})^2)(1 - \tfrac{1}{2}(\sqrt{\alpha}-\sqrt{\beta})^2),$$

$\alpha = \frac{V_0}{E}\lambda(\lambda-1)$, $\beta = \frac{V_0}{E}\kappa(\kappa-1)$. So the motion is possible only if

$$E > \frac{V_0}{2}\left(\sqrt{\lambda(\lambda-1)} + \sqrt{\kappa(\kappa-1)}\right)^2. \tag{9.9}$$

The time evolution of the position is given by

$$x(t) = a \arccos\left[\frac{\alpha-\beta}{2} + \sqrt{\Delta}\cos\left(\sqrt{\frac{2E}{m}}\frac{t}{a}\right)\right], \quad x(0) = x_-. \tag{9.10}$$

Hence, the period is

$$T = 2\pi a\sqrt{\frac{m}{2E}}. \tag{9.11}$$

It is remarkable that the period T does not depend on the strength V_0, nor on λ and κ.

The action variable A satisfies the relation $\frac{dA}{dE} = \frac{T}{2\pi}$, and thus $A = a\sqrt{2mE} +$ const. The constant is determined by the condition that $A = 0$ for $E = V_{\min}$, that is, const. $= -a\sqrt{2mV_{\min}}$. The Pöschl–Teller potential $V(x)$ reaches its minimum at the location x_o defined by

$$\tan^2 \frac{x_o}{2a} = \sqrt{\frac{\kappa(\kappa-1)}{\lambda(\lambda-1)}} . \tag{9.12}$$

So we have, in agreement with (9.9),

$$V_{\min} = V(x_o) = \frac{V_0}{2}\left[\sqrt{\lambda(\lambda-1)} + \sqrt{\kappa(\kappa-1)}\right]^2 , \tag{9.13}$$

and consequently

$$A = a\sqrt{2mE} - a\sqrt{mV_0}[\sqrt{\lambda(\lambda-1)} + \sqrt{\kappa(\kappa-1)}] . \tag{9.14}$$

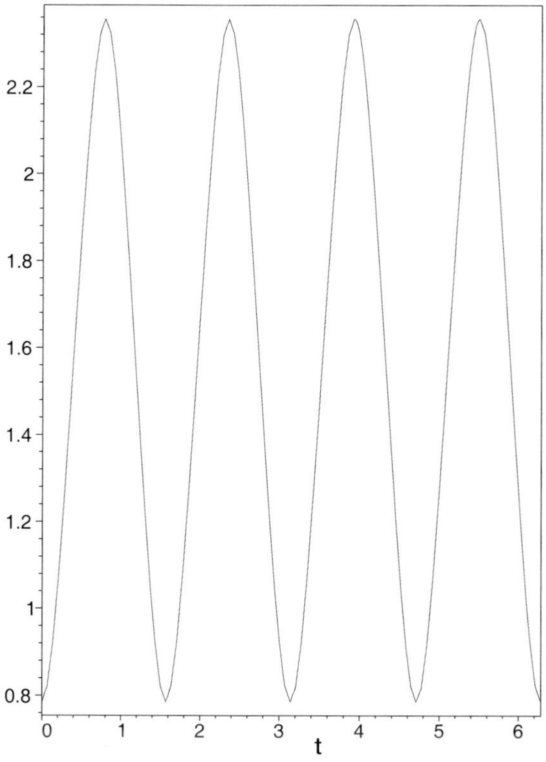

Fig. 9.4 The position $x(t)$ of the particle in the symmetric Pöschl–Teller potential $\lambda = \kappa = 2$ for energy $E = 8V_0$ and period $T = \frac{\pi}{2}$ (cf. Figure 9.2). Source Antoine *et al.* [112] (reprinted with permission from [Antoine, J.-P., Gazeau, J.-P., Monceau, P., Klauder J.R., Penson K.A., Temporally stable coherent states for infinite well, J. Math. Phys., 42, p. 2349, 2001], American Institute of Physics).

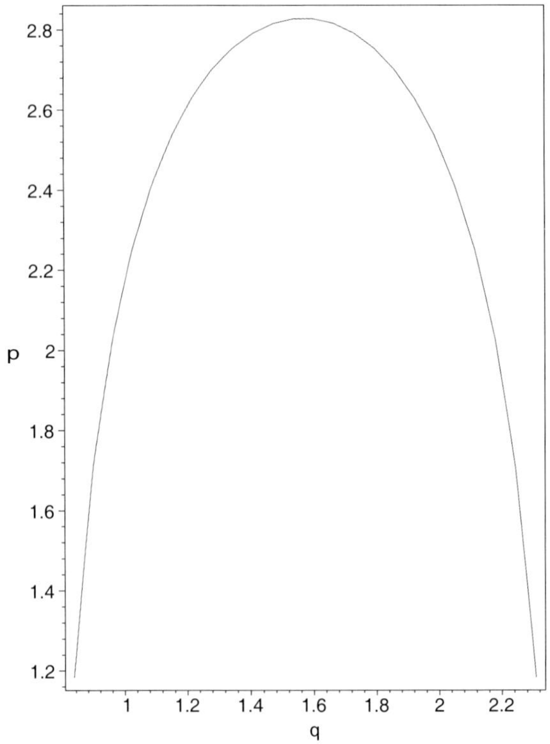

Fig. 9.5 Upper part of the phase trajectory of the particle in the symmetric (2,2) Pöschl–Teller system, for the same values of *E* and *T* as in Figure 9.4 (cf. Figure 9.3). Source Antoine *et al.* [112] (reprinted with permission from [Antoine, J.-P., Gazeau, J.-P., Monceau, P., Klauder J.R., Penson K.A., Temporally stable coherent states for infinite well, J. Math. Phys., 42, p. 2349, 2001], American Institute of Physics).

It is worthwhile comparing (9.11) and (9.14) with their respective infinite-well counterparts (9.2) and (9.7). We should also check that the time behavior (9.10) of $x(t)$ goes into (9.3) at the limits $\alpha, \beta \to 0$. In Figures 9.4 and 9.5 we show the time evolution of the position of the particle trapped in a Pöschl–Teller potential and the corresponding phase trajectory in the plane $(q = x,\ p = mv)$.

Note that, in the general case, the equation for the latter reads (at energy E)

$$p = \pm \frac{\sqrt{2mE}}{\sin \frac{q}{a}} \left[1 - (\alpha + \beta) + (\alpha - \beta) \cos \frac{q}{a} - \cos^2 \frac{q}{a} \right]^{1/2} . \tag{9.15}$$

Finally, let us give the canonical transformation leading to the action–angle variables:

$$\varphi = \arccos \frac{1}{\sqrt{\Delta}} \left[\cos \frac{q}{a} - \frac{\alpha - \beta}{2} \right] A = a \left[p^2 + 2m V(q) \right]^{1/2} . \tag{9.16}$$

9.3 Quantum Motion in the Infinite-Well and Pöschl–Teller Potentials

9.3.1 In the Infinite Well

Any quantum system trapped inside the infinite well $0 \leqslant x \leqslant \pi a$ must have its wave function equal to zero outside the well. It is thus natural to impose on the wave functions the boundary conditions

$$\psi^{\mathrm{iw}}(x) = 0, \quad x \geqslant \pi a \quad \text{and} \quad x \leqslant 0 . \tag{9.17}$$

Since the movement takes place only inside the interval $[0, \pi a]$, we may also ignore the rest of the line and replace the conditions (9.17) by the following ones:

$$\psi^{\mathrm{iw}} \in L^2([0, \pi a], dx) , \quad \psi^{\mathrm{iw}}(0) = \psi^{\mathrm{iw}}(\pi a) = 0 . \tag{9.18}$$

Alternatively, one may consider the periodized well and impose the same periodic boundary conditions, namely, $\psi^{\mathrm{iw}}(n\pi a) = 0$, $\forall n \in \mathbb{Z}$.

In either case, stationary states of the trapped particle of mass m are easily found from the eigenvalue problem for the Schrödinger operator. For reasons to be justified in the sequel, we choose the shifted Hamiltonian:

$$H \equiv H_{\mathrm{w}} = -\frac{\hbar^2}{2m}\frac{d^2}{dx^2} - \frac{\hbar^2}{2ma^2} . \tag{9.19}$$

Then

$$\Psi^{\mathrm{iw}}(x, t) = e^{-\frac{i}{\hbar}Ht}\Psi^{\mathrm{iw}}(x, 0) , \tag{9.20}$$

where $\Psi^{\mathrm{iw}}(x, 0) \equiv \psi^{\mathrm{iw}}(x)$ obeys the eigenvalue equation

$$H\psi^{\mathrm{iw}}(x) = E\psi^{\mathrm{iw}}(x) , \tag{9.21}$$

together with the boundary conditions (9.17). Normalized eigenstates and correspxonding eigenvalues are then given by

$$\psi_n^{\mathrm{iw}}(x) = \sqrt{\frac{2}{\pi a}}\sin(n+1)\frac{x}{a} \equiv \langle x|\psi_n^{\mathrm{iw}}\rangle , \quad 0 \leqslant x \leqslant \pi a , \tag{9.22}$$

$$H|\psi_n^{\mathrm{iw}}\rangle = E_n|\psi_n^{\mathrm{iw}}\rangle , \quad n = 0, 1, \ldots , \tag{9.23}$$

$$E_n = \frac{\hbar^2}{2ma^2}n(n+2) \equiv \hbar\omega x_n , \tag{9.24}$$

with

$$\omega = \frac{\hbar}{2ma^2} \equiv \frac{2\pi}{T_r} \quad \text{and } x_n = n(n+2), \quad n = 0, 1, \ldots ,$$

where T_r is the “revival” time to be compared with the purely classical round-trip time given in (9.2). Now the Bohr–Sommerfeld quantization rule applied to the classical action gives

$$a\sqrt{2mE} = A = (n+1)\hbar\,, \tag{9.25}$$

so

$$E = (n+1)^2 \frac{\hbar^2}{2ma^2} = E_n + \frac{\hbar^2}{2ma^2}\,, \quad n = 0, 1, \ldots \tag{9.26}$$

Thus, here the Bohr–Sommerfeld quantization is exact [17], despite the presence of the extra term $\hbar^2/2ma^2$ that follows from our particular choice of zero in the energy scale (see (9.19)).

9.3.2 In Pöschl–Teller Potentials

Pöschl–Teller potentials were originally introduced in a quantum molecular physics context. The energy eigenvalues and corresponding eigenstates are solutions to the Schrödinger equation

$$\left[-\frac{\hbar^2}{2m}\frac{d^2}{dx^2} + \frac{V_0}{2}\left(\frac{\lambda(\lambda-1)}{\cos^2\frac{x}{2a}} + \frac{\kappa(\kappa-1)}{\sin^2\frac{x}{2a}}\right) - \frac{\hbar^2}{8ma^2}(\lambda+\kappa)^2\right]\psi_n^{\text{pt}}(x) = E\psi_n^{\text{pt}}(x)\,, \tag{9.27}$$

with $0 \leqslant x \leqslant \pi a$ and where we have also shifted the Hamiltonian of the trapped particle of mass m by an amount equal to $-\frac{\hbar^2}{8ma^2}(\lambda+\kappa)^2$. Here too, as for the infinite well, we have the choice of putting the potential equal to infinity outside the interval $[0, \pi a]$, or periodizing the problem, with period $2\pi a$.

Since the potential strength is overdetermined by specifying V_0, λ, and κ simultaneously, we can freely put for convenience, as in [111, 114],

$$V_0 = \frac{\hbar^2}{4ma^2}\,. \tag{9.28}$$

With this choice, and the boundary conditions $\psi^{\text{pt}}(0) = \psi^{\text{pt}}(\pi a) = 0$, the normalized eigenstates, and the corresponding eigenvalues, all of them simple, are given by

$$\begin{aligned}\psi_n^{\text{pt}}(x) &= [c_n(\kappa,\lambda)]^{-\frac{1}{2}}\left(\cos\frac{x}{2a}\right)^{\lambda}\left(\sin\frac{x}{2a}\right)^{\kappa} \\ &\quad\times {}_2F_1\left(-n,\, n+\lambda+\kappa;\, \kappa+\frac{1}{2};\, \sin^2\frac{x}{2a}\right)\,,\end{aligned} \tag{9.29}$$

where $c_n(\kappa,\lambda)$ is a normalization factor that can be given analytically when κ and λ are positive integers, and

$$E_n = \frac{\hbar^2}{2ma^2}\, n(n+\lambda+\kappa) \equiv \hbar\omega x_n\,, \quad n = 0, 1, \ldots\,, \tag{9.30}$$

with

$$\omega = \frac{\hbar}{2ma^2}\,, \qquad x_n = n(n+\lambda+\kappa),\ \lambda, \kappa > 1\,. \tag{9.31}$$

Note that the Bohr–Sommerfeld rule applied to the canonical action (9.14) yields (here we do *not* impose the normalization (9.28))

$$a\sqrt{2mE} - a\sqrt{mV_0}\left[\sqrt{\lambda(\lambda-1)} + \sqrt{\kappa(\kappa-1)}\right] = \hbar(n+\tfrac{1}{2})\,,$$

that is,

$$E_n = \frac{\hbar^2}{2ma^2}(n+\tfrac{1}{2})^2 + \frac{\hbar}{ma}\sqrt{mV_0}(n+\tfrac{1}{2})\left[\sqrt{\lambda(\lambda-1)} + \sqrt{\kappa(\kappa-1)}\right] \tag{9.32}$$

$$+\frac{V_0}{2}\left[\sqrt{\lambda(\lambda-1)} + \sqrt{\kappa(\kappa-1)}\right]^2\,. \tag{9.33}$$

This formula is interesting on two counts at least.

(a) The first term in (9.32) gives, apart from the term $\frac{1}{2}$ in $(n+\frac{1}{2})$, the exact spectrum of the infinite well. More precisely, these values of the energy may be obtained simply by letting $V_0 \to 0$ in $V(x)$ and keeping in mind that $V = \infty$ outside $[0, \pi a]$.

(b) In the limit $V_0 \to \infty$, the first term in (9.32) can be neglected and one is left, up to a global, V_0-dependent, shift, with the spectrum of a harmonic oscillator with elementary quantum

$$\hbar\omega = \hbar\sqrt{\frac{V_0}{ma^2}}\left[\sqrt{\lambda(\lambda-1)} + \sqrt{\kappa(\kappa-1)}\right]\,.$$

Hence, the Pöschl–Teller potential interpolates between the square well and the harmonic oscillator.

9.4 The Dynamical Algebra $\mathfrak{su}(1,1)$

Behind the spectral structure of the infinite well or Pöschl–Teller Hamiltonians, there exists a dynamical algebra generated by lowering and raising operators acting on $|e_n\rangle \equiv |\psi_n^{\mathrm{iw}}\rangle$ or $|\psi_n^{\mathrm{pt}}\rangle$. The latter are defined by

$$a|e_n\rangle = \sqrt{x_n}\,|e_{n-1}\rangle\,, \tag{9.34}$$

$$a^\dagger|e_n\rangle = \sqrt{x_{n+1}}\,|e_{n+1}\rangle\,, \tag{9.35}$$

with

$$x_n = n(n+2)\,, \quad \text{for the infinite well}\,,$$

$$x_n = n(n+\lambda+\kappa)\,, \quad \text{for the Pöschl–Teller potential}\,, \quad n = 0, 1, 2, \ldots$$

Then we observe that the operator $X_N = a^\dagger a$ is diagonal with eigenvalues x_n: $X_N|e_n\rangle = x_n|e_n\rangle$. Note that the number operator N,

$$N|e_n\rangle = n|e_n\rangle\,, \tag{9.36}$$

is given in terms of X_N by

$$N = -\frac{1}{2}(\lambda+\kappa) + \left(X_N + \frac{1}{4}(\lambda+\kappa)^2\right)^{1/2}\,. \tag{9.37}$$

For any diagonal operator Δ with eigenvalues δ_n

$$\Delta|e_n\rangle = \delta_n|e_n\rangle\,, \tag{9.38}$$

we denote its finite difference by Δ'. The latter is defined as the diagonal operator with eigenvalues $\delta'_n \equiv \delta_{n+1} - \delta_n$,

$$\Delta'|e_n\rangle = \delta'_n|e_n\rangle\,. \tag{9.39}$$

More generally, the mth finite difference $\Delta^{(m)}$ will be recursively defined by

$$\Delta^{(m)} = (\Delta^{(m-1)})'\,. \tag{9.40}$$

Now, from the infinite matrix representation (in the basis $\{|e_n\rangle\}$) of the operators a and $a^\dagger$,

$$a = \begin{pmatrix} 0 & \sqrt{x_1} & 0 & 0 & \cdots \\ 0 & 0 & \sqrt{x_2} & 0 & \cdots \\ 0 & 0 & 0 & \sqrt{x_3} & \cdots \\ \cdots & \cdots & \cdots & \cdots & \cdots \end{pmatrix}, \tag{9.41}$$

$$a^\dagger = \begin{pmatrix} 0 & 0 & 0 & 0 & \cdots \\ \sqrt{x_1} & 0 & 0 & 0 & \cdots \\ 0 & \sqrt{x_2} & 0 & 0 & \cdots \\ 0 & 0 & \sqrt{x_3} & 0 & \cdots \\ \cdots & \cdots & \cdots & \cdots & \cdots \end{pmatrix}, \tag{9.42}$$

it is easy to check that

$$[a, a^\dagger] = \begin{pmatrix} x_1 - x_0 & 0 & \cdots & \\ 0 & x_2 - x_1 & \cdots & 0 \\ 0 & 0 & x_3 - x_2 & \end{pmatrix} = X'_N\,, \tag{9.43}$$

$$X'_N|e_n\rangle = x'_n|e_n\rangle, \quad x'_n = x_{n+1} - x_n = 2n + 3\,, \quad \text{or} \quad 2n + 1 + \lambda + \kappa\,. \tag{9.44}$$

We also check that, for any diagonal operator Δ, we have

$$[a, \Delta] = \Delta' a\,, \quad [a^\dagger, \Delta] = -a^\dagger \Delta'\,. \tag{9.45}$$

Therefore,

$$[a, X'_N] = X''_N a \,,$$

with

$$X''_N |e_n\rangle = x''_n |e_n\rangle = (x'_{n+1} - x'_n)|e_n\rangle = 2|e_n\rangle \,. \tag{9.46}$$

So $X''_N = 2I_d$, $X'''_N = 0$, and $[a, X'_N] = 2a$. Similarly, $[a^\dagger, X'_N] = -2a^\dagger$. In summary, there exists a "dynamical" Lie algebra, which is generated by $\{a, a^\dagger, X'_N\}$. Then the commutation rules

$$[a, a^\dagger] = X'_N \,, \quad [a, X'_N] = 2a \,, \quad [a^\dagger, X'_N] = -2a^\dagger \tag{9.47}$$

clearly indicate that it is isomorphic to

$$\mathfrak{su}(1,1) \sim \mathfrak{sl}(2,\mathbb{R}) \sim \mathfrak{so}(2,1) \,. \tag{9.48}$$

A more familiar basis for (9.48) is given (in $\mathfrak{so}(2,1)$ notation) by

$$L^- = \frac{1}{\sqrt{2}} a \,, \quad L^+ = \frac{1}{\sqrt{2}} a^\dagger \,, \quad L_{12} = \frac{1}{2} X'_N \,, \tag{9.49}$$

where L_{12} is the generator of the compact subgroup $SO(2)$, namely,

$$[L^\pm, L_{12}] = \mp L^\pm \,, \quad [L^-, L^+] = L_{12} \,. \tag{9.50}$$

If we add the operator X_N (i.e., the Hamiltonian H) to the set $\{a, a^\dagger, X'_N\}$, we obtain an infinite-dimensional Lie algebra contained in the enveloping algebra. Indeed

$$\begin{aligned}
&[a, X_N] = X'_N a \,, \quad [a^\dagger, X_N] = -a^\dagger X'_N \\
&[a, X'_N a] = 2a^2 \,, \quad [a^\dagger, X'_N a] = -{X'_N}^2 - 2X_N \,, \\
&\text{etc} \ldots
\end{aligned} \tag{9.51}$$

Note also the relation between X_N and X'_N:

$$X_N = \tfrac{1}{4}\left({X'_N}^2 - 2X'_N - 3\right) \,, \text{ or } \tfrac{1}{4}\left({X'_N}^2 - 2X'_N - (\lambda + \kappa + 1)(\lambda + \kappa)\right) \,. \tag{9.52}$$

In the same vein, we note that the condition $X'''_N = 0$ is necessary to obtain a genuine Lie algebra (instead of a subset of the enveloping algebra). Therefore, $\mathfrak{su}(1,1)$ is the *only* dynamical Lie algebra that can arise in such a problem.

It follows from the considerations above that the space $\mathcal{H}$ of states $|e_n\rangle$ carries some representation of $\mathfrak{su}(1,1)$. The latter is found by examining the formulas for the $\mathfrak{su}(1,1)$ discrete series representation [99].

Given $\eta = \frac{1}{2}, 1, \frac{3}{2}, \ldots$, the discrete series unitary irreducible representation U_η is realized on a generic separable Hilbert space $\mathcal{H}_\eta$ with basis $\{|\eta, n\rangle, n \in \mathbb{N}\}$ through the following actions of the Lie algebra elements in the realization (9.49) (slightly different from the Fock–Bargman realization (8.50) introduced in the previous chapter),

$$L_{12}|\eta, n\rangle = (\eta + n)|\eta, n\rangle \,, \tag{9.53}$$

$$L^{-}|\eta, n\rangle = \frac{1}{\sqrt{2}}\sqrt{(2\eta + n - 1)n}\,|\eta, n-1\rangle \,, \tag{9.54}$$

$$L^{+}|\eta, n\rangle = \frac{1}{\sqrt{2}}\sqrt{(2\eta + n)(n+1)}\,|\eta, n+1\rangle \,. \tag{9.55}$$

The representation U_η fixes the Casimir operator $Q = -L_{12}(L_{12} - 1) + 2L^{+}L_{-}$ to the following value: $Q\mathcal{H}_\eta = \eta(\eta - 1)\mathcal{H}_\eta$. Using (9.54) and (9.55), and comparing with (9.34), (9.35), and (9.44), we obtain the specific value of η for the infinite-well problem, namely, $\eta = \frac{3}{2}$, so we can make the identifications $\mathcal{H}_{3/2} \equiv \mathcal{H}$, $|\frac{3}{2}, n\rangle \equiv |\psi_n^{\mathrm{iw}}\rangle$. On the other hand, we obtain a continuous range of values for the Pöschl–Teller potentials,

$$\eta = \frac{\lambda + \kappa + 1}{2} > \frac{3}{2} \,, \tag{9.56}$$

and we shall denote the corresponding Hilbert spaces and states (9.22) by $\mathcal{H}_\eta$ and $|\eta, \psi_n^{\mathrm{pt}}\rangle$, respectively. The relation (9.56) simply means that here we are in the presence of the (abusively called) discrete series representations of the universal covering of $SU(1,1)$, except for the interval $\eta \in \left(\frac{1}{2}, \frac{3}{2}\right)$.

9.5
Sequences of Numbers and Coherent States on the Complex Plane

In a general setting, consider a strictly increasing sequence of positive numbers

$$0 = x_0 < x_1 < x_2 \ldots < x_n < \ldots \,, \tag{9.57}$$

that are eigenvalues of a self-adjoint positive operator X_N in some separable Hilbert space $\mathcal{H}$,

$$X_N|e_n\rangle = x_n|e_n\rangle \,, \tag{9.58}$$

where the set $\{|e_n\rangle, n \in \mathbb{N}\}$ is an orthonormal basis of $\mathcal{H}$. There corresponds to (9.57) a (generically infinite) dynamical Lie algebra with basis $\{a, a^\dagger, X'_N, \ldots\}$, with the notation of the previous section. There also corresponds the sequence of "factorials" $x_n! = x_1 x_2 \ldots x_n$ with $x_0! \stackrel{\mathrm{def}}{=} 1$ and the "exponential"

$$\mathcal{N}(t) = \sum_{n=0}^{+\infty} \frac{t^n}{x_n!} \,, \tag{9.59}$$

with a radius of convergence

$$R = \limsup_{n \to +\infty} \sqrt[n]{x_n!} \tag{9.60}$$

that is assumed to be nonzero, of course. We next suppose that there exists a probability distribution $t \mapsto \pi(t)$ on $[0, R)$ such that

$$x_n! = \int_0^R u^n \, \pi(u) \, du \,. \tag{9.61}$$

Otherwise said, the *Stieltjes moment problem* [115, 116] has a solution for the sequence of factorials $(x_n!)_{n\in\mathbb{N}}$. We know that a necessary and sufficient condition for this is that the two matrices

$$\begin{pmatrix} 1 & x_1! & x_2! & \dots & x_n! \\ x_1! & x_2! & x_3! & \dots & x_{n+1}! \\ x_2! & x_3! & x_4! & \dots & x_{n+2}! \\ \vdots & \vdots & \vdots & \ddots & \vdots \\ x_n! & x_{n+1}! & x_{n+2}! & \dots & x_{2n}! \end{pmatrix} , \tag{9.62}$$

$$\begin{pmatrix} x_1! & x_2! & x_3! & \dots & x_{n+1}! \\ x_2! & x_3! & x_4! & \dots & x_{n+2}! \\ x_3! & x_4! & x_5! & \dots & x_{n+3}! \\ \vdots & \vdots & \vdots & \ddots & \vdots \\ x_{n+1}! & x_{n+2}! & x_{n+3}! & \dots & x_{2n+1}! \end{pmatrix} \tag{9.63}$$

have strictly positive determinants for all n.

Let us consider in the complex plane the open disk $\mathcal{D}_{\sqrt{R}}$ with radius $\sqrt{R}$ and centered at the origin, and the Hilbert space $L^2\left(\mathcal{D}_{\sqrt{R}}, \mu_\pi(d^2z)\right)$ of complex-valued functions that are square-integrable on $\mathcal{D}_{\sqrt{R}}$ with respect to the measure $\mu_\pi(d^2z) \stackrel{\text{def}}{=} \pi(|z|^2)\, d^2z/\pi$. We then choose in this space the set of monomials in $\bar{z}$

$$\phi_n(\bar{z}) = \frac{\bar{z}^n}{\sqrt{x_n!}} \,. \tag{9.64}$$

By construction, this set is orthonormal with respect to the measure μ_π, and it satisfies

$$\sum_{n=0}^{\infty} |\phi_n(z)|^2 = \mathcal{N}(|z|^2) \,.$$

Hence, we are in the Bayesian statistical context described in Section 5.3, that is, the interplay between two probability distributions:

- For each z, the discrete distribution,

$$n \mapsto \frac{|\phi_n(z)|^2}{\mathcal{N}(|z|^2)} \,, \tag{9.65}$$

  which could be considered as concerning experiments performed on the system within some experimental protocol, say, $\mathcal{E}$, to measure the spectral values of X_N.

- For each n, the continuous distribution on $(\mathcal{D}_{\sqrt{R}}, \mu_\pi)$,

$$\mathcal{D}_{\sqrt{R}} \ni z \mapsto |\phi_n(z)|^2 . \tag{9.66}$$

Following our general scheme of coherent state construction, we get in the present situation the family $\{|z\rangle,\ z \in \mathcal{D}_{\sqrt{R}}\}$ of coherent states in the Hilbert space $\mathcal{H}$:

$$|z\rangle = \frac{1}{\sqrt{\mathcal{N}(|z|^2)}} \sum_{n\geqslant 0} \overline{\phi_n(z,\bar{z})}\, |e_n\rangle = \frac{1}{\sqrt{\mathcal{N}(|z|^2)}} \sum_{n\geqslant 0} \frac{z^n}{\sqrt{x_n!}} |e_n\rangle . \tag{9.67}$$

They are normalized, $\langle z|z\rangle = 1$, and resolve the unity in $\mathcal{H}$ with respect to the measure $\mathcal{N}(|z|^2)\,\mu_\pi(d^2 z)$:

$$\int_{\mathcal{D}_{\sqrt{R}}} \frac{d^2 z}{\pi}\, \pi(|z|^2)\, \mathcal{N}(|z|^2)\, |z\rangle\langle z| = I_{\mathcal{H}} . \tag{9.68}$$

Moreover, these coherent states are eigenvectors of the operator a defined in (9.34):

$$a|z\rangle = z|z\rangle . \tag{9.69}$$

Now suppose that X_N is (up to a factor) the Hamiltonian for a quantum system,

$$H = \hbar\omega X_N . \tag{9.70}$$

Then the coherent states (9.67) evolve in time as

$$e^{-\frac{i}{\hbar}Ht} |z\rangle = \frac{1}{\sqrt{\mathcal{N}(|z|^2)}} \sum_{n\geqslant 0} \frac{z^n}{\sqrt{x_n!}} e^{-i\omega x_n t} |e_n\rangle . \tag{9.71}$$

We know that if $x_n \propto n$, that is, in the case of the harmonic oscillator, the temporal evolution of the coherent state $|z\rangle$ reduces to a rotation in the complex plane, namely, $e^{-iHt/\hbar}|z\rangle = |z\, e^{-i\omega t}\rangle$. In general, however, we lose the temporal stability of our family of coherent states (9.67). Hence, to restore it, we must extend our original definitions to the entire family of temporal evolution orbits:

$$\{e^{-i\frac{H}{\hbar}t}|z\rangle,\ z \in \mathcal{D}_{\sqrt{R}},\ t \in I\} . \tag{9.72}$$

The interval I is the whole real line when x_n is generic, whereas it can be restricted to a period, that is, a finite interval $[a, b]$,

$$b - a = \frac{2\pi}{\omega\alpha} \tag{9.73}$$

if $x_n \in \alpha\mathbb{N}$. A straightforward calculation now shows that

$$\langle z|H|z\rangle = \langle z|\hbar\omega X_N|z\rangle = \hbar\omega|z|^2 . \tag{9.74}$$

Therefore, the quantity $|z|^2$ is the average energy evaluated in the elementary quantum unit $\hbar\omega$. Note that

$$\hbar|z|^2 \equiv J \tag{9.75}$$

is simply the action variable in the case where H is the Hamiltonian of the harmonic oscillator and the variable z is given the meaning of a classical state in the phase space $\mathbb{C}$.

On the other hand, introducing the dimensionless number

$$\gamma = \omega t\,, \quad \gamma \in \omega I\,, \tag{9.76}$$

we are naturally led to study the continuous family of states (9.72)

$$|z,\gamma\rangle \stackrel{\text{def}}{=} e^{-i\gamma\frac{H}{\hbar\omega}}|z\rangle = \frac{1}{\sqrt{\mathcal{N}(J/\hbar)}}\sum_{n\geqslant 0}\frac{z^n\, e^{-i\gamma x_n}}{\sqrt{x_n!}}|e_n\rangle\,. \tag{9.77}$$

These states, parameterized by $(z,\gamma) \in \mathcal{D}_{\sqrt{R}} \times I$, may also be called "coherent" for several reasons. First they are, by construction, eigenvectors of the operator

$$a(\gamma) \equiv e^{-i\gamma H/\hbar\omega}\, a\, e^{i\gamma H/\hbar\omega}\,, \tag{9.78}$$

namely,

$$a(\gamma)|z,\gamma\rangle = z|z,\gamma\rangle\,. \tag{9.79}$$

They obey the temporal stability condition

$$e^{-iHt/\hbar}|z,\gamma\rangle = |z,\gamma+\omega t\rangle\,. \tag{9.80}$$

Again, if we consider the harmonic oscillator case, we do not make any distinction between the argument of the complex parameter z and the angle variable γ, since then $x_n = n$ and $z^n e^{-i\gamma n} = (ze^{-i\gamma})^n$, so the only parameters we need are $J = \hbar|z|^2$ and γ. The latter are easily identified with the classical action–angle variables. From now on, we will stick to the minimal parameterization set in the present generalization and shall denote our coherent states by

$$|J,\gamma\rangle = \frac{1}{\sqrt{\mathcal{N}(J)}}\sum_{n\geqslant 0}\frac{J^{n/2}\, e^{-i\gamma x_n}}{\sqrt{x_n!}}|e_n\rangle\,, \tag{9.81}$$

where we have put $\hbar = 1$ for convenience. However, we now have to identify the measure on the observation set $X \equiv \{(J,\gamma),\ J \in [0,R),\ \gamma \in I\}$ for which they resolve the unity, or, up to the factor $\mathcal{N}(J)$, for which the set of functions

$$\chi_n(J,\gamma) = \frac{J^{n/2}\, e^{-i\gamma x_n}}{\sqrt{x_n!}} \tag{9.82}$$

is orthonormal. For a generic real sequence of x_n's and for $I = \mathbb{R}$, the following measure is defined, in the *Bohr's sense* [117], on functions on X by

$$f \mapsto \mu^B(f) = \int_X \mu^B(d\,J\,d\gamma)\, f(J,\gamma) \overset{\text{def}}{=} \lim_{\Gamma\to\infty} \frac{1}{2\Gamma} \int_{-\Gamma}^{\Gamma} d\gamma \int_0^R f(J,\gamma)\pi(J)\, d\,J\,. \tag{9.83}$$

The orthogonality of the functions χ_n derives from the orthogonality of the Fourier exponentials:

$$\lim_{\Gamma\to\infty} \frac{1}{2\Gamma} \int_{-\Gamma}^{\Gamma} e^{i\gamma(x_n - x_{n'})} d\gamma = \lim_{\Gamma\to\infty} \frac{\sin\Gamma\,(x_n - x_{n'})}{\Gamma\,(x_n - x_{n'})} = \delta_{nn'}\,. \tag{9.84}$$

Of course, for a sequence $\{x_n,\ n \in \mathbb{N}\}$ of integers, the measure on $\mathbb{R}$ in the sense of Bohr reduces to the ordinary normalized measure on a period interval, for example, $[0, 2\pi)$.

We thus obtain the resolution of the unity by the coherent states $|\,J,\gamma\rangle$:

$$\int_X \mu^B(d\,J\,d\gamma)\, \mathcal{N}(J) |\,J,\gamma\rangle\langle J,\gamma| = I_{\mathcal{H}}\,. \tag{9.85}$$

In a suitable way [58] (see also the discussion in Section 9.7.4), it is also acceptable to regard the parameterization (J,γ) as "action–angle" variables, and it is convenient to refer to them as such, even when keeping in mind the possibility of extending $\sqrt{J}$ to the complex plane, that is, replacing $\sqrt{J}$ by z.

9.6 Coherent States for Infinite-Well and Pöschl–Teller Potentials

9.6.1 For the Infinite Well

Let us now adapt the material of the previous section to our problem of the infinite well. In that case,

$$x_n! = x_1\, x_2\, \ldots\, x_n = \frac{n!(n+2)!}{2}\,, \tag{9.86}$$

$$|\,J,\gamma\rangle = \frac{1}{\sqrt{\mathcal{N}(J)}} \sum_{n\geqslant 0} \frac{J^{n/2}\, e^{-i\gamma n(n+2)}}{\sqrt{\frac{n!(n+2)!}{2}}} |\psi_n^{\mathrm{iw}}\rangle\,. \tag{9.87}$$

The normalization factor is easily calculated in terms of the modified Bessel function I_ν [18].

$$\mathcal{N}(J) = 2\sum_{n=0}^{+\infty} \frac{J^n}{n!(n+2)!} = \frac{2}{J} I_2(2\sqrt{J})\,. \tag{9.88}$$

The radius of convergence $R = \limsup_{n\to+\infty} \sqrt[n]{\frac{n!(n+2)!}{2}}$ is of course infinite. Moreover, since the x_n's are here natural numbers, the interval of variation of the evolution parameter γ can be chosen as $I = [0, 2\pi]$.

The generalized factorials $x_n!$ arise as moments of a probability distribution $\pi^{\mathrm{iw}}(u)$,

$$x_n! = \int_0^{\infty} u^n \, \pi^{\mathrm{iw}}(u)\, du \,, \tag{9.89}$$

and $\pi^{\mathrm{iw}}(u)$ is explicitly given in terms of the other modified Bessel function K_ν [18],

$$\pi^{\mathrm{iw}}(u) = u K_2(2\sqrt{u}) \,. \tag{9.90}$$

It results from the previous section that the family $\{|J,\gamma\rangle,\ J \in \mathbb{R}^+, \gamma \in [0, 2\pi]\}$ resolves the unity operator,

$$I_d = \int |J,\gamma\rangle\langle J,\gamma|\, \mathcal{N}(J)\, \mu^{\mathrm{iw}}(d J\, d\gamma) \,, \tag{9.91}$$

with

$$\int (\cdot)\mu^{\mathrm{iw}}(d J\, d\gamma) = \frac{1}{2\pi}\int_0^{2\pi} d\gamma \int_0^{+\infty} \pi^{\mathrm{iw}}(J)(\cdot)\, d J \,. \tag{9.92}$$

As is well known, the overlap of two coherent states does not vanish in general. Explicitly, we have

$$\langle J',\gamma'|J,\gamma\rangle = \frac{2}{\sqrt{\mathcal{N}(J)\mathcal{N}(J')}} \sum_{n\geqslant 0} \frac{(J J')^{\frac{n}{2}}}{n!(n+2)!}\, e^{-in(n+2)(\gamma-\gamma')} \,. \tag{9.93}$$

If $\gamma = \gamma'$, we obtain a Bessel function

$$\langle J',\gamma\,|\,J,\gamma\rangle = \frac{2}{(J J')^{\frac{1}{2}}\sqrt{\mathcal{N}(J)\mathcal{N}(J')}}\, I_2\left(2(J J')^{\frac{1}{4}}\right) . \tag{9.94}$$

If $\gamma \neq \gamma'$, we can give an integral representation of (9.93) in terms of a theta function and Bessel functions [18]:

$$\begin{aligned}\langle J',\gamma'|J,\gamma\rangle = {} & \frac{e^{i(\gamma-\gamma')/4}}{i\pi\sqrt{\mathcal{N}(J)\mathcal{N}(J')}} \int_0^{\pi} d\varphi\, \theta_1\left(\frac{\varphi}{\pi}, -\frac{\gamma-\gamma'}{\pi}\right) \\ & \times \left[\frac{-e^{-i(\varphi-\gamma+\gamma')}}{(J J')^{\frac{1}{2}}}\, I_2\left(2(J J')^{\frac{1}{4}}\, e^{i(\varphi-\frac{\gamma-\gamma'}{2}+\frac{\pi}{2})}\right)\right. \\ & \left. + \frac{e^{i(\varphi+\gamma-\gamma')}}{(J J')^{\frac{1}{2}}}\, I_2\left(2(J J')^{\frac{1}{4}}\, e^{-i(\varphi+\frac{\gamma-\gamma'}{2}-\frac{\pi}{2})}\right)\right] .\end{aligned} \tag{9.95}$$

9.6.2
For the Pöschl–Teller Potentials

The relations (9.86) and (9.87) of the previous section are easily generalized to the present case. We shall list them without unnecessary comments.

From the energies $E_n = \hbar\omega\, x_n$ given by (9.30), we get the moments

$$x_n! = x_1\, x_2 \ldots x_n = n!\, \frac{\Gamma(n+\nu+1)}{\Gamma(\nu+1)}\,, \tag{9.96}$$

with $\nu = \lambda + \kappa > 2$.

Thus, the coherent states read as

$$|J,\gamma\rangle = \frac{[\Gamma(\nu+1)]^{1/2}}{\sqrt{\mathcal{N}(J)}} \sum_{n\geq 0} \frac{J^{n/2} e^{-i\gamma n(n+\nu)}}{[n!\,\Gamma(n+\nu+1)]^{\frac{1}{2}}}\, |\psi_n^{\mathrm{pt}}\rangle\,. \tag{9.97}$$

The normalization is then given by

$$\mathcal{N}(J) = \Gamma(\nu+1) \sum_{n\geqslant 0} \frac{J^n}{n!\,\Gamma(n+\nu+1)} = \frac{\Gamma(\nu+1)}{J^{\nu/2}} I_\nu(2\sqrt{J})\,. \tag{9.98}$$

The radius of convergence R is infinite. The interval of variation of the evolution parameter γ is generically the whole real line, unless the parameter ν is an integer.

The numbers $x_n!$ are moments of a probability distribution $\pi^{\mathrm{pt}}(u)$ involving the modified Bessel function K_ν:

$$x_n! = \int_0^\infty u^n\, \pi^{\mathrm{pt}}(u) du\,, \tag{9.99}$$

with (cf. (9.90))

$$\pi^{\mathrm{pt}}(u) = \frac{2}{\Gamma(\nu+1)}\, u^{\nu/2} K_\nu(2\sqrt{u})\,. \tag{9.100}$$

It might be useful to recall here the well-known relation between modified Bessel functions [18],

$$K_\nu(z) = \frac{\pi}{2\sin\pi\nu}[I_{-\nu}(z) - I_\nu(z)]\,, \quad \nu \notin \mathbb{Z}\,. \tag{9.101}$$

The resolution of the unity is then explicitly given by

$$I_d = \int |J,\gamma\rangle\langle J,\gamma|\, \mathcal{N}(J)\, \mu^{\mathrm{pt}}(dJ\, d\gamma)\,, \tag{9.102}$$

with

$$\int (\cdot)\mu^{\mathrm{pt}}(dJ\, d\gamma) = \lim_{\Gamma\to\infty} \frac{1}{2\Gamma} \int_{-\Gamma}^{\Gamma} d\gamma \left[\int_0^{+\infty} \pi^{\mathrm{pt}}(J)\, (\cdot)\, dJ\right]\,. \tag{9.103}$$

Finally, the overlap between two coherent states is given by the series

$$\langle J', \gamma' | J, \gamma \rangle = \frac{\Gamma(\nu+1)}{\sqrt{\mathcal{N}(J)\mathcal{N}(J')}} \sum_{n \geqslant 0} \frac{(J J')^{n/2}}{n!\Gamma(n+\nu+1)} e^{-in(n+\nu)(\gamma-\gamma')} , \tag{9.104}$$

which reduces to a Bessel function for $\gamma = \gamma'$:

$$\langle J', \gamma \,|\, J, \gamma \rangle = \frac{\Gamma(\nu+1)}{\sqrt{\mathcal{N}(J)\mathcal{N}(J')}} \frac{I_\nu(2(J J')^{1/4})}{(J J')^{\nu/4}} . \tag{9.105}$$

At this point, we should emphasize the fact that, when $\gamma = 0$ and J is taken as a complex parameter, our temporally stable families of coherent states (9.87) and (9.97) are nothing other but the temporal evolution orbits of the well-known Barut–Girardello coherent states for $SU(1,1)$ [113].

In addition, we should also quote Nieto and Simmons [118], who considered the infinite square well and the Pöschl–Teller potentials as examples of their construction of coherent states. The latter are required to minimize an uncertainty relation or, equivalently, to be eigenvectors of some "lowering operator" A (à la Barut–Girardello [113]). However, those states have a totally different meaning and should be considered only in the semiclassical limit.

9.7 Physical Aspects of the Coherent States

In this section, we shall review some of the spatial and temporal features of the coherent states described in [112], treating together the infinite-well coherent states (9.81) and the Pöschl–Teller coherent states (9.97), the former being obtained from the latter simply by putting $\nu = \lambda + \kappa = 2$.

9.7.1 Quantum Revivals

As the (infinite) superposition of stationary states which are spatially and temporally periodic for integer values of ν, they should display nonambiguous revivals and fractional revivals. Let us first recall the main definitions concerning the notion of revival, as given in [119]. For other related works, see [120–125, 127] and the recent review by Robinett [128].

A *revival* of a wave function occurs when a wave function evolves in time to a state closely reproducing its initial form. A *fractional revival* occurs when the wave function evolves in time to a state that can be described as a collection of spatially distributed sub-wave functions, each of which closely reproduces the shape of the initial wave function. If a revival corresponds to phase alignments of nearest-neighbor energy eigenstates that constitute the wave function, it can be asserted that a fractional revival corresponds to phase alignments of nonadjacent energy eigenstates that constitute this wave function.

For a general wave packet of the form

$$|\psi(t)\rangle = \sum_{n\geqslant 0} c_n\, e^{-iE_n t/\hbar}|e_n\rangle\,, \tag{9.106}$$

with $\sum_{n\geqslant 0}|c_n|^2 = 1$, the concept of revival arises from the weighting probabilities $|c_n|^2$. Suppose that the expansion (9.106) is strongly weighted around a mean value $\langle n\rangle$ for the number operator N, $N|e_n\rangle = n|e_n\rangle$:

$$\langle\psi|N|\psi\rangle = \sum_{n\geqslant 0} n\,|c_n|^2 \equiv \langle n\rangle\,. \tag{9.107}$$

Let $\bar{n} \in \mathbb{N}$ be the integer closest to $\langle n\rangle$. Assuming that the spread $\sigma \approx \Delta n \equiv \left[\langle n^2\rangle - \langle n\rangle^2\right]^{1/2}$ is small compared with $\langle n\rangle \approx \bar{n}$, we expand the energy E_n in a Taylor series in n around the centrally excited value $\bar{n}$:

$$E_n \simeq E_{\bar{n}} + E'_{\bar{n}}(n-\bar{n}) + \frac{1}{2}E''_{\bar{n}}(n-\bar{n})^2 + \frac{1}{6}E'''_{\bar{n}}(n-\bar{n})^3 + \ldots\,, \tag{9.108}$$

where each prime on $E_{\bar{n}}$ denotes a derivative. These derivatives define distinct time scales [120], namely, the *classical period* $T_{\text{cl}} = 2\pi\hbar/|E'_{\bar{n}}|$; the *revival time* $t_{\text{rev}} = 2\pi\hbar/\frac{1}{2}|E''_{\bar{n}}|$; the *superrevival time* $t_{\text{sr}} = 2\pi\hbar/\frac{1}{6}|E'''_{\bar{n}}|$; and so on. Inserting this expansion into the evolution factor $e^{-iE_n t/\hbar}$ of (9.106) allows us to understand the possible occurrence of a quasiperiodic revival structure of the wave packet (9.106) *according to* the weighting probability $n \mapsto |c_n|^2$. In the present case, we have

$$E_n = \frac{\hbar}{2ma^2}\, n(n+\nu) = \frac{\hbar}{2ma^2}\left[\bar{n}(\bar{n}+\nu) + (2\bar{n}+\nu)(n-\bar{n}) + (n-\bar{n})^2\right]\,. \tag{9.109}$$

So the first characteristic time is the "classical" period

$$T_{\text{cl}} = \frac{2\pi\hbar}{2\bar{n}+\nu}\,\frac{2ma^2}{\hbar^2} = \frac{2\pi ma^2}{\hbar(\bar{n}+\frac{\nu}{2})}\,, \tag{9.110}$$

which should be compared with the actual classical (Bohr–Sommerfeld) counterpart deduced from (9.11) and (9.14),

$$T = \frac{2\pi ma^2}{A + a\sqrt{mV_0}[\sqrt{\lambda(\lambda-1)} + \sqrt{\kappa(\kappa-1)}]}\,. \tag{9.111}$$

The second characteristic time is the revival time

$$t_{\text{rev}} = \frac{4\pi ma^2}{\hbar} = (2\bar{n}+\nu)T_{\text{cl}}\,. \tag{9.112}$$

There is no superrevival time here, because the energy is a quadratic function of n.

With these definitions, the wave packet (9.106) reads in the present situation (up to a global phase factor)

$$|\psi(t)\rangle = \sum_{n\geqslant 0} c_n\, e^{-2\pi i\left[(n-\bar{n})\frac{t}{T_{\text{cl}}} + (n-\bar{n})^2\frac{t}{t_{\text{rev}}}\right]}|e_n\rangle\,. \tag{9.113}$$

Hence, it will undergo motion with the classical period, modulated by the revival phase [126]. Since $T_{\rm cl} \ll t_{\rm rev}$ for large $\bar{n}$, the classical period dominates for small values of t (mod $t_{\rm rev}$), and the motion is then periodic with period $T_{\rm cl}$. As t increases from zero and becomes nonnegligible with respect to $t_{\rm rev}$, the revival term $(n - \bar{n})^2 \frac{t}{t_{\rm rev}}$ in the phase of (9.113) causes the wave packet to spread and collapse. The latter gathers into a series of subsidiary waves, the fractional revivals, which move periodically with a period equal to a rational fraction of $T_{\rm cl}$. Then, a full revival obviously occurs at each multiple of $t_{\rm rev}$.

To put into evidence these revival structures for a given wave packet $\psi(x,t) = \langle x|\psi(t)\rangle$, an efficient method is to calculate its autocorrelation function [126]:

$$\mathrm{A}(t) = \langle \psi(x,0)\,|\psi(x,t)\rangle = \sum_{n\geqslant 0} |c_n|^2\, e^{-iE_n t/\hbar}\,.$$

Numerically, $|\mathrm{A}(t)|^2$ varies between 0 and 1. The maximum $|\mathrm{A}(t)|^2 = 1$ is reached when $\psi(x,t)$ exactly matches the initial wave packet $\psi(x,0)$, and the minimum 0 corresponds to nonoverlapping: $\psi(x,t)$ is far from the initial state. On the other hand, fractional revivals and fractional "superrevivals" appear (in the general case) as periodic peaks in $|\mathrm{A}(t)|^2$ with periods that are rational fractions of the classical round-trip time $T_{\rm cl}$ and the revival time $t_{\rm rev}$.

9.7.2
Mandel Statistical Characterization

Since the weighting distribution $|c_n|^2$ is crucial for understanding the temporal behavior of the wave packet (9.106), it is worthwhile also giving some general precisions of a statistical nature [129–133]. before examining the special case of our coherent states. It is clear that the revival features will be more or less apparent, depending on the value of the deviation $(n - \bar{n})$ (relative to n) that is effectively taken into account in the construction of the wave packet. In this respect, it is interesting to compare $|c_n|^2$ with the Poissonian case $\langle n\rangle^n\, e^{-\langle n\rangle}/n!$ and with the Gaussian case, $(2\pi(\Delta n)^2)^{-1/2}\exp[-(n - \langle n\rangle)^2/2(\Delta n)^2]$.

A quantitative estimate is given by the so-called Mandel parameter Q [129, 131, 132] defined as follows:

$$Q_{\rm M} \overset{\rm def}{=} \frac{(\Delta n)^2}{\langle n\rangle} - 1\,, \tag{9.114}$$

where the average is calculated with respect to the discrete probability distribution $n \mapsto |c_n|^2$. In the Poissonian case, we have $Q_{\rm M} = 0$, that is, $\Delta n = \langle n\rangle^{1/2}$. We say that the weighting distribution is sub-Poissonian (or super-Poissonian) if $Q_{\rm M} < 0$ (or $Q_{\rm M} > 0$). In the super-Poissonian case, that is, $\Delta n > \langle n\rangle^{1/2}$, the set of states $|e_n\rangle$ that contribute significantly to the wave packet can be rather widely spread around $n \simeq \langle n\rangle$, and this may have important consequences for the properties of localization and temporal stability of the wave packet.

When the wave packets are precisely our coherent states (where $|e_n\rangle$ stands for both $|\psi^{\mathrm{iw}}\rangle$ and $|\psi^{\mathrm{pt}}\rangle$),

$$|J,\gamma\rangle = \frac{1}{\sqrt{\mathcal{N}(J)}} \sum_{n\geqslant 0} \frac{J^{n/2}\, e^{-ix_n\gamma}}{\sqrt{\varrho_n}} |e_n\rangle\,, \tag{9.115}$$

the weighting distribution depends on J,

$$|c_n|^2 = \frac{J^n}{\mathcal{N}(J)\,\varrho_n}\,, \tag{9.116}$$

and we can see the interesting statistical interplay with the probability distribution $\pi(J)$ of which the $x_n!$ are the moments; see (9.89).

The following mean values are easily computed, together with their asymptotic values for large J [18]:

$$\begin{aligned}\langle n\rangle &= \frac{J}{\mathcal{N}(J)}\frac{d}{dJ}\mathcal{N}(J) = J\frac{d}{dJ}\ln\mathcal{N}(J)\\ &= \sqrt{J}\frac{I_{\nu+1}(2\sqrt{J})}{I_\nu(2\sqrt{J})} = \sqrt{J} - \frac{\nu}{2} - \frac{1}{4} + O\left(\frac{1}{\sqrt{J}}\right).\end{aligned} \tag{9.117}$$

$$\begin{aligned}\langle n^2\rangle &= \frac{J}{\mathcal{N}(J)}\frac{d}{dJ}J\frac{d}{dJ}\mathcal{N}(J)\\ &= \sqrt{J}\frac{I_{\nu+1}(2\sqrt{J})}{I_\nu(2\sqrt{J})} + J\frac{I_{\nu+2}(2\sqrt{J})}{I_\nu(2\sqrt{J})}\\ &= \langle n\rangle + J\frac{I_{\nu+2}(2\sqrt{J})}{I_\nu(2\sqrt{J})} \approx \sqrt{J}(\sqrt{J}+1) \quad (J \gg 1)\,.\end{aligned} \tag{9.118}$$

So, the dispersion is

$$\begin{aligned}(\Delta n)^2 &= J\,\frac{I_{\nu+2}(2\sqrt{J})}{I_\nu(2\sqrt{J})} + \langle n\rangle - \langle n\rangle^2\\ &= \frac{J}{[I_\nu(2\sqrt{J})]^2}\left(I_{\nu+2}(2\sqrt{J})I_\nu(2\sqrt{J}) - [I_{\nu+1}(2\sqrt{J})]^2\right) + \sqrt{J}\frac{I_{\nu+1}(2\sqrt{J})}{I_\nu(2\sqrt{J})}\\ &\approx \frac{\sqrt{J}}{2}, \quad \text{for } J \text{ large.}\end{aligned} \tag{9.119}$$

Finally, the Mandel parameter is given explicitly by

$$Q_{\mathrm{M}} = J\frac{d}{dJ}\ln\frac{d}{dJ}\ln\mathcal{N}(J) = \sqrt{J}\left[\frac{I_{\nu+2}(2\sqrt{J})}{I_{\nu+1}(2\sqrt{J})} - \frac{I_{\nu+1}(2\sqrt{J})}{I_\nu(2\sqrt{J})}\right]. \tag{9.120}$$

It is easily checked that $(I_{\nu+1}(x))^2 \geqslant I_\nu(x)I_{\nu+2}(x)$, for any $x \geqslant 0$, and, thus, $Q_{\mathrm{M}} \leqslant 0$ for any $J \geqslant 0$. Note that $Q_{\mathrm{M}} \simeq 0$ for large J, while $Q_{\mathrm{M}} \simeq -J$ for small J. Therefore, $|c_n|^2$ is sub-Poissonian in the case of our coherent states, whereas a quasi-Poissonian behavior is restored at high J. This fact is important for understanding the curves presented in Figure 9.6a, which show the distributions

$$D(n,J,\nu) \equiv |c_n|^2 = \frac{1}{n!\,\Gamma(n+\nu+1)}\frac{J^{n+\nu/2}}{I_\nu(2\sqrt{J})} \tag{9.121}$$

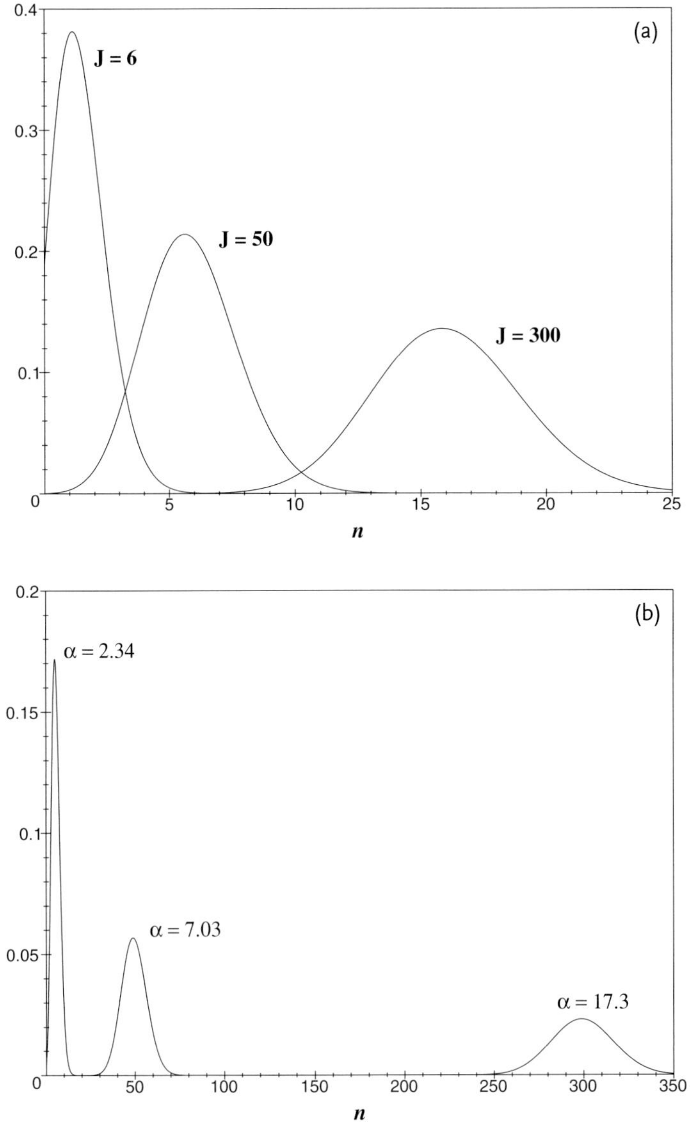

Fig. 9.6 (a) The weighting distribution $|c_n|^2 \equiv D(n, J, \nu)$ given in (9.121) for the infinite square well $\nu = 2$ and different values of J. Note the almost Gaussian shape at $J = 300$, centered at $n = \langle n \rangle = \sqrt{J - \frac{\nu}{2} - \frac{1}{4}} \simeq 16$, a width equal to $2\Delta n = \sqrt{2}J^{1/4} \simeq 5.9$. (b) The same for the harmonic oscillator: $|c_n|^2 = \frac{1}{n!}|\alpha|^{2n}e^{-|\alpha|^2}$. The values of α are chosen so as to get essentially the same mean energy values as in (a): $\alpha = \sqrt{J}$. Source Antoine *et al.* [112] (reprinted with permission from [Antoine, J.-P., Gazeau, J.-P., Monceau, P., Klauder J.R., Penson K.A., Temporally stable coherent states for infinite well, J. Math. Phys., 42, p. 2349, 2001], American Institute of Physics).

for $\nu = 2$ and different values of J. For the sake of comparison, we show in Figure 9.6b the corresponding distribution $|c_n|^2 = \frac{1}{n!}|\alpha|^{2n}e^{-|\alpha|^2}$ for the harmonic oscillator. Exactly as in the latter case, it can be shown easily that the distribution $D(n, J, \nu)$ tends for $J \to \infty$ to a Gaussian distribution. This Gaussian is centered at $\sqrt{J} - \frac{\nu}{2} - \frac{1}{4}$ and has a half-width equal to $\frac{1}{\sqrt{2}}J^{1/4}$:

$$D(n, J, \nu) \approx \frac{1}{\sqrt{\pi\sqrt{J}}} e^{-\left[n-\left(\sqrt{J}-\frac{\nu}{2}-\frac{1}{4}\right)\right]^2/\sqrt{J}} \quad (n \gg 1)\,. \tag{9.122}$$

9.7.3
Temporal Evolution of Symbols

We consider now the probability density $|\langle x|J,\gamma\rangle|^2$ as a function of the evolution parameter $\gamma = \omega t$ for increasing values of J. This evolution is shown in Figure 9.7 in the case of the infinite square well, for $J = 2$, 10, and 50. We can see at $\gamma = \pi = \frac{1}{2}t_{\mathrm{rev}}$ a perfect revival of the initial shape at $\gamma = 0$. This revival takes place near the opposite wall, as expected from the symmetry with respect to the center of the well. On the other hand, the ruling of the wave-packet evolution by the classical period

$$T_{\mathrm{cl}} = \frac{t_{\mathrm{rev}}}{2\bar{n} + \nu} = \frac{\pi}{\bar{n} + 1}$$

becomes more and more apparent as J increases. We also note that, at multiples of the half reversal time $\frac{1}{2}t_{\mathrm{rev}} = \pi$, the probability of localization near the walls increases with the energy J.

In Figure 9.8, we show the squared modulus

$$|\langle J,0|e^{-\frac{i}{\hbar}Ht}\,|\,J,0\rangle|^2 = |\langle J,0\,|\,J,\omega t\rangle|^2 \tag{9.123}$$

$$= \frac{\Gamma(\nu+1)}{N(J)^2}\left|\sum_{n\geqslant 0}\frac{J^n}{n!\Gamma(n+\nu+1)}e^{-i\omega n(n+\nu)t}\right|^2 \tag{9.124}$$

of the autocorrelation versus $\gamma = \omega t$ for the infinite well, for $J = 2$, 10, and 50. Like in Figure 9.7, we draw attention to the large-J regime. Here fractional revivals occur as intermediate peaks at rational multiples of the classical period

$$T_{\mathrm{cl}} = \frac{\pi}{\bar{n}+1} \approx \frac{\pi}{\sqrt{J}}\,, \quad J \gg 1\,,$$

and they tend to diminish as J increases, which is clearly the mark of a quasi-classical behavior. The same quantity is shown in Figure 9.9 for the Pöschl–Teller potential, for $J = 20$ and 40. Note that, in actual calculations like this, one has to choose a finite number of orthonormal eigenstates of the Pöschl–Teller potential, denoted here by $n_{\max}$. Correspondingly, the normalization of the coherent state $|J,\gamma\rangle$ has then to be modified as

$$\sum_{p=0}^{n_{\max}}\frac{J^p}{p!\Gamma(p+\kappa+\lambda)} = \frac{I_{\kappa+\lambda}(2\sqrt{J})}{J^{\frac{1}{2}(\kappa+\lambda)}} - \frac{J^{n_{\max}+1}}{(n_{\max}+1)!\,\Gamma(n_{\max}+\kappa+\lambda+2)} \times {}_1F_2(1; n_{\max}+2, n_{\max}+\kappa+\lambda+2; J)\,, \tag{9.125}$$

where ${}_1F_2$ is the hypergeometrical function.

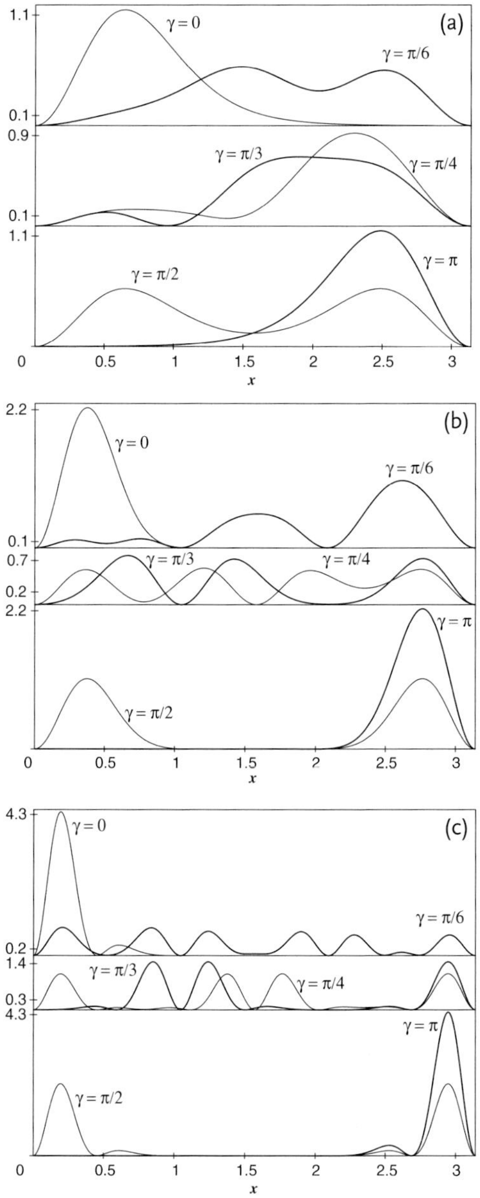

Fig. 9.7 The evolution (versus γ) of the probability density $|\langle x|J,\gamma\rangle|^2$, in the case of the infinite square well for (a) $J = 2$, (b) $J = 10$, and (c) $J = 50$. We note the perfect revival at $\gamma = \pi = \frac{1}{2}t_{\rm rev}$ (in suitable units), symmetrically with respect to the center of the well. Source Antoine *et al.* [112] (reprinted with permission from [Antoine, J.-P., Gazeau, J.-P., Monceau, P., Klauder J.R., Penson K.A., Temporally stable coherent states for infinite well, J. Math. Phys., 42, p. 2349, 2001], American Institute of Physics).

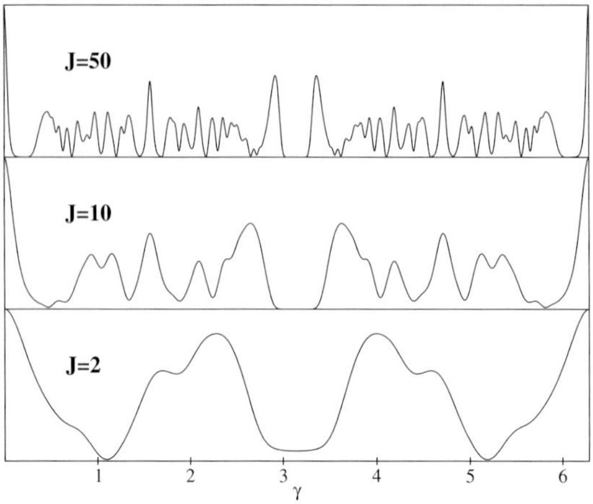

Fig. 9.8 Squared modulus $|\langle J,0 \mid J,\omega t\rangle|^2$ of the autocorrelation versus $\gamma = \omega t$ for the infinite square well, for $J = 2$, 10, and 50. As in Figure 9.7, the large-J regime is characterized by the occurrence of fractional revivals. Source Antoine *et al.* [112] (reprinted with permission from [Antoine, J.-P., Gazeau, J.-P., Monceau, P., Klauder J.R., Penson K.A., Temporally stable coherent states for infinite well, J. Math. Phys., 42, p. 2349, 2001], American Institute of Physics).

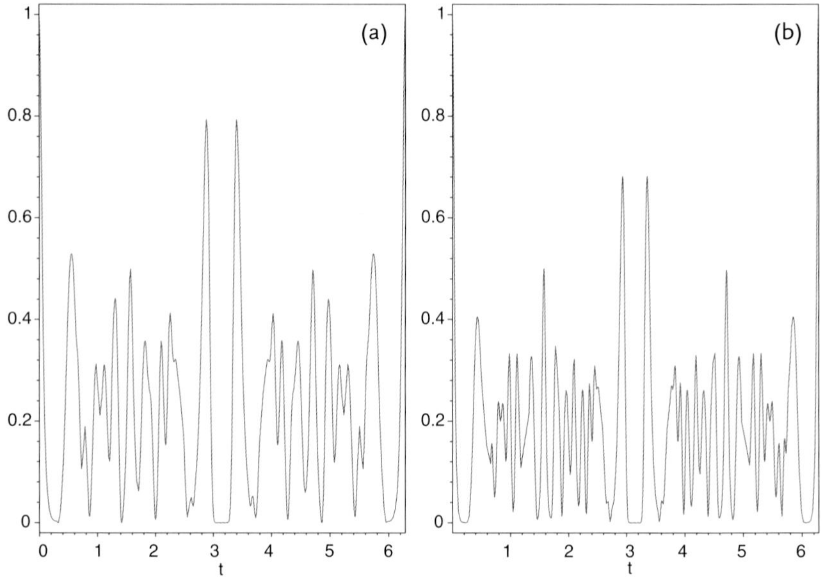

Fig. 9.9 Squared modulus $|\langle J,0 \mid J,\omega t\rangle|^2$ of the autocorrelation for the Pöschl–Teller potential with $n_{\max} = 10$, for (a) $J = 20$ and (b) $J = 40$. Source Antoine *et al.* [112] (reprinted with permission from [Antoine, J.-P., Gazeau, J.-P., Monceau, P., Klauder J.R., Penson K.A., Temporally stable coherent states for infinite well, J. Math. Phys., 42, p. 2349, 2001], American Institute of Physics).

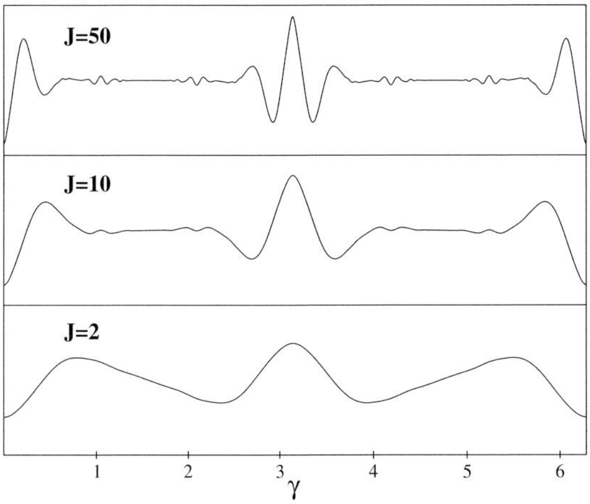

Fig. 9.10 Temporal behavior of the average position of the particle in the infinite square well (in the Heisenberg picture), $\langle J, 0 \mid Q(t) \mid J, 0\rangle = \langle J, \omega t = \gamma \mid Q \mid J, \omega t = \gamma\rangle$, as a function of $\gamma = \omega t$, for $J = 2$, 10, and 50. Source Antoine *et al.* [112] (reprinted with permission from [Antoine, J.-P., Gazeau, J.-P., Monceau, P., Klauder J.R., Penson K.A., Temporally stable coherent states for infinite well, J. Math. Phys., 42, p. 2349, 2001], American Institute of Physics).

Most interesting is the temporal behavior of the average position $\langle Q\rangle$ and more generally of the average of a well-defined operator $A(t)$ in such coherent states:

$$\begin{aligned}\langle J, 0|A(t)|J, 0\rangle &= \langle J, 0|e^{\frac{i}{\hbar}Ht}Ae^{-\frac{i}{\hbar}Ht}|J, 0\rangle \\ &= \langle J, \omega t = \gamma|A|J, \omega t = \gamma\rangle\,. \end{aligned} \tag{9.126}$$

This temporal behavior is shown in Figure 9.10 for the average position in the infinite square well, for $J = 2$, 10, and 50. We note the tendency to stability around the classical mean value $\frac{1}{2}\pi a$, except for strong oscillations of ultrashort duration between the walls near $\gamma = n\pi$. The latter increase with J as expected when one approaches the classical regime. For the sake of comparison, we show in Figure 9.11 the temporal behavior of the average position in the asymmetric Pöschl–Teller potential $(\kappa, \lambda) = (4, 8)$, for $J = 20$ and 50.

As some final information (but not the least important!), we show in Figure 9.12 the temporal behavior of the average position $\langle J, 0 \mid Q(t) \mid J, 0\rangle$ for the infinite square well, for a very high value of $J = 10^6$, near $\gamma = \omega t = 0$. Here the quasi-classical behavior is striking in the range of values considered for γ. These temporal oscillations are clearly governed by $T_{\text{cl}} \simeq \frac{\pi}{\sqrt{J}} = 3 \times 10^{-3}$ and should be compared with their purely classical counterpart in Figure 9.2.

We do not include in this study the temporal behavior of the momentum operator. Indeed, in the case of the infinite well, its tentative expression as an essentially self-adjoint operator raises well-known problems (see, e.g., [112, 134–137]). We will come back to this fundamental question in Chapter 15.

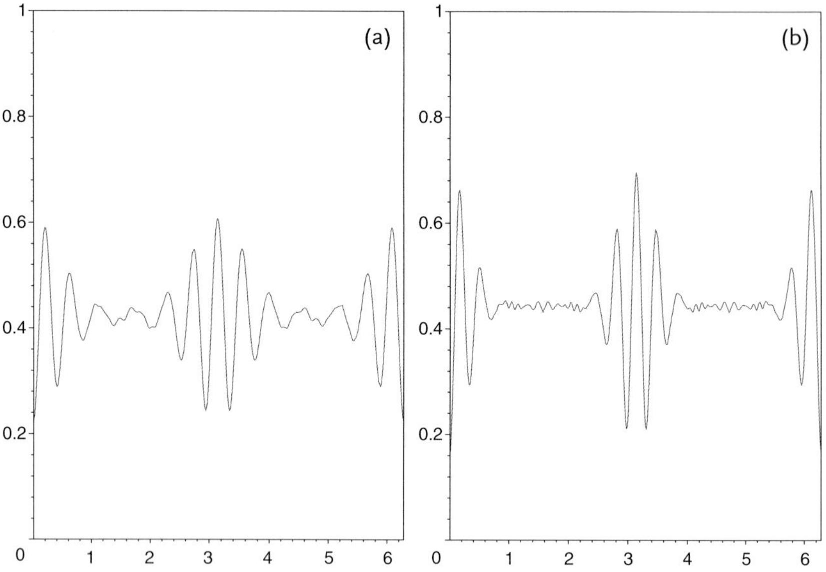

Fig. 9.11 Temporal behavior of the average position for the asymmetric Pöschl–Teller potential $(\lambda, \kappa) = (4, 8)$ with $n_{\max} = 10$, for (a) $J = 20$ and (b) $J = 50$. Source Antoine *et al.* [112] (reprinted with permission from [Antoine, J.-P., Gazeau, J.-P., Monceau, P., Klauder J.R., Pernon K.A., Temporally stable coherent states for infinite well, J. Math. Phys., 42, p. 2349, 2001], American Institute of Physics).

9.7.4
Discussion

The specific choices we have made here for the set of coherent states are based on two additional guiding principles besides continuity and resolution of unity, as was proposed in [58]. The first of these is "temporal stability", which in words asserts that the temporal evolution of any coherent state always remains a coherent state. The second of these, referred to as the "action identity" in [58], chooses variables for the coherent state labels that have as close a connection as possible to classical "action–angle" variables. In particular, for a single degree of freedom, the label pair (J, γ) is used to identify the coherent state $|J, \gamma\rangle$. Temporal stability means that, under the dynamics chosen, temporal evolution proceeds according to $|J, \gamma + \omega t\rangle$, for some fixed parameter ω. To ensure that (J, γ) describes action–angle variables, it is sufficient to require that the symplectic potential induced by the coherent states themselves is of Darboux form, or specifically that

$$i\hbar\langle J, \gamma|\, d\, |J, \gamma\rangle = J\, d\gamma\,,$$

where $d\,|J, \gamma\rangle \equiv |J + dJ, \gamma + d\gamma\rangle - |J, \gamma\rangle$. Temporal stability is what fixes the *phase* behavior of the coherent states, that is, the factor $e^{-i\gamma x_n}$ (cf. (9.77)), while ensuring

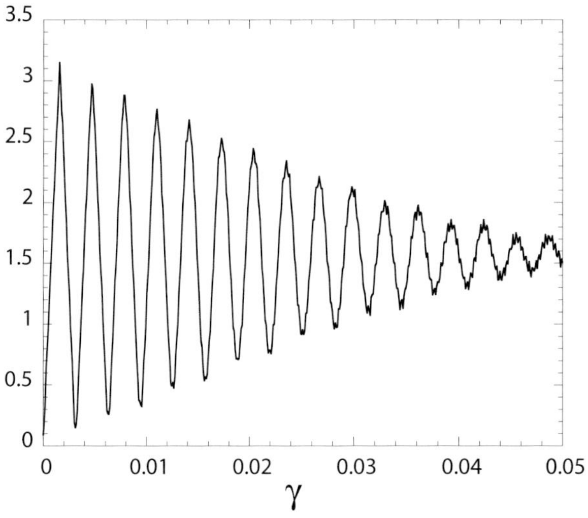

Fig. 9.12 Temporal behavior of the average position in the case of the infinite square well, for a very high value of $J = 10^6$. Source Antoine *et al.* [112] (reprinted with permission from [Antoine, J.-P., Gazeau, J.-P., Monceau, P., Klauder J.R., Penson K.A., Temporally stable coherent states for infinite well, J. Math. Phys., 42, p. 2349, 2001], American Institute of Physics).

that (J, γ) are canonical action–angle variables is what fixes the *amplitude* behavior of the coherent states, that is, $J^n\, e^{-i\gamma x_n}/(\mathcal{N}(J)\sqrt{[x_n]!})$ (cf. (9.81)).

In order for coherent states to interpolate well between quantum and classical mechanics, it is necessary for values of the action $J \gg \hbar$ that the quantum motion be well approximated by the classical motion. In particular, for a classical system with closed, localized trajectories, a suitable wave packet should, if possible, remain "coherent" for a number of classical periods. For the systems under study in this chapter, we have demonstrated the tendency for improved packet coherence with increasing J values within the range studied. For significantly larger values of J, we notice that the packet coherence substantially improves. Interesting results were obtained independently in a related study by Fox and Choi [138], who found a similar packet coherence for 10 or more classical periods for an infinite square well, even though they used a different amplitude prescription for their coherent states. In both approaches, however, the probability distribution shows a Gaussian behavior for large values of J, and this explains the similarity of the results.

It would appear that allowing for generalized phase and amplitude behavior in the definition of coherent states has led us closer to the idealized goal of a set of coherent states adapted to a chosen system and having a large number of properties in common with the associated classical system, despite being fully quantum in their characteristics.

10
Squeezed States and Their $SU(1,1)$ Content

10.1
Introduction

This chapter is devoted to another occurrence of $SU(1,1)$ in the construction of various popular quantum states. The so-called squeezed states, which are to be described here, pertain again to quantum optics and have been raising interest in quantum optics and other fields for the last three decades. Squeezed states, a name given by Hollenhorst [139], were initially introduced in quantum optics [140] for dealing with processes in which emission or absorption of two photons is involved. The formalism underlying the process involves the square of raising and lowering operators, a^2 and ${a^\dagger}^2$, for each mode of the quantized electromagnetic field. In this regard, squeezed states might be viewed as a sort of "two-photon" coherent states. Their mathematical properties had actually been investigated before [141–143]. The experimental evidence for such states was provided by Slusher *et al.* [144]. Interesting applications to quantum nondemolition in view of detecting gravitational waves were envisaged as early as 1981 [145, 146]. In addition, it is interesting to learn from Nieto [147] that these states were already known in 1927, 1 year after the Schrödinger states. For more details on theoretical and experimental aspects of squeezed states, see [25] and the recent report by Dell'Annoa *et al.* [148].

The introduction that is given here to the squeezed coherent states insists more particularly on their relations with the unitary irreducible representations of the symplectic group $Sp(2,\mathbb{R})$, another version of $SU(1,1)$. We will also stress, of course, their importance in quantum optics: they render possible reduction of the uncertainty on one of the two noncommuting observables present in the measurements of the electromagnetic field.

Coherent States in Quantum Physics. Jean-Pierre Gazeau

ISBN: 978-3-527-40709-5

10.2
Squeezed States in Quantum Optics

10.2.1
The Construction within a Physical Context

In Chapter 7, we showed the efficiency of the coherent state formalism in the quantum description of the driven oscillator. An ingoing free oscillator state, $|\Psi_{\text{in}}\rangle$, that experiences during a certain time the action of a driving force $F(t)$ is transformed into an outgoing free oscillator state, $|\Psi_{\text{out}}\rangle$, through the unitary action of an S matrix given by the displacement operator:

$$|\Psi_{\text{out}}\rangle = S^{\dagger}|\Psi_{\text{in}}\rangle\,, \quad S^{\dagger} = D\left(-i\frac{\hat{F}(\omega)}{\sqrt{2m\hbar\omega}}\right)\,, \tag{10.1}$$

with $\hat{F}(\omega) = \int_{-\infty}^{+\infty} e^{i\omega t}\, F(t)\, dt$. Therefore, starting from the ground state, $|\Psi_{\text{in}}\rangle = |0\rangle$, ones ends up with a coherent state $|\alpha\rangle$ with parameter $\alpha = -i\frac{\hat{F}(\omega)}{\sqrt{2m\hbar\omega}}$. For instance, for a constant force with unit duration, $F(t) = F_0\,\chi_{[0,1]}(t)$, where χ_S is the characteristic function of the set S, we obtain

$$|\Psi_{\text{out}}\rangle = \left|-\frac{F_0 e^{-i\frac{\omega}{2}}}{\sqrt{2m\hbar\omega}}\frac{\sin\frac{\omega}{2}}{\frac{\omega}{2}}\right\rangle\,. \tag{10.2}$$

As an example, we can imagine a charged particle, with mass m and charge e, subjected to a harmonic restoring force and acted on during a certain time by a constant electric field E_0 so that $F_0 = eE_0$. In the classical description, the action of this constant force amounts to displacing the potential energy parabola as

$$H_{\text{in}} = \frac{p^2}{2m} + \frac{1}{2}m\omega^2x^2 \to H_{\text{pert}} = \frac{p^2}{2m} + \frac{1}{2}m\omega^2x^2 - F_0x\,. \tag{10.3}$$

The quantum counterpart of this reads as

$$H_{\text{in}} = \hbar\omega\left(a^{\dagger}a + \frac{1}{2}\right) \to H_{\text{pert}} = \hbar\omega\left(a^{\dagger}a + \frac{1}{2}\right) - F_0\sqrt{\frac{\hbar}{2m\omega}}(a + a^{\dagger})\,. \tag{10.4}$$

The coherent states corresponding to the shifted parabola retain a Gaussian shape (e.g., in position representation and up to a phase factor), the same as for the original ground state. Now, one can explore the consequences of adding, at the classical level, a quadratic term to the potential:

$$H_{\text{in}} = \frac{p^2}{2m} + \frac{1}{2}m\omega^2x^2 \to H_{\text{pert}} = \frac{p^2}{2m} + \frac{1}{2}m\omega^2x^2 - F_0(g_1x - g_2x^2)\,. \tag{10.5}$$

This actually amounts, besides the shifting of the parabola, to modifying the spring constant as

$$k \to k' = k + 2g_2F_0\,. \tag{10.6}$$

On the quantum level, the Hamiltonian changes as

$$H_{\text{in}} = \hbar\omega(a^\dagger a + \tfrac{1}{2}) \to H_{\text{pert}} = \hbar\omega(a^\dagger a + \tfrac{1}{2}) + P_2(aa^\dagger)\,, \tag{10.7}$$

where P_2 is quadratic polynomial. Then, the original ground state $|0\rangle$ changes into a kind of "squeezed state," for which we now give a mathematical definition.

Definition 10.1 A squeezed (coherent) state $|\alpha,\xi\rangle$, $\alpha,\xi \in \mathbb{C}$, results from the combined action on the Fock vacuum state of the two following unitary operators:

$$|\alpha,\xi\rangle = D(\alpha)\,S(\xi)|0\rangle\,, \quad D(\alpha) = e^{\alpha a^\dagger - \bar{\alpha} a}\,, \quad S(\xi) \stackrel{\text{def}}{=} e^{\frac{1}{2}(\xi a^{\dagger 2} - \bar{\xi} a^2)}\,, \tag{10.8}$$

with $|0,0\rangle \equiv |0\rangle$.

The squeezing operator $S(\xi)$ is unitary since it is the exponential of the anti-Hermitian combination $\xi a^{\dagger 2} - \bar{\xi} a^2$. Note that its inverse is given, like for the displacement operator $D(\alpha)$, by $S^{-1}(\xi) = S^\dagger(\xi) = S(-\xi)$.

Let us examine more closely the interest in such states before pursuing the development of the formalism. The elliptic deformation of the Gaussian results from the construction of the squeezed states. There follows a decreasing value for one of the two variance factors in the Heisenberg uncertainty inequalities. This fact can be exploited in the detection of weak signals; see, for instance, [139]. Examples drawn from quantum optics are standard in regard to this notion of squeezing; as a matter of fact, squeezed light can be generated from light in a coherent state or a vacuum state by using certain optical nonlinear interactions. For the sake of simplicity, let us consider a one-mode electromagnetic field propagating in one-dimensional volume L, with vector potential

$$A(x,t) = -i\left(\frac{4\pi\hbar c^2}{2V\omega}\right)^{1/2}\left(a e^{i(kx-\omega t)} - a^\dagger e^{-i(kx-\omega t)}\right)\,.$$

Neglecting spatial dependence, which is not important for our purpose, the electric field operator reads as

$$E(x,t) = -\frac{1}{c}\frac{\partial}{\partial t}A(x,t) = \mathcal{E}\left(a e^{-i\omega t} + a^\dagger e^{i\omega t}\right)\,,$$

where $\mathcal{E} = \left(\frac{4\pi\hbar\omega}{2L}\right)^{1/2}$.

Let us now introduce the following two self-adjoint observables

$$X = \tfrac{1}{2}(a + a^\dagger)\,, \quad Y = \frac{1}{2i}(a - a^\dagger)\,, \tag{10.9}$$

which would stand, in the context of one-particle quantum mechanics and up to the factor $\frac{1}{\sqrt{2}}$ and appropriate units, for position and momentum operator, respectively. Here, they describe the two components in phase quadrature of the field. They can, in principle, be measured separately through homodyne detection [43], but not simultaneously, of course. Indeed, the commutation rule and the Heisenberg inequality read, respectively,

$$[X, Y] = \frac{i}{2} I_d \,, \quad \Delta X \, \Delta Y \geq \frac{1}{4} \,. \tag{10.10}$$

In optical interferometry, the term *homodyne* signifies that the reference radiation (the "local oscillator") is derived from the same source as the signal before the modulating process. For example, in a laser scattering measurement, the laser beam is split into two parts. One is the local oscillator and the other is sent to the system to be probed. The scattered light is then mixed with the local oscillator at the detector.

In terms of these "quadratures," the electric field reads

$$E(t) = 2\mathcal{E}(X \cos \omega t + Y \sin \omega t) \,. \tag{10.11}$$

Its mean value and variance in a certain state are given by

$$\langle E(t) \rangle = 2\mathcal{E}(\langle X \rangle \cos \omega t + \langle Y \rangle \sin \omega t) \,, \tag{10.12}$$

$$(\Delta E(t))^2 = (2\mathcal{E})^2 \begin{pmatrix} \cos \omega t & \sin \omega t \end{pmatrix} \begin{pmatrix} (\Delta X)^2 & \Delta\{X\,Y\} \\ \Delta\{X\,Y\} & (\Delta Y)^2 \end{pmatrix} \begin{pmatrix} \cos \omega t \\ \sin \omega t \end{pmatrix} \,, \tag{10.13}$$

where $\Delta\{X\,Y\} = \langle \frac{1}{2}(XY + YX) \rangle - \langle X \rangle \langle Y \rangle$. The Hermitian square matrix in (10.13) is called a *covariance matrix*. It is an important object for the statistical description of the quantum field. Let us now compare these expressions when we choose a coherent state, a pure squeezed state, or a mixed (coherent + squeezed) state.

10.2.1.1
In a Coherent State

The mean value of the electric field in a coherent state $|\alpha\rangle = D(\alpha)|0\rangle$ is given by

$$\langle E(t) \rangle / 2\mathcal{E} = \Re \left(\alpha e^{-i\omega t} \right) = \begin{pmatrix} \Re(\alpha) & \Im(\alpha) \end{pmatrix} \begin{pmatrix} \cos \omega t \\ \sin \omega t \end{pmatrix} \,. \tag{10.14}$$

The corresponding covariance matrix is equal to

$$\begin{pmatrix} (\Delta X)^2 & \Delta\{X\,Y\} \\ \Delta\{X\,Y\} & (\Delta Y)^2 \end{pmatrix} = \frac{1}{4} \begin{pmatrix} 1 & 0 \\ 0 & 1 \end{pmatrix} \,, \tag{10.15}$$

with $(\Delta E)^2 = \mathcal{E}^2$.

10.2.1.2
In a "Pure" Squeezed State

A pure squeezed state is defined as produced by the sole action of the unitary operator $S(\xi)$ on the vacuum $|0\rangle$:

$$|0, \xi\rangle \overset{\text{def}}{=} S(\xi)|0\rangle = e^{\frac{1}{2}(\xi a^{\dagger 2} - \bar{\xi} a^2)}|0\rangle \,. \tag{10.16}$$

In view of the computation of the mean value and the covariance matrix for the electric field, we need to determine the unitary transport of the lowering and raising operators, $S^\dagger(\xi)aS(\xi)$ and $S^\dagger(\xi)a^\dagger S(\xi)$. By applying the formula

$$e^A B e^{-A} = B + [A, B] + \frac{1}{2!}[A, [A, B]] + \cdots ,$$

and, with $\xi = re^{i\theta}$, one finds

$$S^\dagger(\xi)aS(\xi) = a\cosh r + a^\dagger e^{i\theta}\sinh r\,, \quad S^\dagger(\xi)a^\dagger S(\xi) = a^\dagger\cosh r + ae^{-i\theta}\sinh r\,.$$

Hence, with $U + iV \equiv (X + iY)e^{-i\theta/2}$, we have

$$S^\dagger(\xi)(U + iV)S(\xi) = Ue^r + iVe^{-r}\,.$$

There follows for the field mean value

$$\langle 0, \xi|E(t)|0, \xi\rangle = 0\,, \tag{10.17}$$

and for the covariance matrix

$$\begin{pmatrix} (\Delta X)^2 & \Delta\{X\,Y\} \\ \Delta\{X\,Y\} & (\Delta Y)^2 \end{pmatrix} = \frac{1}{4}\left(\cosh 2r I_2 + \sinh 2r \begin{pmatrix} \cos\theta & \sin\theta \\ \sin\theta & \cos\theta \end{pmatrix}\right)\,. \tag{10.18}$$

In particular, we have for the product of variances

$$(\Delta X)^2\,(\Delta Y)^2 = \frac{1}{16}(1 + \sinh^2 2r\,\sin^2\theta)\,,$$

or equivalently

$$\Delta U\,\Delta V = \frac{1}{4}\,. \tag{10.19}$$

In consequence, with pure squeezed states, there is saturation of the Heisenberg inequalities, but, contrarily to the coherent states, with $\Delta U \neq \Delta V$.

10.2.1.3
In a General Squeezed State

In a general squeezed state, $|\alpha, \xi\rangle = D(\alpha)S(\xi)|0\rangle$, the average value of the field is the same as in a coherent state:

$$\langle\alpha, \xi|E(t)|\alpha, \xi\rangle = \langle\alpha, 0|E(t)|\alpha, 0\rangle = 2\mathcal{E}\,\Re\left(\alpha e^{-i\omega t}\right) = \begin{pmatrix}\Re(\alpha) & \Im(\alpha)\end{pmatrix}\begin{pmatrix}\cos\omega t \\ \sin\omega t\end{pmatrix}.$$

On the other hand, the variance is the same as for a pure squeezed state:

$$(\Delta E)^2 = \langle\alpha, \xi|E^2 - \langle E\rangle^2|\alpha, \xi\rangle = \langle 0, \xi|E^2 - \langle E\rangle^2|0, \xi\rangle\,.$$

The same holds for the covariance matrix.

Note that these results are independent of the manner in which one would define the squeezed states, $|\alpha, \xi\rangle \overset{\text{def}}{=} D(\alpha)\,S(\xi)|0\rangle$ or $|\xi, \alpha\rangle \overset{\text{def}}{=} S(\xi)\,D(\alpha)|0\rangle$.

10.2.1.4
A General Definition of Squeezing

One can now give a general definition of squeezing with respect to a pair of quantum observables A and B with commutator $[A, B] = iC \neq 0$. When variances are calculated in a generic state, one obtains from Cauchy–Schwarz inequality $\Delta A\,\Delta B \geq \frac{1}{2}|\langle C\rangle|$. Now a state will be called *squeezed* (with respect to the pair (A, B)) if $(\Delta A)^2$ (or $(\Delta B)^2$) $< \frac{1}{2}|\langle C\rangle|$. A state will be called *ideally squeezed* (with respect to the pair (A, B)) if the equality $\Delta A\,\Delta B = \frac{1}{2}|\langle C\rangle|$ is reached together with $(\Delta A)^2$ (or $(\Delta B)^2$) $< \frac{1}{2}|\langle C\rangle|$. Hence, the class of ideally squeezed states contains the set of pure squeezed states.

10.2.1.5
Squeezing the Uncertainties

To conclude this section, let us briefly explain the interest in squeezed states in various applications, particularly in the coding and transmission of information through optical devices. From the above study of the electric field, Glauber (standard) optical coherent states have, in the phase quadrature plane, circularly symmetric uncertainty regions, so the uncertainty relation dictates some minimum noise amplitudes, for instance, for the amplitude and phase (Figure 10.1). A further reduction in amplitude noise is possible only by "squeezing" the uncertainty region, reducing its width in the amplitude, that is, radial, direction while increasing it in the orthogonal, that is, angular or phase, direction, so that the phase uncertainty is increased. Such light is called amplitude-squeezed (Figure 10.2). Conversely, phase-squeezed light (Figure 10.3) has decreased phase fluctuations at the expense of increased amplitude fluctuations. There is also the so-called squeezed vacuum, where the center of the uncertainty region (corresponding to the average amplitude) is at the origin of the coordinate system, and the fluctuations are reduced in some direction. The mean photon number is larger than zero in this case; a squeezed vacuum is a "vacuum" only in the sense that the average amplitude (but

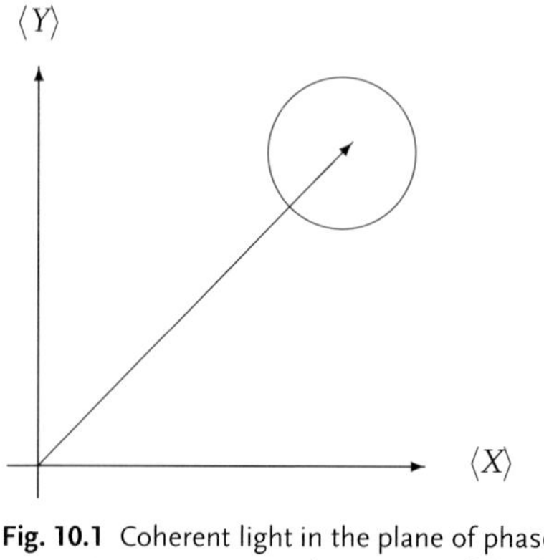

Fig. 10.1 Coherent light in the plane of phase quadratures. The uncertainty region around one point of the plane is circularly symmetric.

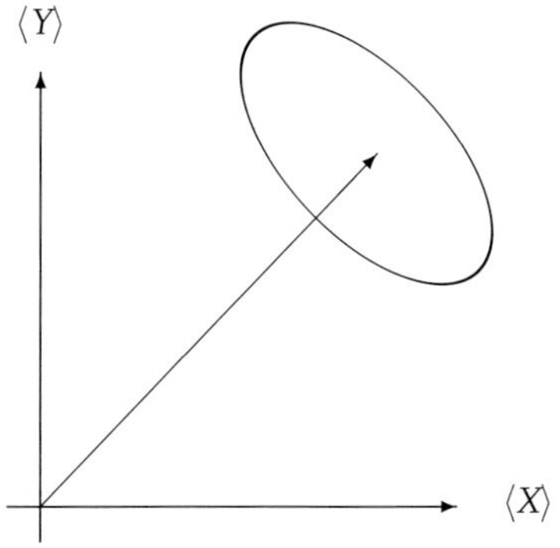

Fig. 10.2 Amplitude-squeezed light in the plane of phase quadratures. The uncertainty ellipse is stretched in the angular direction.

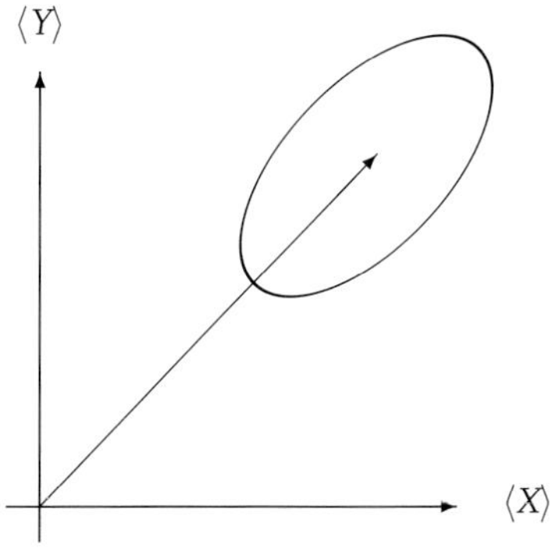

Fig. 10.3 Phase-squeezed light in the plane of phase quadratures. The uncertainty ellipse is stretched in the radial direction.

not the average photon number) is zero. For example, an optical parametric amplifier with a vacuum input can generate a squeezed vacuum with a reduction in the noise of one quadrature components on the order of 10 dB.

In summary, the main feature of the squeezed states lies in the fact that they offer the opportunity of deforming (squeezing!) in a certain direction the circle of "quantum uncertainty" that so becomes an uncertainty ellipse. Hence, they provide a way of reducing the quantum noise for one of the two quadratures.

10.2.2
Algebraic ($\mathfrak{su}(1,1)$) Content of Squeezed States

We have seen that squeezed states are obtained by forcing two-photon processes with time-dependent classical sources (from this the appellation *two-photon coherent states*). The most general form of the corresponding Hamiltonian reads

$$H = \hbar\omega\left(a^\dagger a + \frac{1}{2}\right) + f_2(t)\, a^{\dagger 2} + \overline{f_2(t)}\, a^2 + f_1(t)\, a^\dagger + \overline{f_1(t)}\, a\,. \tag{10.20}$$

The production of coherent states or of squeezed states will depend on the respective importance granted to the factors $f_1(t)$ and $f_2(t)$. Now, the Hamiltonian in (10.20) has clearly the form of an element in the "two-photon" Lie algebra, denoted by $\mathfrak{h}_6$, generated by the set of operators $\{a, a^\dagger, I_d, N = a^\dagger a, a^2, a^{\dagger 2}\}$, and already mentioned in Section 2.2.4. From the (nontrivial) commutation rules,

$$\begin{aligned} &[a, a^\dagger] = I_d\,, \quad [a, N] = a\,, \quad [a^\dagger, N] = -a^\dagger\,, \\ &[a^2, a^\dagger] = 2a\,, \quad [a^{\dagger 2}, a] = -2a^\dagger\,, [a^2, N] = 2a^2\,, \quad [a^{\dagger 2}, N] = -2a^{\dagger 2}\,, \end{aligned} \tag{10.21}$$

we see that $\mathfrak{h}_6$ is a representation of the semidirect sum of $\mathfrak{su}(1, 1)$ with the Weyl–Heisenberg algebra. The corresponding group, denoted by H_6, is the semidirect product $H_6 = W \rtimes SU(1, 1)$.

The existence of such an algebraic tool in the construction of squeezed states is very useful in solving problems involving Hamiltonians such as (10.20). Suppose we have to deal with an evolution equation of the type

$$i\frac{\partial}{\partial t} U(t, t_0) = H(t)\, U(t, t_0)\,, \quad U(t_0, t_0) = I_d\,, \tag{10.22}$$

where the mathematical objects have a "$\mathfrak{h}_6$" nature: the time-dependent Hamiltonian H is an element of the Lie algebra $\mathfrak{h}_6$, whereas the evolution operator $U(t, t_0)$, as a solution to (10.22), should be an element of a unitary representation of the Lie group H_6. The trick is of disentangling nature [9], like in (8.7.4). It amounts to solving the equation by choosing among linear faithful representations of H_6 or $\mathfrak{h}_6$ the simplest one, namely, the four-dimensional one in which group and algebra elements are realized as 4×4 matrices. Let us associate with a generic element X in $\mathfrak{h}_6$, written as

$$X = \eta\left(N + \tfrac{1}{2}\right) + \delta I_d + R a^{\dagger 2} + L a^2 + r a^\dagger + l a\,, \quad \eta, \delta, R, L, r, l \in \mathbb{C}\,, \tag{10.23}$$

the following matrix in $\mathrm{M}(4, \mathbb{C})$:

$$M(X) = \begin{pmatrix} 0 & 0 & 0 & 0 \\ r & \eta & 2R & 0 \\ -l & -2L & -\eta & 0 \\ -2\delta & -l & -r & 0 \end{pmatrix}. \tag{10.24}$$

This representation of X is made possible because of the Lie algebra isomorphism between basic operators defining $\mathfrak{h}_6$ and elementary projectors E_{ij} with matrix elements $\left(E_{ij}\right)_{kl} = \delta_{ik}\,\delta_{jl}$ generating the Lie algebra $\mathrm{M}(4, \mathbb{C})$:

$$\begin{aligned} &N + \frac{1}{2} \mapsto E_{22} - E_{33}\,, \quad I_d \mapsto -2E_{41}\,, \\ &a^{\dagger 2} \mapsto 2E_{23}\,, \quad a^\dagger \mapsto E_{21} - E_{43}\,, \\ &a^2 \mapsto -2E_{32}\,, \quad a^\dagger \mapsto -E_{31} - E_{42}\,. \end{aligned} \tag{10.25}$$

Owing to the general form (10.20) of the Hamiltonian, (10.22) is equivalent to a dynamical system in the space of time-dependent parameters $(\eta, \delta, R, L, r, l)$. This dynamical system is easily made explicit thanks to the map $M : X \mapsto M(X)$, which permits one to write (10.22) in its 4×4-matrix representation:

$$i \frac{\partial}{\partial t} M(U) = M(H)\, M(U)\,, \quad M(U_0) = I_4\,, \tag{10.26}$$

where $U = e^X$. Now, the exponential of a matrix such as (10.24) has the general form

$$\begin{pmatrix} 1 & 0 & 0 & 0 \\ & & & 0 \\ & P & & 0 \\ & & & 1 \end{pmatrix},$$

where P is a 3×3 matrix. Therefore, the solution $M(U)$ to (10.26) should be of this type, and its parameters $(\eta, \delta, R, L, r, l)$ should be, in general, determined numerically. Inverting the map M allows one to find the unitary operator $U(t, t_0)$ and eventually the resulting S matrix for a specific physical process. Note that there exist various forms for $U = e^X$ to be used as an ansatz in dealing with (10.26), for instance,

$$\begin{aligned} U &= \exp\left[\eta\left(N + \tfrac{1}{2}\right) + \delta I_d + R a^{\dagger 2} + L a^2 + r a^\dagger + l a\right] \\ &= \exp[R' a^{\dagger 2} + r' a^\dagger] \exp\left[\eta'\left(N + \tfrac{1}{2}\right) + \delta' I_d\right] \exp[L' a^2 + l' a] \\ &= \exp[r'' a^\dagger + l'' a] \exp[R'' a^{\dagger 2} + L'' a^2] \exp\left[\eta''\left(N + \tfrac{1}{2}\right) + \delta'' I_d\right]. \end{aligned}$$

We give below the matrix P appearing through the map M for each of these expressions:

$$\begin{aligned} P &= \begin{pmatrix} \dfrac{e^G - I_2}{G}\begin{pmatrix} r \\ -l \end{pmatrix} & e^G \\ -2\delta + \begin{pmatrix} -l & -r \end{pmatrix} \dfrac{e^G - I_2 - G}{G}\begin{pmatrix} r \\ -l \end{pmatrix} & \begin{pmatrix} -l & -r \end{pmatrix} \dfrac{e^G - I_2}{G} \end{pmatrix} \\ &= \begin{pmatrix} -2R'l'e^{-\eta'} & e^{\eta'} - 4L'R'e^{-\eta'} & 2R'e^{-\eta'} \\ -l'e^{-\eta'} & -2L'e^{-\eta'} & e^{-\eta'} \\ -2\delta' + r'l'e^{-\eta'} & 2r'L'e^{-\eta'} & -r'e^{-\eta'} \end{pmatrix} \\ &= \begin{pmatrix} r'' & \cosh\theta e^{\eta''} & 2R''\dfrac{\sinh\theta}{\theta}e^{-\eta''} \\ -l'' & -2L''\dfrac{\sinh\theta}{\theta}e^{\eta''} & \cosh\theta e^{-\eta''} \\ -2\delta'' & -l''\cosh\theta e^{\eta''} - 2L''r''\dfrac{\sinh\theta}{\theta} & -r''\cosh\theta e^{-\eta''} - 2R''l''\dfrac{\sinh\theta}{\theta} \end{pmatrix}, \end{aligned}$$

with

$$G = \begin{pmatrix} \eta & 2R' \\ -2L & -\eta \end{pmatrix},$$

and $e^G = \cosh\theta I_2 + G\frac{\sinh\theta}{\theta}$, $\theta^2 = \eta^2 - 4LR$. Such results, obtained from tractable *disentangling* factorizations of 4×4 matrices, permit us to calculate easily mean values and variances of arbitrary powers of elements of $\mathfrak{h}_6$ or H_6. These disentangling manipulations are an easy alternative to the Baker–Campbell–Hausdorff–Zassenhaus formulas mentioned in Section 6.5. Let us start with the mean value $\langle X\rangle$ of $X \in \mathfrak{h}_6$ in the state $|\Psi(t)\rangle = U(t, t_0)|0\rangle$. We can write

$$\langle X\rangle = \langle 0|U^\dagger X\, U|0\rangle = \frac{\partial}{\partial\gamma}\langle 0|U^\dagger e^{\gamma X}\, U|0\rangle\bigg|_{\gamma=0} .$$

Now, the expression $U^\dagger e^{\gamma X}\, U$, as a product of operator-valued exponentials, is precisely computed by using the finite-dimensional representation M:

$$U^\dagger e^{\gamma X}\, U \overset{M}{\to} (M(U))^{-1} M\left(e^{\gamma X}\right) M(U) .$$

The appropriate disentangling relation is that which leads to $X = \eta(N + 1/2) + \delta\mathbb{I}_d$, $U = e^{La^2+la}$, and

$$M\left(e^{La^{\dagger 2}+ra^\dagger}\right) M\left(e^{\eta(N+1/2)+\delta\mathbb{I}_d}\right) M\left(e^{La^2+ra}\right) ,$$

where η and δ are functions of γ such that $\eta(0) = 0 = \delta(0)$. Then, by inverting the map M, we get the trivial calculation of the mean value:

$$\langle 0|e^{La^{\dagger 2}+ra^\dagger}\, e^{\eta(N+1/2)+\delta\mathbb{I}_d}\, e^{La^2+ra}|0\rangle = e^{\eta(\gamma)/2+\delta(\gamma)} .$$

The successive moments of the observable X are then given by

$$\langle X^n\rangle = \left(\frac{\partial}{\partial\gamma}\right)^n e^{\eta(\gamma)/2+\delta(\gamma)}\bigg|_{\gamma=0} . \tag{10.27}$$

For the first two, one gets

$$\langle X\rangle = \tfrac{1}{2}\eta'(0) + \delta'(0) , \quad \langle X^2\rangle = \left(\tfrac{1}{2}\eta + \delta\right)''\Big|_{\gamma=0} + (\langle X\rangle)^2 . \tag{10.28}$$

Therefore, the variance is given by

$$(\Delta X)^2 = \langle X^2\rangle - \langle X\rangle^2 = \left(\tfrac{1}{2}\eta + \delta\right)''\Big|_{\gamma=0} . \tag{10.29}$$

As a complementary expression to these formulas, let us consider the case in which we calculate the mean value of the operator $e^{\gamma a^\dagger + \bar{\gamma} a}$ in a squeezed state $|\Psi\rangle$:

$$F(\bar{\gamma}, \gamma) \overset{\text{def}}{=} \left\langle e^{\gamma a^\dagger + \bar{\gamma} a}\right\rangle .$$

(i) *First expression of squeezed states.* If

$$|\Psi\rangle = D(\alpha)S(\xi)|0\rangle ,$$

then

$$F(\bar{\gamma}, \gamma) = \left(\bar{\gamma}\alpha + \tfrac{1}{2}\bar{\gamma}^2 e^{i\theta}\sinh\varrho\right) + (\text{c.c.}) + \tfrac{1}{2}|\gamma|^2\cosh 2\varrho ,$$

with $\xi \equiv \varrho e^{i\theta}$.

(ii) *Second expression of squeezed states.* If

$$|\Psi\rangle = S(\xi)D(\alpha)|0\rangle ,$$

then

$$F(\overline{\gamma}, \gamma) = \left(\overline{\gamma}(\alpha \cosh \varrho + \overline{\alpha} e^{i\theta} \sinh \varrho) + \tfrac{1}{2}\overline{\gamma}^2 e^{i\theta} \sinh \varrho\right) + (\text{c.c}) .$$

(iii) *Most general expression of squeezed states.* If

$$|\Psi\rangle = e^{Ra^{\dagger 2} + ra^{\dagger} - \bar{R}a^2 - \bar{r}a}|0\rangle ,$$

then

$$F(\overline{\gamma}, \gamma) = \left(\frac{\overline{\gamma} r + 2\bar{r}R}{1 - 4\bar{R}R} + \tfrac{1}{2}\overline{\gamma}^2 \frac{2R}{1 - 4|R|^2}\right) + \text{c.c.} .$$

10.2.3 Using Squeezed States in Molecular Dynamics

Let us give a nice illustration (see [9] and references therein for more details) of the use of the previous formulas in the domain of molecular dynamics. Let us consider a diatomic molecule (e.g., O_2, NO, N_2) hit by an inert atom (e.g., He). The Hamiltonian of the system reads as

$$H = H_{\text{target}} + H_{\text{projectile}} + H_{\text{int}} ,$$

where

$$H_{\text{int}}(\mathcal{R}, q) = V(\mathcal{R}, q) - V(\infty, q) ,$$

$\mathcal{R}$ being the relative position of the projectile with respect to the center of mass of the molecule, whereas $q = q(t)$ is the position operator corresponding to the displacement of the intermolecular relative position with respect to the equilibrium. Within the usual framework of the semiclassical approach, namely, the Born–Oppenheimer approximation, the trajectory of the projectile $\mathcal{R} = \mathcal{R}(t)$ is considered as classical, and we can deal with the series expansion of H_{int} in terms of the small displacement q:

$$H_{\text{int}} = h_0(t) + h_1(t)q + \frac{1}{2}h_2(t)q^2 + \cdots .$$

By putting $q = \sqrt{\frac{\hbar}{m\omega}}(a+a^{\dagger})/\sqrt{2}$ and neglecting terms beyond the quadratic ones, we get the following expression for the Hamiltonian:

$$H = \frac{1}{2}M\dot{\mathcal{R}}^2 + \left(\hbar\omega + \frac{\hbar}{m\omega}B(t)\right)\left(a^{\dagger}a + \frac{1}{2}\right) + \frac{\hbar}{2m\omega}A(t)(a + a^{\dagger}) + \frac{\hbar}{2m\omega}B(t)(a^2 + a^{\dagger 2}) . \tag{10.30}$$

From this (approximate) Hamiltonian it becomes possible to determine, for instance, the probability of the target of changing its vibrational ground state to an excited state $|n\rangle$ after the collision. The calculation goes through the S-matrix formalism:

$$|\Psi_{\text{out}}\rangle \equiv |\Psi(t=-\infty)\rangle = S^{\dagger}|\Psi_{\text{in}}\rangle \equiv S^{\dagger}|\Psi(t=-\infty)\rangle\,. \tag{10.31}$$

Thus the S matrix is the limit (in a certain functional sense)

$$S^{\dagger} = \lim_{\substack{t_0 \to -\infty \\ t \to +\infty}} U(t, t_0)$$

of the evolution operator associated with the Hamiltonian through the equation

$$i\hbar \frac{\partial}{\partial t} U(t, t_0) = H(t) U(t, t_0)\,, \qquad U(t_0, t_0) = I_d\,. \tag{10.32}$$

In dealing with this equation, one should consider the constraint resulting from the conservation of the classical energy:

$$\tfrac{1}{2} M \left(\dot{\mathcal{R}}^2(\infty) - \dot{\mathcal{R}}^2(-\infty)\right) + \Delta V = 0\,. \tag{10.33}$$

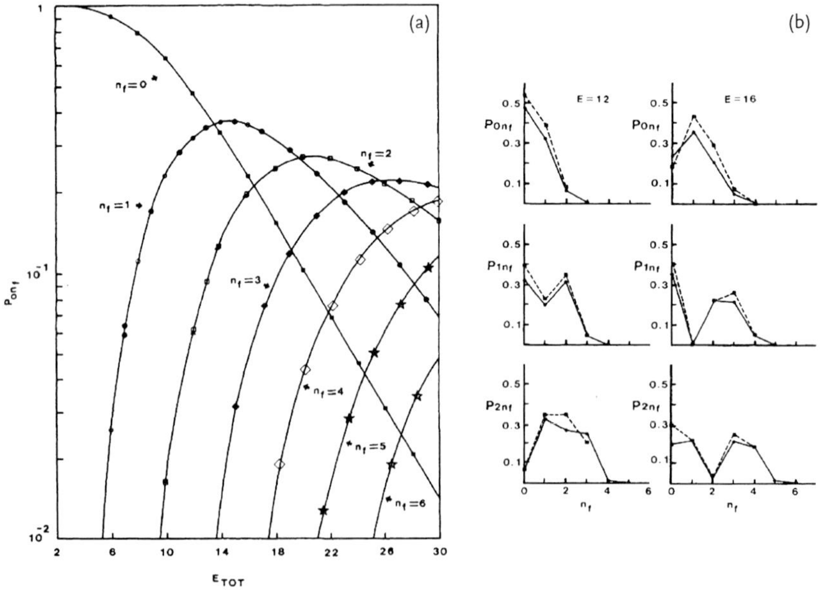

Fig. 10.4 Transition probabilities for excitations of H_2 by He scattering. (a) Transition probabilities $P_{0\to n_f}$ ($n_f = 0$ to 6) as a function of the total energy E_{tot} of the He + H_2 system. All curves are drawn smoothly through the data points of Gilmore and Yuan [149, 150]. • and ◇ denote results obtained by Gazdy and Micha [151, 152]. (b) Transition probabilities $P_{n_i \to n_f}$ as a function of n_f for $E = 12$ and 16. The three plots on the left are for $E = 12$ and those on the right are for $E = 16$ [149, 150] (reprinted with permission from [Zang, W.-M., Feng, D.-H., and Gilmore, R., Rev. Mod. Phys., 26, 867, 1990], American Institute of Physics).

The function $\mathcal{R}(t)$ is determined classically by taking into account a loss of kinetic energy suitably smoothed during the collision. Once $\mathcal{R}(t)$ is known, one proceeds to the numerical integration of the 4×4-matrix version of (10.32):

$$i\hbar \frac{\partial}{\partial t} M[U(t, t_0)] = M[H(t)]\, M[U(t, t_0)]\,, \quad M[U(t_0, t_0)] = I_4\,. \tag{10.34}$$

Taking the limits $t \longrightarrow +\infty$, $t_0 \longrightarrow -\infty$, one obtains the parameters $(R, r, \eta, \delta, L, l)$ of the S matrix, as they were introduced in (10.23):

$$S = e^{Ra^{\dagger 2} + ra^{\dagger}}\, e^{\eta(n+1/2)+\delta I_d}\, e^{La^2 + la}\,. \tag{10.35}$$

Its transition $0 \longrightarrow n$ matrix element is then given by

$$\langle n|S|0\rangle = \sum_{s=0}^{\lfloor \frac{n}{2} \rfloor} \frac{r^{n-2s} R^s \sqrt{n!}}{(n-2s)!s!}\, e^{\eta/2+\delta}\,. \tag{10.36}$$

Some examples of curves giving the probability $P_{0n} = |\langle n|S|0\rangle|^2$ are shown in Figure 10.4 for the collision $He + H_2 \longrightarrow He + {H_2}^{\star}$.

11
Fermionic Coherent States

11.1
Introduction

Quantum elementary systems can be divided into two classes, bosons and fermions, according to the integral or half-integral value of their spins. This division rules many fundamental physical evidences, such as the possibility of particle collisions or nuclear reactions or the stability of matter. Fermions are particles that obey Fermi–Dirac statistics (*identical particles are forbidden from being in the same state*), while bosons are particles that obey Bose–Einstein statistics (*identical particles can be in the same state*). One can state that the standard coherent states are bosonic, in the sense that they are superpositions of number states for which there is no limitation on the number of excitations involved.

We present here the so-called fermionic coherent states, which are, on their side, superpositions of number states $|n\rangle$ where the number n assumes two values only, $n = 0$ or $n = 1$. After their algebraic description, we will give some insight into their utilization in the study of many-fermion systems (e.g., the Hartree–Fock–Bogoliubov approach).

Fermionic models are essentially encountered in atomic and nuclear physics. Typically, they account for assemblies of about 100 particles. Therefore, it is necessary to proceed to approximations such as variational methods. The criteria that are retained for acceptable test functions are simplicity and optimal consideration of quantum correlations, criteria that are met by coherent states.

11.2
Coherent States for One Fermionic Mode

The Pauli exclusion principle requires the creation and annihilation operators for a fermionic mode to obey the following anticommutation rules:

$$[a, a^\dagger]_+ \stackrel{\text{def}}{=} aa^\dagger + a^\dagger a = I_d\,, \quad [a, a]_+ = 0 = [a^\dagger, a^\dagger]_+\,. \tag{11.1}$$

Any system having such commutation rules has the group $SU(2)$ as a "dynamical" group, that is, a group whose elements connect states with different energies.

Coherent States in Quantum Physics. Jean-Pierre Gazeau

ISBN: 978-3-527-40709-5

The generators of this group are in the present context [2]$\{a^\dagger, a, a^\dagger a - 1/2\}$, with the commutation rules

$$[a^\dagger, a] = 2\left(a^\dagger a - \tfrac{1}{2}\right) \quad \left[a^\dagger a - \tfrac{1}{2}, a\right] = -a \quad \left[a^\dagger a - \tfrac{1}{2}, a^\dagger\right] = a^\dagger\,. \tag{11.2}$$

It is important to note that these commutation rules are derived *from* the anticommutation rules (11.1) and *not* from the usual canonical commutation rules.

Let us now proceed to the construction of the corresponding coherent states. Since ${a^\dagger}^2 = 0 = a^2$, the only pertinent Hermitian space is the two-dimensional Hermitian one, $\mathcal{H}^{j=1/2} \cong \mathbb{C}^2$. Thus, it is enough to repeat the construction of the standard coherent states, but keeping the constraint ${a^\dagger}^2 = 0 = a^2$, or the construction of the $SU(2)$ coherent states with spin generators $J_\pm$ and $j = 1/2$. Hence, we associate with $\xi \in \mathbb{C}$ the states

$$|\xi\rangle_\pm = e^{\xi a^\dagger - \bar{\xi} a}\,|\pm\rangle\,, \tag{11.3}$$

where $|\pm\rangle$ ($\equiv |\frac{1}{2} \pm \frac{1}{2}\rangle$ in $SU(2)$ notation) are such that $a^\dagger|-\rangle = |+\rangle$, $a|+\rangle = |-\rangle$. Since $(\xi a^\dagger - \bar{\xi} a)^2 = -|\xi|^2(a^\dagger a + a a^\dagger) = -|\xi|^2$, we have

$$e^{\xi a^\dagger - \bar{\xi} a} = \cos|\xi|\, I_d + \frac{\xi}{|\xi|}\sin|\xi|\, a^\dagger - \frac{\bar{\xi}}{|\xi|}\sin|\xi|\, a\,.$$

Putting $\xi = -\frac{\theta}{2}\,e^{-i\phi}$, we get the two expressions

$$|\xi\rangle_+ = \cos\frac{\theta}{2}\,|+\rangle + \sin\frac{\theta}{2}\,e^{i\phi}|-\rangle\,, \quad |\xi\rangle_- = \sin\frac{\theta}{2}\,e^{-i\phi}|+\rangle + \cos\frac{\theta}{2}\,|-\rangle\,, \tag{11.4}$$

in which we recognize the spin or Bloch coherent states for $j = 1/2$. Therefore, we inherit the properties of spin coherent states in the present fermionic context, like the overlap relation (6.31), the way they behave under the action of an $SU(2)$ transformation (6.28), and the resolution of the identity (6.10). Note that we could also use the other complex parameterization $\zeta = \tan\frac{\theta}{2}\,e^{i\phi}$, a notation that we will recover on a more general level in the next sections.

11.3
Coherent States for Systems of Identical Fermions

11.3.1
Fermionic Symmetry *SU*(*r*)

We now consider a system of r fermionic states or modes occupied by at most one fermion. To each mode corresponds one pair $(a_i^\dagger, a_i)$ of operators obeying the anticommutation rules (11.1):

$$[a_i, a_j^\dagger]_+ = \delta_{ij}\,, \quad [a_i, a_j]_+ = 0 = [a_i^\dagger, a_j^\dagger]_+\,, \quad 1 \le i, j \le r\,. \tag{11.5}$$

Table 11.1 A short glossary of matrix groups and algebras.

Groups		
Symbol	**Definition**	**Real dimension**
$M(r, \mathbb{C})$	$r \times r$ matrices with complex entries	$2r^2$
$GL(r, \mathbb{C}) \subset M(r, \mathbb{C})$	Invertible matrices	$2r^2$
$SL(r, \mathbb{C}) \subset GL(r, \mathbb{C})$	Unit determinant	$2r^2 - 2$
$U(r) \subset GL(r, \mathbb{C})$	Unitary matrices $mm^\dagger = I_d$	r^2
$SU(r) = U(r) \bigcap SL(r, \mathbb{C})$	Unitary and unit determinant	$r^2 - 1$

Lie algebras		
Symbol	**Definition**	**Real dimension**
$\mathfrak{gl}(r, \mathbb{C}) \cong M(r, \mathbb{C})$	$r \times r$ matrices with complex entries	$2r^2$
$\mathfrak{sl}(r, \mathbb{C})$	Null trace	$2r^2 - 2$
$\mathfrak{u}(r)$	Anti-Hermitian matrices $m^\dagger = -m$	r^2
$\mathfrak{su}(r)$	Anti-Hermitian matrices with null trace	$r^2 - 1$

Several Lie algebras (see Table 11.1) can be built from the set of operators $a_i, a_i^\dagger$. The most immediate one is the algebra $\mathfrak{u}(r)$, linear span of the r^2 quadratic operators

$$a_i^\dagger a_j \,, \quad 1 \leq i, j \leq r \,. \tag{11.6}$$

They obey the commutation rules

$$[a_i^\dagger a_j, a_k^\dagger a_l] = \delta_{jk}\, a_i^\dagger a_l - \delta_{il}\, a_k^\dagger a_j \,. \tag{11.7}$$

The r elements $H_i = a_i^\dagger a_i$ span a maximal Abelian subalgebra of $\mathfrak{u}(r)$. The way this subalgebra acts on the other elements of $\mathfrak{u}(r)$ is described by the commutation relations

$$[H_i, a_j^\dagger a_k] = \left(\delta_{ij} - \delta_{ik}\right) a_j^\dagger a_k \,. \tag{11.8}$$

This algebra acts on the Hermitian spaces describing r fermionic states, more precisely that which carries a completely antisymmetric fundamental representation of $\mathfrak{u}(r)$, denoted $\Lambda = \{\lambda_1, \lambda_2, \ldots, \lambda_r\}$, with $\lambda_i \in \{0, 1\}$. The latter are the values assumed by the r fundamental weights of the Lie algebra $\mathfrak{u}(r)$ (see Appendix B). Thus, when $p \leq r$ identical fermions are involved,

$$\Lambda = \{\lambda_1, \lambda_2, \ldots, \lambda_r\} = \{1, 1, \ldots, 1, 0, 0 \ldots, 0\} \equiv \{1^p, 0^{r-p}\} \,, \quad p \leq r \,. \tag{11.9}$$

This space of states has its dimension equal to $\binom{r}{p}$, and a suitable basis is formed by the kets $|n_1, n_2, \ldots, n_r\rangle$, where $n_i = 0$ or 1 denotes the occupation number of the ith level, and the constraint $\sum_{i=1}^{r} n_i = p$ expresses the conservation of the total number of fermions. These basis states are common eigenstates of the complete set $\{H_i, 1 \le i \le r\}$ of commuting observables. With respect to this basis, the matrix representative E_{ij} of the generator $a_i^\dagger a_j$ of $\mathrm{u}(r)$ has all its entries equal to 0 with the exception of entry (i, j), which is equal to 1:

$$a_i^\dagger a_j \longleftrightarrow E_{ij} \quad \text{with} \quad \left(E_{ij}\right)_{kl} = \delta_{ik}\,\delta_{jl} , \tag{11.10}$$

a notation previously used in (10.25). Let us consider the following "extremal" state for the representation $\Lambda = \{1^p, 0^{r-p}\}$:

$$|\mathrm{extr}\rangle = |\overbrace{1, 1, \ldots, 1}^{p}, \overbrace{0, \ldots, 0}^{r-p}\rangle . \tag{11.11}$$

It is actually the nonperturbed ground state of a p-fermion Hamiltonian expressed in a mean-field approximation:

$$H = \sum_{i,j=1}^{p} k_{ij}\, a_i^\dagger a_j + \sum_{i,j,l,m=1}^{p} V_{ijlm}\, a_i^\dagger a_j^\dagger a_m a_l . \tag{11.12}$$

The mean-field approach is necessary when we deal with a many-body system with interactions, since in general there is no way to solve exactly the model, except for extremely simple cases. A major drawback (e.g., when computing the partition function of the system) is the treatment of combinatorics generated by the interaction terms in the Hamiltonian when summing over all states. The goal of mean-field theory (also known as self-consistent-field theory) is precisely to resolve these combinatorial problems. The main idea of mean-field theory is to replace all interactions with any one body with an average or effective interaction. This reduces any multibody problem into an effective one-body problem.

State (11.11) is extremal in the sense that it is annihilated by all elements as $a_i^\dagger a_j |\mathrm{extr}\rangle = 0$ for all i, j, such that $1 \le i \neq j \le p$ or $p + 1 \le i, j \le r$. Together with the H_i's, $1 \le i \le p$, these generators span the subalgebra $\mathrm{u}(p) \oplus \mathrm{u}(r-p)$. The corresponding group, $U(p) \otimes U(r-p)$, is the subgroup of elements in $U(r)$ leaving invariant, up to a phase, the extremal state $|\mathrm{extr}\rangle$, that is, it is the *stability* group of the latter. One then defines a family of coherent states for the group $U(r)$ by

$$|\xi \equiv (\xi_{ij})\rangle = \exp\left[\sum_{\substack{1 \le j \le p \\ p+1 \le i \le r}} (\xi_{ij} a_i^\dagger a_j - \bar{\xi}_{ij} a_j^\dagger a_i)\right] |\mathrm{extr}\rangle . \tag{11.13}$$

These states are in one-to-one correspondence with the points of the coset $U(r)/(U(p) \bigotimes U(r-p))$ [153]. They represent in a certain sense a "quantum" version of this compact manifold. Note that for a system of bosonic states, the manifold to be considered would be the coset $U(r)/(U(1) \otimes U(r-1))$.

Given the set of complex parameters $\{\xi_{ij}, 1 \le j \le p, p + 1 \le i \le r\}$ viewed as the entries of a $(r-p) \times p$ matrix ξ, let us introduce two other $(r-p) \times p$ matrices:

$$Z = \xi \frac{\sin\sqrt{\xi^\dagger \xi}}{\sqrt{\xi^\dagger \xi}} , \quad \sin\sqrt{\xi^\dagger \xi} = \sqrt{\xi^\dagger \xi} - \frac{(\sqrt{\xi^\dagger \xi})^3}{3!} + \ldots , \tag{11.14}$$

$$\zeta = Z\,(I_p - Z^\dagger Z)^{-\frac{1}{2}}\,. \tag{11.15}$$

The matrix ζ is well defined because $I_p - Z^\dagger Z$ is by construction a positive matrix. It can be coordinatized by $p(r-p)$ complex parameters, denoted ζ_α, $p(r-p)$ being the complex dimension of the coset $U(r)/(U(p) \otimes U(r-p))$. The ζ_α' represent a kind of projective coordinates for this manifold. With this notation, if one considers the matrix representation $a_i^\dagger a_j \longleftrightarrow E_{ij}$ of the $\mathfrak{u}(r)$ basis elements, we find that the displacement operator

$$\exp\left[\sum_{\substack{1\le j\le p\\ p+1\le i\le r}} (\xi_{ij} a_i^\dagger a_j - \bar{\xi}_{ij} a_j^\dagger a_i)\right]$$

or equivalently the coherent state $|\xi\rangle \equiv |\zeta\rangle$, is represented by the following unitary $r \times r$ matrix:

$$\begin{pmatrix} \sqrt{I_{r-p} - ZZ^\dagger} & Z \\ -Z^\dagger & \sqrt{I_p - Z^\dagger Z} \end{pmatrix} \equiv \gamma(Z) \in U(r)\,. \tag{11.16}$$

The notation ζ is particularly suitable for describing the action of the group $U(r)$ on its coset manifold $U(r)/(U(p) \otimes U(r-p))$. Let us write an element of $U(r)$ in the following block matrix form:

$$U(r) \ni g = \begin{pmatrix} \underset{(r-p)\times(r-p)}{A} & B \\ C & \underset{(p)\times(p)}{D} \end{pmatrix}\,. \tag{11.17}$$

Its action on the elements ζ of the coset reads as a projective one:

$$\zeta \overset{g}{\mapsto} \zeta' \equiv g \cdot \zeta = (A\zeta + B)(C\zeta + D)^{-1}\,. \tag{11.18}$$

11.3.1.1
Remark
This action of the group on one of its cosets is not mysterious. We already described an example of it in (8.36). On a general level, let G be a group and H be a subgroup of G. The right coset G/H is the set of equivalence classes "modulo H" defined by the relation $g \underset{H}{\equiv} g'$ if and only if there exists $h \in H$ such that $g' = gh$. Thus, $G/H = \{gH \overset{\text{def}}{=} \{gh,\, h \in H\},\, g \in G\}$. Now, let $g_0 \in G$ and g_0H be its equivalence class. An element $g \in G$ "transports" this class by left multiplication:

$$g_0 H \overset{g}{\mapsto} g(g_0 H) = (g g_0) H\,.$$

In the present case, it is enough to note that any element

$$g = \begin{pmatrix} A & B \\ C & D \end{pmatrix} \in U(r)$$

with nonsingular D can be factorized as $g = \gamma(Z_g)h_g$ with $\zeta_g = BD^{-1}$, $\gamma(Z_g)$ like in (11.16) and $h_g \in U(p) \otimes U(r-p)$. So, by simple matrix multiplication, we have $g\gamma(Z) = \gamma(Z')h' \underset{H}{\equiv} \gamma(Z')$, where ζ' is like in (11.18).

The compact manifold $U(r)/(U(p) \otimes U(r-p))$ has a rich structure. It is, like $\mathbb{C}$ or the sphere S^2 or the Poincaré half-plane, a Kähler manifold: its Riemannian symplectic structure is simply encoded by the Kähler potential $F = F(\zeta, \bar{\zeta})$ given by

$$F(\zeta, \bar{\zeta}) = \ln \det(I_p + \zeta^\dagger \zeta) . \tag{11.19}$$

The metric element and the 2-form derive from it as follows:

$$ds^2 = \sum_{\alpha,\beta} g_{\alpha\beta}\, d\zeta_\alpha\, d\bar{\zeta}_\beta , \quad g_{\alpha\beta} = \frac{\partial^2}{\partial\zeta_\alpha \partial\bar{\zeta}_\beta} F(\zeta, \bar{\zeta}) , \tag{11.20a}$$

$$\omega = \frac{i}{2} \sum_{\alpha,\beta} g_{\alpha\beta}\, d\zeta_\alpha \wedge d\bar{\zeta}_\beta . \tag{11.20b}$$

The volume element is given by

$$d\mu(\zeta, \bar{\zeta}) = \frac{\dim V^\Lambda}{\mathrm{Vol}\left(U(r)/U(r-p)\right)} \left(\det\left(I_p + \zeta^\dagger\zeta\right)\right)^{-r} \prod_\alpha d\zeta_\alpha\, d\bar{\zeta}_\beta . \tag{11.21}$$

In the normalization factor, V^Λ denotes the finite-dimensional space carrying the unitary irreducible representation Λ of $U(r)$. The explicit form of the coherent states is established thanks to the existence of a Baker–Campbell–Hausdorff–Zassenhaus factorization formula for the group $U(r)$. This formula expresses the displacement operator appearing in (11.13) in terms of the variables ζ_{ij}:

$$\begin{aligned}
&\exp\left[\sum_{\substack{1\le j\le p\\ p+1\le i\le r}} \left(\xi_{ij} a_i^\dagger a_j - \bar{\xi}_{ij} a_j^\dagger a_i\right)\right] = \\
&\exp\left[\sum_{\substack{1\le j\le p\\ p+1\le i\le r}} \zeta_{ij} a_i^\dagger a_j\right] \exp\left[\sum_{\substack{1\le i,j\le p\\ p+1\le i,j\le r}} \lambda_{ij} a_i^\dagger a_j\right] \exp\left[\sum_{\substack{1\le i\le p\\ p+1\le j\le r}} -\bar{\zeta}_{ij} a_i^\dagger a_j\right] .
\end{aligned} \tag{11.22}$$

Equation (11.22) is proven through the use of the following factorization in the finite matrix representation of the operators involved:

$$\begin{pmatrix} \sqrt{I_{r-p} - ZZ^\dagger} & Z \\ -Z^\dagger & \sqrt{I_p - Z^\dagger Z} \end{pmatrix} = \begin{pmatrix} I_{r-p} & \zeta \\ 0 & I_p \end{pmatrix} \begin{pmatrix} e^{\lambda_1} & 0 \\ 0 & e^{\lambda_2} \end{pmatrix} \begin{pmatrix} I_{r-p} & 0 \\ -\zeta^\dagger & I_p \end{pmatrix} , \tag{11.23}$$

with $e^{\lambda_1} = (I_{r-p} - ZZ^\dagger)^{-1/2}$ and $e^{\lambda_2} = (I_p - Z^\dagger Z)^{1/2}$.

By using the factorization (11.22) and the fact that the two right factors stabilize the extremal state

$$a_i^\dagger a_j |\text{extr}\rangle = 0 \quad \text{for} \quad 1 \le i \ne j \le p \quad \text{or} \quad p+1 \le i, j \le r ,$$

one can write the coherent states as

$$|\zeta\rangle = \frac{1}{(\mathcal{N}(\zeta, \bar{\zeta}))^{1/2}} \exp\left[\sum_{\substack{1\le j\le p \\ p+1\le i\le r}} \zeta_{ij} a_i^\dagger a_j\right] |\text{extr}\rangle , \tag{11.24}$$

with $\mathcal{N}(\zeta, \bar{\zeta}) = \det\left(I_{p\times p} - \zeta^\dagger \zeta\right)$. The scalar product (overlap) between two coherent states is then given in terms of the Kähler potential by

$$\langle \zeta' | \zeta \rangle = \frac{e^{F(\zeta, \bar{\zeta}')}}{(\mathcal{N}(\zeta, \bar{\zeta}))^{1/2} (\mathcal{N}(\zeta', \bar{\zeta}'))^{1/2}} . \tag{11.25}$$

These coherent states, as an overcomplete family of vectors in V^Λ, resolve the identity:

$$\int_{U(r)/(U(p)\otimes U(r-p))} d\mu(\zeta, \bar{\zeta}) \, |\zeta\rangle\langle\zeta| = I_d . \tag{11.26}$$

This implies that any state $|\Psi\rangle$ in the Hermitian space of fermionic levels can be expanded in terms of those states,

$$|\Psi\rangle = \int d\mu(\zeta, \bar{\zeta}) \, \Psi(\zeta) \, |\zeta\rangle$$

where the "contravariant" symbol of $|\Psi\rangle$ is given by

$$\Psi(\zeta) \stackrel{\text{def}}{=} \langle \bar{\zeta} | \Psi \rangle \equiv (\mathcal{N}(\zeta, \bar{\zeta}))^{-1/2} f(\zeta) .$$

Here, $f(\zeta)$ is an analytical function of ζ, thus providing the space V^Λ with a realization of the Fock–Bargmann type.

11.3.2
Fermionic Symmetry *SO*(2*r*)

A second relevant algebra, namely, $\mathfrak{so}(2r)$, is constructed from the operators $a_i^\dagger$ and a_j. Its generators are the $r(2r-1)$ operators

$$a_i^\dagger a_j - \frac{1}{2}\delta_{ij} , \quad 1 \le i, j \le r \tag{11.27}$$

$$a_i a_j \quad \text{and} \quad a_i^\dagger a_j^\dagger , \quad 1 \le i \ne j \le r . \tag{11.28}$$

The maximal Abelian subalgebra, that is, Cartan subalgebra, is generated by the r operators $H_i = a_i^\dagger a_i - 1/2$. The relevant Hermitian spaces, denoted $V^{\Lambda_\pm}$, are those which carry the two fundamental spinor representations of $\mathfrak{so}(2r)$, denoted by $\Lambda_\pm = \{\frac{1}{2}, \frac{1}{2}, \ldots, \pm\frac{1}{2}\}$. The sign $\pm$ corresponds to the parity of the number $n = \sum_{i=1}^{r} n_i$, where the components n_i are the respective eigenvalues of the generators H_i. They assume their values in $\{0, 1\}$ and are used to label the basis states $|n_1, n_2, \ldots, n_r\rangle$. Thus, the dimension of $V^{\Lambda_\pm}$ is equal to $2^{r-1} = \frac{1}{2}\sum_p \binom{r}{p}$: here, the number of existing fermions varies from 0 to r.

The matrix representation of the generators of $\mathfrak{so}(2r)$ is as follows:

$$\begin{aligned} a_i^\dagger a_j - \tfrac{1}{2}\delta_{ij} &\leftrightarrow E_{ij} - E_{r+jr+i}\,, \\ a_i^\dagger a_j^\dagger &\leftrightarrow E_{ir+j} - E_{jr+i} \\ a_i a_j &\leftrightarrow E_{r+ij} - E_{r+ji}\,, \end{aligned}$$

where the matrices E_{ij} were defined in (11.10), and with $1 \le i, j \le r$. Extremal states can be chosen as

$$|\text{extr}\rangle = \begin{cases} |0, 0, \ldots, 0\rangle \equiv |0\rangle & \text{for } \Lambda_+ \\ |1, 0, \ldots, 0\rangle \equiv |1\rangle & \text{for } \Lambda_-\,. \end{cases} \tag{11.29}$$

These extremal states are nonperturbed ground states of an r fermionic state Hamiltonian of the type (11.12), already considered for $\mathfrak{su}(r)$, with the restriction that the parity of the total number of fermions now be fixed to be even (for V^{Λ_+}) or odd (for V^{Λ_-}). These two extremal states are annihilated or left invariant in the following way:

$$a_i^\dagger a_j|0\rangle = 0 \quad \text{for} \quad 1 \le i, j \le r\,, \quad \text{for the } \Lambda_+ \text{ ground state}\,,$$

$$\begin{cases} a_1^\dagger a_1|1\rangle = |1\rangle\,, & \\ a_i^\dagger a_j|1\rangle = 0 & \text{for} \quad 2 \le i, j \le r\,, \\ a_1^\dagger a_i^\dagger|1\rangle = 0 & \text{for} \quad 2 \le i \le r\,, \\ a_i a_1|1\rangle = 0 & \text{for} \quad 2 \le i \le r\,, \end{cases} \quad \text{for the } \Lambda_- \text{ ground state}\,.$$

Thus, the stability subgroup of the state $|0\rangle$ is generated by the operators $a_i^\dagger a_j - \frac{1}{2}\delta_{ij}$, $1 \le i, j \le r$ and so is isomorphic to $U(r)$. Similarly, the subgroup that stabilizes the state $|1\rangle$ is generated by the operators $a_1^\dagger a_1 - 1/2$, $a_i^\dagger a_j - \frac{1}{2}\delta_{ij}$, $2 \le i, j \le r$, $a_1 a_i$, and $a_1^\dagger a_i^\dagger$, $2 \le i \le r$, and so is also isomorphic to $U(r)$. Both are subgroups of $\text{Spin}(2r)$, the double covering of $SO(2r)$. In consequence, the family of coherent states involved is in one-to-one correspondence with the points of the coset $\text{Spin}(2r)/U(r)$:

$$\text{Spin}(2r)/U(r) \ni \xi \mapsto |\xi\rangle = \exp\left(\sum_{1 \le i \ne j} \xi_{ij} a_i^\dagger a_j^\dagger - \text{h.c.}\right)|\text{extr}\rangle\,. \tag{11.30}$$

In what follows, we choose as the extremal state the even one, $|0\rangle$. By using the matrix representation, one easily checks that a coherent state corresponds to the following point in the coset (11.30):

$$|\xi\rangle \leftrightarrow \begin{pmatrix} \sqrt{I_r - ZZ^\dagger} & Z \\ -Z^\dagger & \sqrt{I_r - Z^\dagger Z} \end{pmatrix}, \quad Z \overset{\text{def}}{=} \xi \frac{\sin\sqrt{\xi^\dagger \xi}}{\sqrt{\xi^\dagger \xi}}, \tag{11.31}$$

where ξ and Z are $r \times r$ antisymmetric matrices, $\xi_{ij} = -\xi_{ji}$. As for $U(r)$, we introduce the projective matrix variable,

$$\zeta = Z\,(I_p - Z^\dagger Z)^{-\frac{1}{2}} \equiv (\zeta_\alpha)\,, \tag{11.32}$$

and we will denote coherent states as well by $|\zeta\rangle \equiv |\xi\rangle$.

Here too, the coset $\mathrm{Spin}(2r)/U(r)$ has the geometrical structure of a Kähler manifold with potential equal to $F(\zeta, \bar{\zeta}) = \frac{1}{2} \ln \det(I_r + \zeta^\dagger \zeta)$. From this potential the Riemannian metric,

$$ds^2 = \sum_{\alpha,\beta} g_{\alpha,\beta}\, d\zeta_\alpha\, d\bar{\zeta}_\beta\,, \quad g_{\alpha,\beta} = \frac{\partial^2}{\partial \zeta_\alpha \partial \bar{\zeta}_\beta} F(\zeta, \bar{\zeta})\,,$$

and the 2-form,

$$\omega = \frac{1}{2} \sum_{\alpha,\beta} g_{\alpha,\beta} d\zeta_\alpha \wedge d\bar{\zeta}_\alpha\,,$$

are derived. The volume element is given by

$$d\mu(\zeta, \bar{\zeta}) = \frac{\dim V^\Lambda}{\mathrm{Vol}\left(\mathrm{Spin}(2r)/U(r)\right)} \left(\det\left(I_{r\times r} + \zeta^\dagger \zeta\right)\right)^{-\frac{1}{2}} \prod_\alpha d\zeta_\alpha\, d\bar{\zeta}_\alpha\,.$$

As expected, we have the resolution of the identity,

$$\int d\mu(\zeta, \bar{\zeta})\, |\zeta\rangle\langle\zeta| = I_d\,,$$

and the resulting analytical Fock–Bargmann realization of the space of states V^Λ:

$$V^\Lambda \ni \Psi \mapsto \Psi(\zeta) \overset{\text{def}}{=} \langle \bar{\zeta} | \Psi \rangle\,.$$

11.3.3
Fermionic Symmetry $SO(2r+1)$

With the adding of the operators $a_i, a_j^\dagger$, $1 \le i, j \le r$, to the generators of $\mathfrak{so}(2r)$, one obtains the algebra $\mathfrak{so}(2r+1)$. The commutations rules for $\mathfrak{so}(2r)$ are now completed by

$$[a_i^\dagger, a_j] = 2(a_i^\dagger a_j - \tfrac{1}{2}\delta_{ij})\,, \tag{11.33}$$

$$[a_i^\dagger a_j - \tfrac{1}{2}\delta_{ij}, a_k] = -\delta_{ik} a_j\,, \tag{11.34}$$

$$[a_i^\dagger a_j - \tfrac{1}{2}\delta_{ij}, a_k^\dagger] = \delta_{jk} a_i^\dagger \,. \tag{11.35}$$

The Cartan subalgebra is generated by the r operators

$$H_i = a_i^\dagger a_i - \frac{1}{2} \quad i = 1, 2, \ldots, r \,. \tag{11.36}$$

The apposite Hermitian space V^Λ of states is now the one that carries the spinorial representation $\Lambda \equiv \left[\frac{1}{2}, \frac{1}{2}, \ldots \frac{1}{2}\right]$, for which the basis states are $|n_1, n_2, \ldots, n_r\rangle$, with $n_i = 0$ or 1. There are 2^r such states. The matrix representation is expressed in terms of matrices E_{ij} of order $2r + 1$, for example,

$$a_i^\dagger \leftrightarrow E_{i0} - E_{0r+i} \,, \qquad \text{etc} \,.$$

By choosing as the extremal state $|0\rangle \equiv |0, 0, \ldots, 0\rangle$, the stability group of the latter is a subgroup isomorphic to $U(r)$ and is generated by the operators

$$a_i^\dagger a_j - \frac{1}{2}\delta_{ij} \,, \quad 1 \le i, j \le r \,.$$

Hence, the coherent states built from the state $|0\rangle$ are in one-to-one correspondence with the points ζ in the coset $\text{Spin}(2r + 1)/U(r)$.

11.3.4
Graphic Summary

In Table 11.2 we summarize the chain of dynamical groups for the Hamiltonian

$$H = \sum_{i,j} \varepsilon_{ij} a_i^\dagger a_j + \sum_{i,j,l,m} V_{ijlm} a_i^\dagger a_j^\dagger a_l a_m$$

for a system of r identical fermions in the mean field approximation.

Table 11.2 Dynamical groups for systems of identical fermions.

Chain of groups and representation carrier spaces			
Group	**Carrier space**	**Dimension**	**Conserved quantity**
$U(r)$	V^{Λ_k}	$\binom{r}{k}$	Number k of fermions
$\downarrow$			
$SO(2r)$	$V^{\Lambda_+} \oplus V^{\Lambda_-} = \oplus_{k=0}^r V^{\Lambda_k}$	$2^{r-1} + 2^{r-1} = 2^r$	Parity of the total number of occupied levels
$\downarrow$			
$SO(2r + 1)$	$\oplus_{k=0}^r V^{\Lambda_k}$	2^r	

11.4 Application to the Hartree–Fock–Bogoliubov Theory

For most of the N fermion systems (e.g., nuclei, atoms, molecules), the shell structure emerges in a systematic way from approximations of the mean-field type. Hence, by restricting ourselves to one- or two-body interactions, we consider Hamiltonians of a form that is a particular case of (11.12):

$$H = \sum_{j=1}^{r} \varepsilon_j a_j^\dagger a_j + \tfrac{1}{4} \sum_{i,j,k,l} V_{ijkl}\, a_i^\dagger a_j^\dagger a_k a_l \equiv H_0 + H_{\text{int}} . \tag{11.37}$$

The maximal dynamical symmetry group is $SO(2r+1)$. If one excludes from the scope of our analysis the transfer of one sole particle, then the dynamical symmetry group reduces to $SO(2r)$. If the pairing is also neglected, then the dynamical symmetry group becomes $U(r)$.

Let us consider the case $SO(2r)$ with its fundamental spinorial representation $\Lambda_+ = \left[\frac{1}{2}, \frac{1}{2}, \ldots, \frac{1}{2}\right]$. We have defined the coherent states by

$$|\zeta\rangle \equiv |\xi\rangle = T\,|0\rangle , \quad \text{where} \quad T = \exp\left(\sum_{j<k} \left(\bar{\xi}_{jk} a_j^\dagger a_k^\dagger - \xi_{jk} a_j\, a_k \right) \right) . \tag{11.38}$$

The "displacement" operator T acting on the vacuum $|0\rangle = |0, 0, \ldots, 0\rangle$ is the most general possible unitary transformation since $U(r)$ is the stability subgroup of this extremal state and any element of $SO(2r)$ can be written as the product of the element corresponding to T with an element of $U(r)$.

To the transformation T there corresponds the following transformation of the one-particle annihilation and creation operators a_j and $a_j^\dagger$:

$$\begin{pmatrix} \boldsymbol{\alpha} \\ \boldsymbol{\alpha}^\dagger \end{pmatrix} = T \begin{pmatrix} \mathbf{a} \\ \mathbf{a}^\dagger \end{pmatrix} T^{-1} , \quad \mathbf{a} \equiv \begin{pmatrix} a_1 \\ a_2 \\ \vdots \\ a_r \end{pmatrix} \equiv (a_j) , \quad \text{and so on} . \tag{11.39}$$

This is precisely the most general one among the so-called Hartree–Fock–Bogoliubov transformations particle $\mapsto$ "quasiparticle". Within the present framework, this transformation is rewritten in matrix-block operatorial form as

$$\begin{pmatrix} \boldsymbol{\alpha} \\ \boldsymbol{\alpha}^\dagger \end{pmatrix} = \mathbb{T}^\dagger \begin{pmatrix} \mathbf{a} \\ \mathbf{a}^\dagger \end{pmatrix} , \quad \text{with} \quad \mathbb{T}^\dagger = \begin{pmatrix} U & -V^\dagger \\ V & U^t \end{pmatrix} , \tag{11.40}$$

and

$$U = \cos\sqrt{\xi^\dagger \xi} = U^\dagger , \quad \overline{V} = -\xi \frac{\sin\sqrt{\xi^\dagger \xi}}{\sqrt{\xi^\dagger \xi}} = -V^\dagger .$$

It follows that the new vacuum is precisely the coherent state $|\zeta\rangle$. The quasiparticles are then obtained from the resolution of the variational equation:

$$\delta \langle\zeta|H - \lambda N|\zeta\rangle = 0\,, \tag{11.41}$$

where N is the operator number of particles and λ is a constraint parameter called the chemical potential. Condition (11.41) leads, within the framework of the Hartree–Fock–Bogoliubov approximation, to the system

$$\begin{pmatrix} \nu & \Delta \\ -\bar{\Delta} & -\bar{\nu} \end{pmatrix} \begin{pmatrix} U_i \\ V_i \end{pmatrix} = E_i \begin{pmatrix} U_i \\ V_i \end{pmatrix}\,, \tag{11.42}$$

where ν is the so-called potential, Δ is the pairing potential,

$$\nu_{ij} = \varepsilon_{ij} - \lambda\delta_{ij} + \Gamma_{ij} \qquad \Delta_{ij} = \frac{1}{2}\sum_{i',j'} V_{iji'j'}\,\kappa_{i'j'}\,,$$

and Γ is the "deformation potential,"

$$\Gamma_{ij} = \sum_{i'j'} V_{ii'jj'}\,\varrho_{i'j'}\,.$$

In these definitions two matrices have been introduced. Their elements are lower symbols with respect to the coherent states $|\zeta\rangle$:

$$\begin{aligned} \varrho_{ij} &= \langle\zeta|a_i^\dagger a_j|\zeta\rangle \quad \text{"matrix density,"} \\ \kappa_{ij} &= \langle\zeta|a_i a_j|\zeta\rangle \quad \text{"pairing tensor."} \end{aligned}$$

If the pairing is neglected, then the relevant dynamical group is $U(r)$, ζ is a point in the coset $U(r)/(U(p) \otimes U(r-p))$, and the above procedure is equivalent to the so-called mean-field Hartree–Fock theory.

Part Two Coherent State Quantization

12
Standard Coherent State Quantization: the Klauder–Berezin Approach

12.1
Introduction

The second part of this book is devoted to the quantization of sets, more precisely measure spaces, through coherent states. To introduce that other important aspect of these states, we explain in this chapter how standard coherent states allow a natural quantization of the complex plane viewed as the phase space of the particle motion on the line. We show how they offer a classical-like representation of the evolution of quantum observables. They also help to set Heisenberg inequalities concerning "phase operator" and number operator for the oscillator Fock states. By restricting the formalism to the finite dimension, we present new quantum inequalities concerning the respective spectra of "position" and "momentum" matrices that result from such a coherent state quantization scheme for the motion on the line.

12.2
The Berezin–Klauder Quantization of the Motion of a Particle on the Line

Let us go back to the quantum harmonic oscillator, and more generally the quantum version of the particle motion on the real line. Let us first recall some of the features of this basic model. On the classical level, the corresponding phase space is $X = \mathbb{R}^2 \simeq \mathbb{C} = \{z = \frac{1}{\sqrt{2}}(q + i p)\}$ (in complex notation and with suitable physical units). This phase space is equipped with the ordinary Lebesgue measure on a plane that coincides with the symplectic 2-form: $\mu(dz\, d\bar{z}) \equiv \frac{1}{\pi}\, d^2 z$ where $d^2 z = d\Re z\, d\Im z$. Strictly included in the Hilbert space $L^2(\mathbb{C}, \mu(dz\, d\bar{z}))$ of all complex-valued functions on the complex plane that are square-integrable with respect to this measure, there is the Fock–Bargmann Hilbert subspace $\mathcal{FB}$ of all square-integrable functions that are of the form $\phi(z) = e^{-\frac{|z|^2}{2}}\, g(\bar{z})$, where $g(z)$ is analytically entire. As an orthonormal basis of this subspace we have chosen the normalized powers of the conjugate of the complex variable z weighted by the Gaussian, that

Coherent States in Quantum Physics. Jean-Pierre Gazeau

ISBN: 978-3-527-40709-5

is, $\phi_n(z) \equiv e^{-\frac{|z|^2}{2}} \frac{\bar{z}^n}{\sqrt{n!}}$ with $n \in \mathbb{N}$. Standard coherent states stem from the general construction described at length in Chapter 5:

$$|z\rangle = \frac{1}{\sqrt{\mathcal{N}(z)}} \sum_n \overline{\phi_n(z)}|n\rangle = e^{-\frac{|z|^2}{2}} \sum_{n\in\mathbb{N}} \frac{z^n}{\sqrt{n!}}|n\rangle . \tag{12.1}$$

For the normalized states (12.1), the resolution of the unity results:

$$\frac{1}{\pi}\int_{\mathbb{C}} |z\rangle\langle z|\, d^2z = \mathbb{I}_{\mathcal{H}} . \tag{12.2}$$

The property (12.2) is crucial for our purpose in setting the bridge between the classical and the quantum worlds. It encodes the quality of standard coherent states of being *canonical quantizers* along a guideline established by Klauder [15, 154] and Berezin [20]. This *Berezin–Klauder coherent state quantization*, also named *anti-Wick quantization* or *Toeplitz quantization* [156][12] by many authors, consists in associating with any classical observable f that is a function of phase space variables (q, p) or equivalently of $(z, \bar{z})$, the operator-valued integral

$$\frac{1}{\pi}\int_{\mathbb{C}} f(z,\bar{z})\, |z\rangle\langle z|\, d^2z = A_f . \tag{12.3}$$

The function f is usually supposed be be smooth, but we will not retain in the sequel this too restrictive attribute. The resulting operator A_f, if it exists, at least in a weak sense, acts on the Hilbert space $\mathcal{H}$ of quantum states for which the set of Fock (or number) states $|n\rangle$ is an orthonormal basis. It is worthwhile being more explicit about what we mean by "weak sense": the integral

$$\langle\psi|A_f|\psi\rangle = \int_{\mathbb{C}} f(z,\bar{z})|\langle\psi|z\rangle|^2 \frac{d^2z}{\pi} \tag{12.4}$$

should be finite for any $|\psi\rangle$ in some dense subset in $\mathcal{H}$. One should note that if ψ is normalized, then (12.4) represents the mean value of the function f with respect to the ψ-dependent probability distribution $z \mapsto |\langle\psi|z\rangle|^2$ on the phase space.

More mathematical rigor is necessary here, and we will adopt the following acceptance criteria for a function (or distribution) to belong to the class of quantizable classical observables.

Definition 12.1 A function $\mathbb{C} \ni z \mapsto f(z,\bar{z}) \in \mathbb{C}$ and more generally a distribution $T \in \mathcal{D}'(\mathbb{R}^2)$ is a *CS quantizable classical observable* along the map $f \mapsto A_f$ defined by (12.3), and more generally by $T \mapsto A_T$:

- if the map $\mathbb{C} \ni z = \frac{1}{\sqrt{2}}(q + ip) \equiv (q, p) \mapsto \langle z|A_f|z\rangle$ (or $\mathbb{C} \ni z \mapsto \langle z|A_T|z\rangle$) is a smooth (i.e., $\in C^\infty$) function with respect to the (q, p) coordinates of the phase plane.

12) More generally, the Berezin–Toeplitz quantization is for symplectic manifolds that admit a Kähler structure.

- and if we restore the dependence on $\hbar$ through $z \to \frac{z}{\sqrt{\hbar}}$, we must get the right semiclassical limit, which means that $\langle \frac{z}{\sqrt{\hbar}}|A_f|\frac{z}{\sqrt{\hbar}}\rangle \approx f(\frac{z}{\sqrt{\hbar}}, \frac{\bar{z}}{\sqrt{\hbar}})$ as $\hbar \to 0$. The same asymptotic behavior must hold in a distributional sense if we are quantizing distributions.

The function f (or the distribution T) is an *upper* or *contravariant* symbol of the operator A_f (or A_T), and the mean value $\langle z|A_f|z\rangle$ (or $\langle z|A_T|z\rangle$) is the *lower* or *covariant* symbol of the operator A_f (or A_T). The map $f \mapsto A_f$ is linear and associates with the function $f(z, \bar{z}) = 1$ the identity operator in $\mathcal{H}$. Note that the lower symbol of the operator A_f is the Gaussian convolution of the function $f(z, \bar{z})$:

$$\left\langle \frac{z}{\sqrt{\hbar}} \middle| A_f \middle| \frac{z}{\sqrt{\hbar}} \right\rangle = \int \frac{d^2 z'}{\pi\hbar} e^{-\frac{|z-z'|^2}{\hbar}} f\left(\frac{z'}{\sqrt{\hbar}}, \frac{\bar{z}'}{\sqrt{\hbar}}\right). \tag{12.5}$$

This expression is of great importance and is actually the reason behind the robustness of coherent state quantization, since it is well defined for a large class of non-smooth functions and even for a class of distributions comprising the tempered ones. Equation (12.5) illustrates nicely the regularizing role of quantum mechanics compared with classical singularities. Note also that the Gaussian convolution helps us carry out the semiclassical limit, since the latter can be extracted by using a saddle point approximation. For regular functions for which A_f exists, the application of the saddle point approximation is trivial and we have

$$\left\langle \frac{z}{\sqrt{\hbar}} \middle| A_f \middle| \frac{z}{\sqrt{\hbar}} \right\rangle \approx f\left(\frac{z}{\sqrt{\hbar}}, \frac{\bar{z}}{\sqrt{\hbar}}\right) \quad \text{as } \hbar \to 0\,, \tag{12.6}$$

as expected.

At this point, we should mention the "reverse" of the coherent state quantization, that is, the *coherent state dequantization*. This problem consists in finding, or, at least, proving the existence of, an upper symbol f_A for an operator A in $\mathcal{H}$ along the equation

$$A = \frac{1}{\pi} \int_{\mathbb{C}^2} d^2 z \, f_A(z, \bar{z}) \, |z\rangle\langle z| \,. \tag{12.7}$$

This question was also examined by Sudarshan in his seminal paper [8] and by many others, mainly within the context of quantum optics [6, 7, 157–159]. Our approach should be viewed as conceptually different. It raises different mathematical problems because we are not subjected to the same constraints. More details are given in [160].

Expanding bras and kets in (12.3) in terms of the Fock states yields the expression of the operator A_f in terms of its infinite matrix elements $(A_f)_{nn'} \stackrel{\text{def}}{=} \langle n|A_f|n'\rangle$:

$$A_f = \sum_{n,n' \geq 0} (A_f)_{nn'} |n\rangle\langle n'| \,, \quad (A_f)_{nn'} = \frac{1}{\sqrt{n!n'!}} \int_{\mathbb{C}} \frac{d^2 z}{\pi} e^{-|z|^2} z^n \bar{z}^{n'} f(z, \bar{z}) \,. \tag{12.8}$$

Alternatively, if the classical observable is "isotropic", that is, $f(z,\bar{z}) \equiv h(|z|^2)$, then A_f is diagonal, with matrix elements given by a kind of gamma transform:

$$(A_f)_{nn'} = \delta_{nn'} \frac{1}{n!} \int_0^{\infty} du\, e^{-u}\, u^n\, h(u)\,. \tag{12.9}$$

In the case where the classical observable is purely angular-dependent, that is, $f(z,\bar{z}) = g(\theta)$ for $z = |z|\, e^{i\theta}$, the matrix elements $(A_f)_{nn'}$ are given by

$$(A_f)_{nn'} = \frac{\Gamma(\frac{n+n'}{2} + 1)}{\sqrt{n!n'!}}\, c_{n'-n}(g)\,, \tag{12.10}$$

where $c_m(g) \stackrel{\text{def}}{=} \frac{1}{2\pi} \int_0^{2\pi} g(\theta)\, e^{-im\theta}\, d\theta$ is the Fourier coefficient of the 2π-periodic function g.

Let us explore what the coherent state quantization map (12.3) produces, starting with some elementary functions f. We have for the most basic one

$$f(z) = z \mapsto \int_{\mathbb{C}} z\, |z\rangle\langle z|\, \frac{d^2z}{\pi} = \sum_n \sqrt{n+1}|n\rangle\langle n+1| \equiv a\,, \tag{12.11}$$

which is the lowering operator, $a|n\rangle = \sqrt{n}|n-1\rangle$. The adjoint $a^{\dagger}$ is obtained by replacing z by $\bar{z}$ in (12.11).

From $q = \frac{1}{\sqrt{2}}(z+\bar{z})$ and $p = \frac{1}{\sqrt{2}i}(z-\bar{z})$, one easily infers by linearity that the canonical position q and momentum p map to the quantum observables $\frac{1}{\sqrt{2}}(a+a^{\dagger}) \equiv Q$ and $\frac{1}{\sqrt{2}i}(a-a^{\dagger}) \equiv P$, respectively. In consequence, the self-adjoint operators Q and P obtained in this way obey the canonical commutation rule $[Q,P] = i\mathbb{I}_{\mathcal{H}}$, and for this reason fully deserve the name of position and momentum operators of the usual (Galilean) quantum mechanics, together with all localization properties specific to the latter, as they were listed at length in Chapter 1.

12.3 Canonical Quantization Rules

At this point, it is worth recalling what *quantization of classical mechanics* means in a commonly accepted sense (for a recent and complete review, see [155]). In this context, a classical observable f is supposed to be a smooth function with respect to the canonical variables. Here we reintroduce again the Planck constant since it parameterizes the link between classical and quantum mechanics.

12.3.1 Van Hove Canonical Quantization Rules [161]

Given a phase space with canonical coordinates $(\boldsymbol{q},\boldsymbol{p})$

(i) to the classical observable $f(\boldsymbol{q},\boldsymbol{p}) = 1$ corresponds the identity operator in the (projective) Hilbert space $\mathcal{H}$ of quantum states,

(ii) the correspondence that assigns to a classical observable $f(\boldsymbol{q}, \boldsymbol{p})$, supposed be to infinitely differentiable, a (essentially) self-adjoint operator A_f in $\mathcal{H}$ is a linear map,

(iii) to the classical Poisson bracket corresponds, *at least at the order* $\hbar$, the quantum commutator, multiplied by $i\hbar$:

$$\begin{aligned} &\text{with} \quad f_j(\boldsymbol{q}, \boldsymbol{p}) \mapsto A_{f_j} \quad \text{for} \quad j = 1, 2, 3 \\ &\text{we have} \quad \{f_1, f_2\} = f_3 \mapsto [A_{f_1}, A_{f_2}] = i\hbar A_{f_3} + o(\hbar)\,, \end{aligned}$$

(iv) some conditions of minimality on the resulting observable algebra.

The last point can give rise to technical and interpretational difficulties.

It is clear that points (i) and (ii) are fulfilled with the coherent state quantization, the second one at least for observables obeying fairly mild conditions. To see better the "asymptotic" meaning of condition (iii), let us quantize higher-degree monomials, starting with $|z|^2 = \frac{p^2+q^2}{2}$, the classical harmonic oscillator Hamiltonian. For the latter, we get immediately from (12.9)

$$|z|^2 \mapsto A_{|z|^2} = \sum_{n \geq 0} (n+1)|n\rangle\langle n| = N + I_d\,, \tag{12.12}$$

where $N = a^{\dagger}a$ is the number operator. Since $N = \frac{P^2+Q^2}{2} - \frac{1}{2} I_d$, we see from this elementary example that the coherent state quantization does not fit exactly with the canonical one, which consists in just replacing q by Q and p by P in the expressions of the observables $f(q, p)$ and next proceeding to a symmetrization to comply with self-adjointness. In the present case, there is a shift by $1/2$, more precisely $\hbar/2$ after restoring physical dimensions, of the quantum harmonic oscillator spectrum, that is, both quantizations differ by an amount on the order of $\hbar$. Actually, it seems that no physical experiment can discriminate between those two spectra that differ from each other by a simple shift; see [162] for a careful discussion of this point.

12.4
More Upper and Lower Symbols: the Angle Operator

Since we do not retain in our quantization scheme the smoothness condition on the classical observables, we feel free to quantize with standard coherent states another elementary classical object, namely, the argument $\theta \in [0, 2\pi)$ of the complex variable $z = r\,e^{i\theta}$. Computing its quantum counterpart, say, $A_{\arg}$ from (12.3) is straightforward and yields the infinite matrix:

$$A_{\arg} = \pi I_d + i \sum_{n \neq n'} \frac{\Gamma\left(\frac{n+n'}{2} + 1\right)}{\sqrt{n!n'!}} \frac{1}{n' - n} |n\rangle\langle n'|\,. \tag{12.13}$$

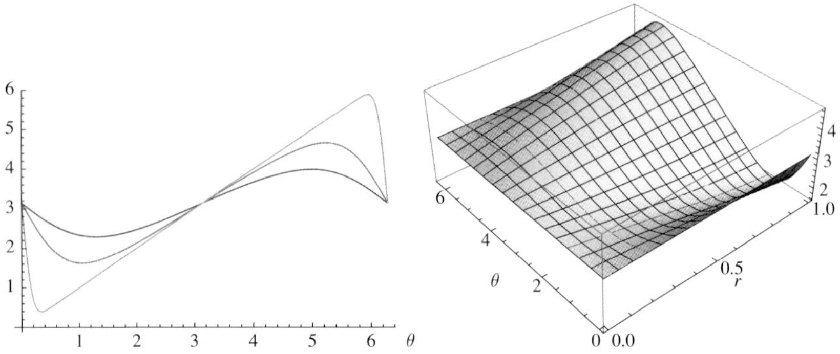

Fig. 12.1 Lower symbol of the angle operator for $r = \{0.5, 1, 5\}$ and $\theta \in [0, 2\pi)$ and for $(r, \theta) \in [0, 1] \times [0, 2\pi)$.

The corresponding lower symbol reads as the sine Fourier series:

$$\langle z|A_{\text{arg}}|z\rangle \equiv \langle (r,\theta)|A_{\text{arg}}|(r,\theta)\rangle = \pi + i\, e^{-|z|^2} \sum_{n \neq n'} \frac{\Gamma\left(\frac{n+n'}{2} + 1\right)}{n!n'!} \frac{z^{n'} \bar{z}^n}{n' - n}$$

$$= \pi - 2 \sum_{q=1}^{\infty} \frac{c_q(r)}{q} \sin q\theta\,, \tag{12.14}$$

where

$$c_q(r) = \frac{2}{q} e^{-r^2} r^q \frac{\Gamma(\frac{q}{2}+1)}{\Gamma(q+1)}\, {}_1F_1\left(\frac{q}{2}+1; q+1; r^2\right).$$

One can easily prove that this symbol is infinitely differentiable in the variables r and θ [160].

The behavior of the lower symbol (12.14) is shown in Figure 12.1.

It is interesting to evaluate the asymptotic behaviors of the function (12.14) at small and large r, respectively. At small r, it oscillates around its average value π with amplitude equal to $\sqrt{\pi}r$:

$$\langle (r,\theta)|A_{\text{arg}}|(r,\theta)\rangle \approx \pi - \sqrt{\pi}r\, \sin\theta\,.$$

At large r, we recover the Fourier series of the 2π-periodic angle function:

$$\langle (r,\theta)|A_{\text{arg}}|(r,\theta)\rangle \approx \pi - 2\sum_{q=1}^{\infty} \frac{1}{q} \sin q\theta = \theta \quad \text{for} \quad \theta \in [0, 2\pi)\,.$$

The latter result can be equally understood in terms of the classical limit of these quantum objects. Indeed, by reinjecting into our formulas physical dimensions, we know that the quantity $|z|^2 = r^2$ acquires the dimension of an action and should appear in the formulas divided by the Planck constant $\hbar$. Hence, the limit $r \to \infty$ in our previous expressions can also be considered as the classical limit $\hbar \to$

0. Since we have at our disposal the number operator $N = a^\dagger a$, which is, up to a constant shift, the quantization of the classical action, and an angle operator, we can examine their commutator and its lower symbol to measure the extent to which we get something close to the expected canonical value, namely, $i\,\mathbb{I}_{\mathcal{H}}$. The commutator reads as

$$[A_{\arg}, N] = i \sum_{n \neq n'} \frac{\Gamma\left(\frac{n+n'}{2} + 1\right)}{\sqrt{n! n'!}} |n\rangle\langle n'| . \tag{12.15}$$

Its lower symbol is then given by

$$\langle (r,\theta)|[A_{\arg}, N]|(r,\theta)\rangle = i \sum_{q=1}^{\infty} q c_q(r) \cos q\theta \equiv i\, C(r,\theta) , \tag{12.16}$$

with the same $c_q(r)$ as in (12.14).

At small r, the function $C(r,\theta)$ oscillates around 0 with amplitude equal to $\sqrt{\pi}r$:

$$C(r,\theta) \approx \sqrt{\pi} r \ \cos\theta .$$

At large r, the function $C(r,\theta)$ tends to the Fourier series $2\sum_{q=1}^{\infty} \cos q\theta$, whose convergence has to be understood in the sense of distributions. Applying the Poisson summation formula, we get at $r \to \infty$ (or $\hbar \to 0$) the expected "canonical" behavior for $\theta \in (0, 2\pi)$. More precisely, we obtain in this quasi-classical regime the following asymptotic behavior:

$$\langle (r,\theta)|[A_{\arg}, N]|(r,\theta)\rangle \approx -i + 2\pi i \sum_{n\in\mathbb{Z}} \delta(\theta - 2\pi n) . \tag{12.17}$$

The fact that this commutator is not exactly canonical was expected since we know from Dirac [163] the impossibility of getting canonical commutation rules for the quantum versions of the classical action–angle pair. One can observe that the commutator symbol becomes canonical for $\theta \neq 2\pi n$, $n \in \mathbb{Z}$. Dirac singularities are located at the discontinuity points of the 2π-periodic extension of the linear function $f(\theta) = \theta$ for $\theta \in [0, 2\pi)$.

12.5 Quantization of Distributions: Dirac and Others

It is commonly accepted that a "coherent state diagonal" representation of the type (12.3) is possible only for a restricted class of operators in $\mathcal{H}$. The reason is that we usually impose too strong conditions on the upper symbol $f(z,\bar{z})$ viewed as a classical observable on the phase space, and so subjected to belonging to the space of infinitely differentiable functions on $\mathbb{R}^2$. We already noticed that a "reasonable" phase or angle operator is easily built starting from the classical discontinuous periodic angle function. The application of the same quantization method to the free particle "time" q/p, which is obviously singular at $p = 0$, was carried out in [160] and yields a fairly reasonable time operator for the free motion on the line.

Owing to the general expression (12.8) for matrix elements of the quantized version of an observable f, one can immediately think of extending the method to tempered distributions. In particular, we will show that *any* simple operator $\Pi_{nn'} \stackrel{\text{def}}{=} |n\rangle\langle n'|$ also has a coherent state diagonal representation by including distributions on $\mathbb{R}^2$ in the class of classical observables.

In (12.8), the functions

$$(z, \bar{z}) \mapsto e^{-|z|^2} z^n \bar{z}^{n'} \tag{12.18}$$

are rapidly decreasing C^∞ functions on the plane with respect to the canonical coordinates (q, p), or equivalently with respect to the coordinates $(z, \bar{z})$: they belong to the Schwartz space $\mathcal{S}(\mathbb{R}^2)$. The use of complex coordinates is clearly more adapted to the present context, and we adopt the following definitions and notations for tempered distributions. Firstly, any function $f(z, \bar{z})$ that is "slowly increasing" and locally integrable with respect to the Lebesgue measure d^2z on the plane defines a regular tempered distribution T_f, that is, a continuous linear form on the vector space $\mathcal{S}(\mathbb{R}^2)$ equipped with the usual topology of uniform convergence at each order of partial derivatives multiplied by polynomial of arbitrary degree [164]. This definition rests on the map

$$\mathcal{S}(\mathbb{R}^2) \ni \psi \mapsto \langle T_f, \psi \rangle \stackrel{\text{def}}{=} \int_{\mathbb{C}} d^2z\, f(z, \bar{z})\, \psi(z, \bar{z}) \,, \tag{12.19}$$

and the notation is the same for all tempered distributions T. In view of (12.18), we can extend (12.19) to locally integrable functions $f(z, \bar{z})$ that increase like $e^{\eta|z|^2} p(z, \bar{z})$ for some $\eta < 1$ and some polynomial p, and to all distributions that are derivatives (in the distributional sense) of such functions. We recall here that partial derivatives of distributions are given by

$$\left\langle \frac{\partial^r}{\partial z^r} \frac{\partial^s}{\partial \bar{z}^s} T, \psi \right\rangle = (-1)^{r+s} \left\langle T, \frac{\partial^r}{\partial z^r} \frac{\partial^s}{\partial \bar{z}^s} \psi \right\rangle . \tag{12.20}$$

We also recall that the multiplication of distributions T by smooth functions $\alpha(z, \bar{z}) \in C^\infty(\mathbb{R}^2)$ is understood through

$$\mathcal{S}(\mathbb{R}^2) \ni \psi \mapsto \langle \alpha T, \psi \rangle \stackrel{\text{def}}{=} \langle T, \alpha\, \psi \rangle \,. \tag{12.21}$$

Now, equipped with the above distributional material, we consider as "acceptable" observables those distributions in $\mathcal{D}'(\mathbb{R}^2)$ that obey the following condition.

Definition 12.2 (Coherent state quantizable observable) A distribution $T \in \mathcal{D}'(\mathbb{R}^2)$ is a *coherent state quantizable classical observable* if there exists $\eta < 1$ such that the product $e^{-\eta|z|^2}\, T \in \mathcal{S}'(\mathbb{R}^2)$, that is, is a tempered distribution.

Of course, all compactly supported distributions such as the Dirac distribution and its derivatives are tempered and so are *coherent state quantizable classical observable*. The Dirac distribution supported by the origin of the complex plane is denoted as usual by δ (and abusively in the present context by $\delta(z,\bar{z})$):

$$C^{\infty}(\mathbb{R}^2) \ni \psi \mapsto \langle \delta, \psi \rangle \equiv \int_{\mathbb{C}} d^2z\, \delta(z,\bar{z})\, \psi(z,\bar{z}) \stackrel{\text{def}}{=} \psi(0,0)\,. \tag{12.22}$$

Let us now coherent state quantize the Dirac distribution using the recipe provided by (12.3) and (12.8):

$$\begin{aligned}\frac{1}{\pi}\int_{\mathbb{C}} \delta(z,\bar{z})\, |z\rangle\langle z|\, d^2z &= \sum_{n,n'\geq 0} \frac{1}{\sqrt{n!n'!}} \int_{\mathbb{C}} \frac{d^2z}{\pi}\, e^{-|z|^2}\, z^n\, \bar{z}^{n'}\, \delta(z,\bar{z})\, |n\rangle\langle n'| \\ &= \frac{1}{\pi}\, |z=0\rangle\langle z=0| \;=\; \frac{1}{\pi}\, \Pi_{00}\,.\end{aligned} \tag{12.23}$$

We thus find that the ground state (as a projector) is the quantized version of the Dirac distribution supported by the origin of the phase space. The obtaining of all possible diagonal projectors $\Pi_{nn} = |n\rangle\langle n|$ or even all possible operators $\Pi_{nn'} = |n\rangle\langle n'|$ is based on the quantization of partial derivatives of the δ distribution. First, let us quantize the various derivatives of the Dirac distribution:

$$\begin{aligned} U_{a,b} &= \int_{\mathbb{C}} \left[\frac{\partial^b}{\partial z^b}\frac{\partial^a}{\partial \bar{z}^a}\delta(z,\bar{z})\right] |z\rangle\langle z|\, d^2z \\ &= \sum_{n,n'\geq 0} (-1)^{n+a} \frac{b!\,a!}{(b-n)!}\frac{1}{\sqrt{n!n'!}}\, \delta_{n-b,n'-a}\, \Pi_{nn'}\,. \end{aligned} \tag{12.24}$$

Once this quantity $U_{a,b}$ is at hand, one can invert the formula to get the operator $\Pi_{r+s,r} = |r+s\rangle\langle r|$ as

$$\Pi_{r+s,r} = \sqrt{r!(r+s)!}(-1)^s \sum_{p=0}^{r} \frac{1}{p!(s+p)!(r-p)!}\, U_{p,p+s}\,, \tag{12.25}$$

and so its upper symbol is given by the distribution supported by the origin:

$$f_{r+s,r}(z,\bar{z}) = \sqrt{r!(r+s)!}(-1)^s \sum_{p=0}^{r} \frac{1}{p!(s+p)!(r-p)!} \left[\frac{\partial^{p+s}}{\partial z^{p+s}}\frac{\partial^p}{\partial \bar{z}^p}\delta(z,\bar{z})\right]. \tag{12.26}$$

Note that this distribution, as is well known, can be approached, in the sense of the topology on $\mathcal{D}'(\mathbb{R}^2)$, by smooth functions, such as linear combinations of derivatives of Gaussians. The projectors $\Pi_{r,r}$ are then obtained trivially by setting $s = 0$ in (12.25) to get

$$\Pi_{r,r} = \sum_{p=0}^{r} \frac{1}{p!}\binom{r}{p} U_{p,p}\,. \tag{12.27}$$

This "phase space formulation" of quantum mechanics enables us to mimic at the level of functions and distributions the algebraic manipulations on operators within the quantum context. By carrying out the coherent state quantization of Cartesian powers of planes, we could obtain an interesting "functional portrait" in terms of a "star" product on distributions for the quantum logic based on manipulations of tensor products of quantum states.

12.6 Finite-Dimensional Canonical Case

The idea of exploring various aspects of quantum mechanics by restricting the Hilbertian framework to finite-dimensional space is not new, and has been intensively used in the last decade, mainly in the context of quantum optics [165–167], but also in the perspective of noncommutative geometry and "fuzzy" geometrical objects [168], or in matrix model approaches in problems such as the quantum Hall effect [169]. For quantum optics, a comprehensive review (mainly devoted to the Wigner function) is provided in [170]. In [166, 167], the authors defined normalized finite-dimensional coherent states by truncating the Fock expansion of the standard coherent states. Let us see through the approach presented in this chapter how we recover their coherent states [171]. We just restrict the choice of the orthonormal set $\{\phi_n\}$ to a finite subset of it, more precisely to the first N elements:

$$\phi_n(z) = e^{-\frac{|z|^2}{2}} \frac{\bar{z}^n}{\sqrt{n!}}, \quad n = 0, 1, \ldots N-1 < \infty . \tag{12.28}$$

The coherent states then read

$$|z\rangle = \frac{e^{-\frac{|z|^2}{2}}}{\sqrt{\mathcal{N}(|z|^2)}} \sum_{n=0}^{N-1} \frac{z^n}{\sqrt{n!}} |n\rangle , \tag{12.29}$$

with

$$\mathcal{N}(|z|^2) = e^{-|z|^2} \sum_{n=0}^{N-1} \frac{|z|^{2n}}{n!} . \tag{12.30}$$

The quantization procedure (12.3) yields for the classical observables q and p the position operator, Q_N, and the momentum operator, P_N, or phase quadratures in quantum optics, acting in N-dimensional Hermitian space. They read as the $N \times N$ matrices

$$Q_N = \begin{pmatrix} 0 & \frac{1}{\sqrt{2}} & 0 & \cdots & 0 \\ \frac{1}{\sqrt{2}} & 0 & 1 & \cdots & 0 \\ 0 & 1 & \ddots & \ddots & \vdots \\ \vdots & \cdots & \ddots & 0 & \sqrt{\frac{N-1}{2}} \\ 0 & 0 & \cdots & \sqrt{\frac{N-1}{2}} & 0 \end{pmatrix} , \tag{12.31}$$

$$P_N = -i \begin{pmatrix} 0 & \frac{1}{\sqrt{2}} & 0 & \dots & 0 \\ -\frac{1}{\sqrt{2}} & 0 & 1 & \dots & 0 \\ 0 & -1 & \ddots & \ddots & \vdots \\ \vdots & \dots & \ddots & 0 & \sqrt{\frac{N-1}{2}} \\ 0 & 0 & \dots & -\sqrt{\frac{N-1}{2}} & 0 \end{pmatrix} . \tag{12.32}$$

Their commutator is "almost" canonical:

$$[Q_N, P_N] = i I_N - i N E_N , \tag{12.33}$$

where E_N is the orthogonal projector on the last basis element:

$$E_N = \begin{pmatrix} 0 & \dots & 0 \\ \vdots & \ddots & \vdots \\ 0 & \dots & 1 \end{pmatrix} .$$

The presence of such a projector in (12.33) is clearly a consequence of the truncation at the Nth level.

A very interesting finding comes from this finite-dimensional quantization of the classical phase space. Let us examine the spectral values of the position operator, that is, the allowed or experimentally measurable quantum positions. They *are* just the zeros of the Hermite polynomials $\mathrm{H}_N(\lambda)$, and the same result holds for the momentum operator. We know that $\mathrm{H}_N(0) = 0$ if and only if N is odd and that the other zeros form a set symmetric with respect to the origin. Let us order the nonnull roots of the Hermite polynomial $\mathrm{H}_N(\lambda)$ as

$$\left\{-\lambda_{\lfloor \frac{N}{2} \rfloor}(N), -\lambda_{\lfloor \frac{N}{2} \rfloor -1}(N), \dots, -\lambda_1(N), \lambda_1(N), \dots, \lambda_{\lfloor \frac{N}{2} \rfloor -1}(N), \lambda_{\lfloor \frac{N}{2} \rfloor}(N)\right\} . \tag{12.34}$$

It is a well-known property of the Hermite polynomials that $\lambda_{i+1}(N) - \lambda_i(N) > \lambda_1(N)$ for all $i \geq 1$ if N is odd, whereas $\lambda_{i+1}(N) - \lambda_i(N) > 2\lambda_1(N)$ for all $i \geq 1$ if N is even, and that the zeros of the Hermite polynomials H_N and H_{N+1} intertwine.

Let us denote by $\lambda_M(N) = \lambda_{\lfloor \frac{N}{2} \rfloor}(N)$ (or $\lambda_m(N) = \lambda_1(N)$) the largest root (or the smallest nonzero root in absolute value) of H_N. A numerical study of the product

$$\pi_N = \lambda_m(N)\lambda_M(N) \tag{12.35}$$

was carried out in [171]. It was found that π_N goes asymptotically to π for large even N and to 2π for large odd N.

This result (which can be rigorously proved by using the Wigner semicircle law for the asymptotic distribution of zeros of the Hermite polynomials [172, 173]) is

Table 12.1 Values of $\sigma_N = \delta_N(Q)\Delta_N(Q)$ up to $N = 10^6$. Compare with the value of 2π.

N	$\delta_N(Q)\Delta_N(Q)$	2π
10	4.713 054	
55	5.774 856	
100	5.941 534	
551	6.173 778	
1000	6.209 670	
5555	6.259 760	
10 000	6.267 356	
55 255	6.278 122	
100 000	6.279 776	
500 555	6.282 020	
1 000 000	6.282 450	6.283 185 3

important with regard to its physical implications in terms of correlation between small and large distances. Define by

- $\Delta_N(Q) = 2\lambda_M(N)$ the "size" of the "universe" accessible to exploration by the quantum system
- $\delta_N(Q) = \lambda_m(N)$ (resp. $\delta_N(Q) = 2\lambda_m(N)$) for odd (or even) N, the "size" of the smallest "cell" for which exploration is forbidden by the same system,

and introduce the product

$$\delta_N(Q)\Delta_N(Q) \equiv \sigma_N = \begin{cases} 4\lambda_m(N)\lambda_M(N) & \text{for } N \text{ even,} \\ 2\lambda_m(N)\lambda_M(N) & \text{for } N \text{ odd,} \end{cases} \tag{12.36}$$

then σ_N, as a function of N, is strictly increasing and goes asymptotically to 2π (Table 12.1).

Hence, we can assert the new inequalities concerning the quantum position and momentum:

$$\delta_N(Q)\Delta_N(Q) \le 2\pi\,, \quad \delta_N(P)\Delta_N(P) \le 2\pi \quad \forall N\,. \tag{12.37}$$

In order to fully perceive the physical meaning of such inequalities, it is necessary to reintegrate into them physical constants or scales appropriate for the physical system considered, that is, characteristic length l_c and momentum p_c:

$$\delta_N(Q)\Delta_N(Q) \le 2\pi l_c^2\,, \quad \delta_N(P)\Delta_N(P) \le 2\pi p_c^2 \quad \forall N\,, \tag{12.38}$$

where $\delta_N(Q)$ and $\Delta_N(Q)$ are now expressed in unit l_c.

Realistically, in any physical situation N cannot be infinite: there is an obvious limitation on the frequencies or energies accessible to observation/experimentation. So it is natural to work with a finite although large value of N, which need not be determinate. In consequence, there exist irreducible limitations, namely, $\delta_N(Q)$ and $\Delta_N(Q)$, in the exploration of small and large distances, and both limitations have the correlation $\delta_N(Q)\Delta_N(Q) \leq 2\pi l_c^2$.

Suppose that there exists, for theoretical reasons, a fundamental or "universal" minimal length, say, l_m, which could be the Planck length $l_{\text{Pl}} = \sqrt{\hbar G/c^3} \approx 10^{-35}$ m or something else, depending on the experimental or observational context, or, equivalently, a universal ratio $\varrho_u = l_c/l_m \geq 1$. Then, from $\delta_N(Q) \geq l_m$, we infer that there exists a universal maximal length l_M given by

$$l_M \approx 2\pi\varrho_u l_c\,. \tag{12.39}$$

Of course, if we choose $l_m = l_c$, then the size of the "universe" is $l_M \approx 2\pi l_m$, a fact that leaves no room for the observer and observed things! Now, if we choose a characteristic length appropriate for atomic physics, such as the Bohr radius, $l_c \approx 10^{-10}$ m, and for the minimal length the Planck length, $l_m = l_{\text{Pl}} \approx 10^{-35}$ m, we find for the maximal size the astronomical quantity $l_M \approx 10^{16}$ m $\approx$ 1 light year, which is also of the order of one parsec. On the other hand, if we consider the (controversial) estimated size of our present universe $L_u = cT_u$, with $T_u \approx 13 \times 10^9$ years, we get from $l_p\, L_u \approx 2\pi l_c^2$ a characteristic length $l_c \approx 10^{-5}$ m, that is, a wavelength in the infrared region of the electromagnetic spectrum.

Another interesting outcome of the monotonic increase of the product

$$l_m\, l_M \underset{N\to\infty}{\overset{<}{\to}} 2\pi l_c^2$$

is that the reasoning leading to (12.39) can be reversed. Suppose that there exists an absolute confinement, of size l_M, for the system considered. Then, at large N, there exists as well an impassable "core" of size

$$l_m \approx 2\pi \frac{l_c^2}{l_M}\,. \tag{12.40}$$

Then, the allowed range of values of the characteristic length l_c is $l_m \leq l_c \leq l_M/\sqrt{\pi}$ since, for the upper limit $l_c = l_M/\sqrt{\pi}$, we have $l_m \approx l_M$, whereas at the lowest limit we recover $l_M \approx 2\pi l_m$.

Let us turn to another example, which might be viewed as more concrete, namely, the quantum Hall effect in its matrix model version [169]. The planar coordinates X_1 and X_2 of quantum particles in the lowest Landau level of a constant magnetic field do not commute:

$$[X_1, X_2] = i\theta\,, \tag{12.41}$$

where θ represents a minimal area. We recall that the average density of $N \to \infty$ electrons is related to θ by $\varrho_o = 1/2\pi\theta$ and the filling fraction is $\nu = 2\pi\varrho_0/B$. The quantity $l_m = \sqrt{\theta}$ can be considered as a minimal length. The model deals with a finite number N of electrons:

$$[X_{1,N}, X_{2,N}] = i\theta(1 - N|N-1\rangle\langle N-1|)\,. \tag{12.42}$$

In this context, our inequalities read as

$$\delta_N(X_i)\Delta_N(X_i) \le 2\pi l_c^2\,, \quad i = 1, 2\,, \tag{12.43}$$

where l_c corresponds to a choice of experimental unit. Since $l_m = \sqrt{\theta}$ affords an irreducible lower limit in this problem, we can assert that the *maximal linear size* L_M of the sample should satisfy

$$l_M \le 2\pi \frac{l_c}{\sqrt{\theta}} l_c \tag{12.44}$$

for *any* finite N.

The experimental interpretation of such a result certainly deserves a deeper investigation.

As a final comment concerning the inequalities (12.37), we would like to insist on the fact they are not just an outcome of finite approximations Q_N and P_N (or $X_{1,N}$ and $X_{2,N}$) to the canonical position and momentum operators (or to X_1 and X_2) in infinite-dimensional Hilbert space of quantum states. They hold however large the dimension N is, as long as it is finite. Furthermore, let us advocate the idea that a quantization of the classical phase space results from the choice of a specific (reproducing) Hilbert subspace $\mathcal{H}$ in $L^2(\mathbb{C}, \mu(dz\, d\bar{z}))$ in which coherent states provide a frame resolving the identity. This frame corresponds to a certain point of view in dealing with the classical phase space, and this point of view yields the quantum versions Q_N and P_N (or $X_{1,N}$ and $X_{2,N}$) of the classical coordinates q and p (or x_1 and x_2).

13
Coherent State or Frame Quantization

13.1
Introduction

Physics is part of the natural sciences and its prime object is what we call "nature", or rather, in a more restrictive sense "time", "space", "matter", "energy", and "interaction", which appear at a certain moment of the process in the form of "significant" data, such as position, speed, and frequencies. So the question arises how to process those data, and this raises the question of a selected point of view or *frame*. Faced with a set of "raw" collected data encoded into a certain mathematical form and provided by a measure, that is, a function which attributes a weight of importance to subsets of data, we give in addition more or less importance to different aspects of those data by choosing, opportunistically, the most appropriate frame of analysis.

We include in this general scheme the *quantum processing*, that is, the way of considering objects from a quantum point of view, exactly like we quantize the classical phase space in quantum mechanics. To a certain extent, quantization pertains to a larger discipline than just restricting ourselves to specific domains of physics such as mechanics or field theory. The aim of this chapter is to provide a generalization of the Berezin–Klauder–Toeplitz quantization illustrated in Chapter 12, precisely the quantization of a measure space X once we are given a family of coherent states or frame constructed along the lines given in Chapter 5. We also develop the notion of lower and upper symbols resulting from such a quantization scheme, and finally discuss the probabilistic content of the construction. A quite elementary example, the quantization of the circle with 2×2 matrices, is presented as an immediate illustration of the formalism.

13.2
Some Ideas on Quantization

Many reviews exist on the quantization problem and the variety of approaches for solving it; see, for instance, the recent ones by Ali and Englis [155] and by Landsman [174]. For simplification, let us say that a quantization is a procedure that

Coherent States in Quantum Physics. Jean-Pierre Gazeau

ISBN: 978-3-527-40709-5

associates with an algebra $\mathcal{A}_{\mathrm{cl}}$ of classical observables an algebra $\mathcal{A}_{\mathrm{q}}$ of quantum observables. The algebra $\mathcal{A}_{\mathrm{cl}}$ is usually realized as a commutative Poisson algebra[13)] of derivable functions on a symplectic (or phase) space X. The algebra $\mathcal{A}_{\mathrm{q}}$ is, however, noncommutative in general and the quantization procedure must provide a correspondence $\mathcal{A}_{\mathrm{cl}} \mapsto \mathcal{A}_{\mathrm{q}} : f \mapsto A_f$. Various procedures of quantization exist, and minimally require the following conditions, which loosely parallel those listed in Section 12.3:

- With the constant function 1 is associated the unity of $\mathcal{A}_{\mathrm{q}}$,
- The commutation relations of $\mathcal{A}_{\mathrm{q}}$ reproduce the Poisson relations of $\mathcal{A}_{\mathrm{cl}}$. Moreover, they offer a realization of the Heisenberg algebra.
- $\mathcal{A}_{\mathrm{q}}$ is realized as an algebra of operators acting in some Hilbert space.

Most physical quantum theories may be obtained as the result of a canonical quantization procedure. However, the prescriptions for the latter appear quite arbitrary. Moreover, it is difficult, if not impossible, to implement it covariantly. It is thus difficult to generalize this procedure to many systems. Geometrical quantization [175] exploits fully the symplectic structure of the phase space, but generally requires more structure, such as a symplectic potential, for example, the Legendre form on the cotangent bundle of a configuration space. In this regard, the deformation quantization appears more general in the sense that it is based on the symplectic structure only and it preserves symmetries (symplectomorphisms) [176, 177].

The coherent state quantization that is presented here is by far more universal since it does not even require a symplectic or Poisson structure. The only structure that a space X must possess is a measure. This procedure can be considered from different viewpoints:

- It is mostly genuine in the sense that it verifies all the requirements above, including those relative to the Poisson structure when the latter is present.
- It may also be viewed as a "fuzzyfication" of X: an algebra $\mathcal{A}_{\mathrm{cl}}$ of functions on X is replaced by an algebra $\mathcal{A}_{\mathrm{q}}$ of operators, which may be considered as the "coordinates" of a fuzzy version of X, even though the original X is not equipped with a manifold structure. The term "fuzzy" manifold, where points have noncommutative coordinates, was proposed by Madore [178] in his presentation of noncommutative geometry of simple manifolds such as the sphere. As a matter of fact, ordinary quantum mechanics may also be viewed as a noncommutative version of the geometry of the phase space, where position and momentum operators do not commute [179]. In this regard, the quantization of a "set of data" makes a fuzzy or noncommutative geometry emerge.

13) A Poisson algebra $\mathcal{A}$ is an associative algebra together with a Lie bracket $X, Y \mapsto \{X, Y\}$ that also satisfies the Leibniz law: for any $X \in \mathcal{A}$ the action $D_X(Y) \stackrel{\mathrm{def}}{=} \{X, Y\}$ on $\mathcal{A}$ is a *derivation* $\{X, YZ\} = \{X, Y\}Z + Y\{X, Z\}$. As a matter of fact, the space of real-valued smooth functions over a symplectic manifold, when equipped with the Poisson bracket, forms a Poisson algebra.

- Finally, this procedure is, to a certain extent, a change of point of view in considering the "system" X, not necessarily a path to quantum physics. In this sense, it could be called a discretization or a regularization [180]. It shows a similitude with standard procedures pertaining to signal processing, for instance, those involving wavelets, which are coherent states for the affine group transforming the time-scale half-plane into itself. In many respects, the choice of a quantization appears as the choice of a resolution in looking at the system.

13.3
One more Coherent State Construction

In Section 5.3 we drew a parallel between quantum mechanics and signal analysis: the object considered is an observation set X of data or parameters equipped with a measure μ defined on a σ-algebra $\mathcal{F}$. For both theories, the natural framework of investigation is the Hilbert space $L^2_{\mathbb{K}}(X,\mu)$ ($\mathbb{K} = \mathbb{R}$ or $\mathbb{C}$) of square-integrable real or complex functions $f(x)$ on the observation set X: $\int_X |f(x)|^2 \mu(dx) < \infty$. We then observed that "quantum processing" of X differs from signal processing since, in particular, not all square-integrable functions are eligible as quantum states. We meet, precisely at this point, the quantization problem: how do we select quantum states among simple signals? In other words, how do we select the true (projective) Hilbert space of quantum states, denoted by $\mathcal{K}$, that is, a closed subspace of $L^2_{\mathbb{K}}(X,\mu)$, or equivalently the corresponding orthogonal projector $I_{\mathcal{K}}$?

The only requirements on $\mathcal{K}$, in addition to being a Hilbert space, amount to the following technical conditions:

Condition 13.1 (Quantum states)

(i) For all $\phi \in \mathcal{K}$ and all x, $\phi(x)$ is well defined (this is, of course, the case whenever X is a topological space and the elements of $\mathcal{K}$ are continuous functions).

(ii) The linear map ("evaluation map")

$$\begin{aligned} \delta_x : \mathcal{K} &\to \mathbb{K} \\ \phi &\mapsto \phi(x) \end{aligned} \tag{13.1}$$

is continuous with respect to the topology of $\mathcal{K}$, for almost all x.

The last condition is realized as soon as the space $\mathcal{K}$ is finite-dimensional since all the linear forms are continuous in this case. We see below that some other examples can be found.

As a consequence, using the Riesz theorem[14], there exists, for almost all x, a unique element $p_x \in \mathcal{K}$ (a function) such that

$$\langle p_x | \phi \rangle = \phi(x) . \tag{13.2}$$

We define the *coherent states* as the *normalized* vectors corresponding to p_x, written in Dirac notation as

$$|x\rangle \equiv \frac{|p_x\rangle}{[\mathcal{N}(x)]^{\frac{1}{2}}} \quad \text{where} \quad \mathcal{N}(x) \equiv \langle p_x | p_x \rangle . \tag{13.3}$$

One can see at once that for any $\phi \in \mathcal{K}$

$$\phi(x) = \left[\mathcal{N}(x)\right]^{\frac{1}{2}} \langle x | \phi \rangle. \tag{13.4}$$

One obtains the following resolution of the identity of $\mathcal{K}$ that is at the basis of the whole construction:

$$I_{\mathcal{K}} = \int_X \mu(dx)\, \mathcal{N}(x)\, |x\rangle\langle x| . \tag{13.5}$$

This equation is a direct consequence of the following equalities:

$$\langle \phi_1 | \int_X \mu(dx)\, \mathcal{N}(x)\, |x\rangle\langle x || \phi_2\rangle = \int_X \mu(dx)\, \mathcal{N}(x) \langle \phi_1 | x\rangle\langle x | \phi_2 \rangle$$
$$= \int_X \mu(dx)\, \overline{\phi_1(x)} \phi_2(x) = \langle \phi_1 | \phi_2 \rangle ,$$

which hold for any $\phi_1, \phi_2 \in \mathcal{K}$.

Note that

$$\phi(x) = \int_X \mu(dx')\, \sqrt{\mathcal{N}(x)\mathcal{N}(x')} \langle x | x' \rangle \phi(x') , \quad \forall \phi \in \mathcal{K} . \tag{13.6}$$

Hence, $\mathcal{K}$ is a reproducing Hilbert space with kernel

$$K(x, x') = \sqrt{\mathcal{N}(x)\mathcal{N}(x')} \langle x | x' \rangle, \tag{13.7}$$

and the latter assumes finite diagonal values (almost everywhere), $K(x, x) = \mathcal{N}(x)$, by construction. Note that this construction yields an embedding of X into $\mathcal{K}$ and one could interpret $|x\rangle$ as a state localized at x once a notion of localization has been properly defined on X.

In view of (13.5), the set $\{|x\rangle\}$ is called a *frame* for $\mathcal{K}$. This frame is said to be overcomplete when the vectors $\{|x\rangle\}$ are not linearly independent.

14) *In a Hilbert space $\mathcal{H}$ with inner product $\langle v_1 | v_2 \rangle$, to any continuous linear form $\varphi : \mathcal{H} \mapsto \mathbb{K}$ there corresponds one and only one vector v_φ such that $\varphi(v) = \langle v_\varphi | v \rangle$ for all $v \in \mathcal{H}$.*

The technical conditions and the definition of coherent states can be easily expressed when we have an orthonormal basis of $\mathcal{K}$. Let $\{\phi_n,\ n \in I\}$ be such a basis; the technical condition is equivalent to

$$\sum_n |\phi_n(x)|^2 < \infty \quad \text{a.e.} \tag{13.8}$$

The coherent state is then defined by

$$|x\rangle = \frac{1}{(\mathcal{N}(x))^{\frac{1}{2}}} \sum_n \overline{\phi_n(x)}\phi_n \quad \text{with} \quad \mathcal{N}(x) = \sum_n |\phi_n(x)|^2 .$$

We thus recover the original construction of coherent states described in Chapter 5. We have just adopted here a more "functional" approach. We also remember that the Hilbert space of "quantum states" can be chosen a priori, independently of the Hilbert space $L^2_{\mathbb{K}}(X,\mu)$ of "signals." Given a separable Hilbert space $\mathcal{H}$ and an orthonormal basis $(|e_n\rangle)_{n\in\mathbb{N}}$ in one-to-one correspondence with the ϕ_n's, one defines *in* $\mathcal{H}$ the family of coherent states

$$|x\rangle \equiv \frac{1}{\sqrt{\mathcal{N}(x)}} \sum_n \overline{\phi_n(x)}|e_n\rangle . \tag{13.9}$$

Consistently,

$$\phi_n(x) \equiv \sqrt{\mathcal{N}(x)}\langle x\,|e_n\rangle . \tag{13.10}$$

These states resolve the unity in $\mathcal{H}$

$$\int_X |x\rangle\langle x|\,\nu(dx) = I_{\mathcal{H}} , \tag{13.11}$$

where $\nu(dx) = \mathcal{N}(x)\,\mu(dx)$.

13.4
Coherent State Quantization

A *classical* observable is a function $f(x)$ on X having specific properties with respect to some supplementary structure allocated to X, such as topology, geometry, or something else. It could be a distribution if the topological structure assigned to X allows the existence of such an object. More precisely, within the framework of our approach, the actual definition of classical observable will depend on the coherent state quantization scheme. Inspired by Chapter 12, we define the quantization of a classical observable $f(x)$ *with respect to* the family of states $\{|x\rangle,\ x \in X\}$ as the linear map associating with the function $f(x)$ the operator in $\mathcal{H}$ given by

$$f \mapsto A_f = \int_X f(x)|x\rangle\langle x|\,\nu(dx) . \tag{13.12}$$

Note that the function $f(x) = 1$ goes to the identity operator $I_{\mathcal{H}}$. Borrowing from Berezin and Lieb their terminology, one names the function $f(x) \equiv \widehat{A}_f(x)$ the contravariant (Berezin) or the upper symbol (Lieb) of the operator A_f, whereas one names the mean value $\langle x|A_f|x\rangle \equiv \check{A}_f(x)$ the covariant or the lower symbol of A_f.

The terminology "lower" and "upper" comes from the following inequalities involving symbols of self-adjoint operators. Let $g(\cdot)$ be a real-valued convex function

$$g\left(\sum_i a_i \alpha_i\right) \leq \sum_i a_i \, g(\alpha_i) \quad \text{for all set } \{a_i \geq 0\} \text{ with } \sum_i a_i = 1 \tag{13.13}$$

and let A be a self-adjoint operator on $\mathcal{H}$ that has well-defined lower and upper symbols with respect to the coherent state frame $\{|x\rangle\}_{x\in X}$. Then the *Berezin–Lieb inequalities*

$$\int_X g(\check{A}(x)) \, \nu(dx) \leq \operatorname{Tr} g(A) \leq \int_X g(\hat{A}(x)) \, \nu(dx) \tag{13.14}$$

hold depending on suitable properties of A. A particular case of these inequalities was proved and applied to quantum spin systems in Chapter 7.

Now, in agreement with Chapter 12, we should make more precise, although still loose, our definition of a *classical observable* in terms of lower symbols.

Definition 13.1 Given a frame $\{|x\rangle, x \in X\}$, a function $X \ni x \mapsto f(x) \in \mathbb{K}$, possibly understood in a distributional sense, is a *coherent state quantizable classical observable* along the map $f \mapsto A_f$ defined by (13.12)

(i) if the map $X \ni x \mapsto \langle x|A_f|x\rangle$ is a regular function with respect to some additional structure allocated to X.

(ii) we must get the right classical limit, which means that $\langle x|A_f|x\rangle \approx f(x)$ as a certain parameter goes to 0.

One can say that, according to this approach, a quantization of the observation set X is in one-to-one correspondence with the choice of a frame in the sense of (13.11). This term of *frame* [61] is more appropriate for designating the total family $\{|x\rangle\}_{x\in X}$. To a certain extent, the coherent state quantization scheme consists in adopting a certain point of view in dealing with X. This frame can be discrete or continuous, depending on the topology additionally allocated to the set X, and it can be overcomplete, of course. The validity of a precise frame choice is determined by comparing spectral characteristics of quantum observables A_f with experimental data. Of course, a quantization scheme associated with a specific frame is intrinsically limited to all those classical observables for which the expansion (13.12) is mathematically justified within the theory of operators in Hilbert space (e.g., weak convergence). However, it is well known that limitations hold for *any* quantization scheme.

The advantage of the coherent state quantization illustrated here is that it requires a minimal significant structure on X, namely, the existence of a measure $\mu(dx)$, together with a σ-algebra of measurable subsets, and some additional structure to be defined according to the context. The construction of the Hilbert space $\mathcal{H}$ (or $\mathcal{K}$) is equivalent to the choice of a class of eligible quantum states, together with a technical condition of continuity. A correspondence between classical and quantum observables is then provided through a suitable generalization of the standard coherent states.

An interesting question to be addressed concerns the emergence or not of noncommutativity through the quantum processing of the set X. Suppose the "large" signal space $L^2_{\mathbb{K}}(X,\mu)$ has an orthonormal *basis* $\{\phi_n(x),\ n \in \mathbb{N}\}$ that satisfies the required condition

$$\mathcal{N}(x) \equiv \sum_n |\phi_n(x)|^2 < \infty \quad \text{a.e.}$$

Then the coherent state quantization with the corresponding coherent states

$$|x\rangle \equiv \frac{1}{\sqrt{\mathcal{N}(x)}} \sum_n \overline{\phi_n(x)} |\phi_n\rangle$$

provides a commutative algebra of operators! Indeed, in this case, to $f(x)$ there corresponds the multiplication operator

$$A_f \phi(x) = f(x)\phi(x) \quad \text{a.e.}$$

This means that we reach through that coherent state quantization only this class of operators, and other operators cannot be expressed in such a "diagonal form" with respect to the coherent state family. The essence of the noncommutative "quantum" reading of the observation set X lies in a strict inclusion of the space $\mathcal{K}$ of quantum states into the space $L^2_{\mathbb{K}}(X,\mu)$ of "signals."

Another possible approach to quantization would consist first in proceeding to the most demanding sampling, namely, in choosing the "continuous orthogonal" basis of $L^2_{\mathbb{K}}(X,\mu)$ made up of the "Dirac functions" $\{\delta_y(x)\,,\ y \in X\}$, operationally defined on a suitable dense subspace $\mathcal{V}$ of functions in $L^2_{\mathbb{K}}(X,\mu)$ by

$$\int_X \mu(dx)\, \delta_y(x)\, v(x) \equiv \langle \delta_y | v \rangle = v(y) \quad \forall v \in \mathcal{V}\,.$$

Here too we have a resolution of the identity at least on $\mathcal{V}$:

$$\int_X \mu(dx)\, |\delta_x\rangle\langle\delta_x| = I_d\,.$$

Then a quantization with this Dirac basis would also provide a commutative algebra of operators. Indeed, in this case, to a function $f(x)$ there corresponds the multiplication operator

$$A_f \phi(x) = f(x)\phi(x) \quad \text{a.e.}$$

13.5
A Quantization of the Circle by 2 × 2 Real Matrices

In Section 3.2.3, we illustrated the notion of resolution of the unity by considering the set of unit vectors as a continuous frame for the Euclidean plane:

$$\frac{1}{\pi}\int_0^{2\pi} d\theta\, |\theta\rangle\langle\theta| = I_d\,. \tag{13.15}$$

Let us now employ this set of "coherent states" to describe the corresponding quantization along the lines illustrated in the previous section.

13.5.1
Quantization and Symbol Calculus

The existence of the set $\{|\theta\rangle\}$ offers the possibility of proceeding to the following *quantization* of the circle that associates with a function $f(\theta)$ the linear operator A_f in the plane:

$$f \mapsto A_f = \frac{1}{\pi}\int_0^{2\pi} d\theta\, f(\theta)|\theta\rangle\langle\theta|\,. \tag{13.16}$$

For instance, let us choose the angle function $f(\theta) = \theta$. Its quantized version is equal to the matrix

$$A_\theta = \begin{pmatrix} \pi & -\frac{1}{2} \\ -\frac{1}{2} & \pi \end{pmatrix}, \tag{13.17}$$

with eigenvalues $\pi \pm \frac{1}{2}$.

The frame $\{|\theta\rangle\}$ allows one to carry out a symbol calculus à la Berezin and Lieb. With any *self-adjoint* linear operator A, that is, a real symmetric matrix with respect to some orthonormal basis, one associates the two types of *symbol* functions $\check{A}(\theta)$ and $\hat{A}(\theta)$, respectively, defined on the unit circle by

$$\check{A}(\theta) = \langle\theta|A|\theta\rangle \ :\ \text{lower or covariant symbol} \tag{13.18}$$

and

$$A = \frac{1}{\pi}\int_0^{2\pi} d\theta\, \hat{A}(\theta)|\theta\rangle\langle\theta|\,. \tag{13.19}$$

The upper or contravariant symbol $\hat{A}(\theta)$ that appears in this operator-valued integral is highly nonunique, but will be chosen as the simplest one.

Three basic matrices generate the Jordan algebra[15] of all real symmetric 2 × 2 matrices. They are the identity matrix I_d, the symbol of which is trivially the

15) A Jordan algebra is an algebra (not necessarily associative) over a field whose multiplication is commutative and satisfies the *Jordan identity*: $(xy)x^2 = x(yx^2)$.

function 1, and the two real Pauli matrices,

$$\sigma_1 = \begin{pmatrix} 0 & 1 \\ 1 & 0 \end{pmatrix}, \quad \sigma_3 = \begin{pmatrix} 1 & 0 \\ 0 & -1 \end{pmatrix}. \tag{13.20}$$

Any element A of the algebra decomposes as

$$A \equiv \begin{pmatrix} a & b \\ b & d \end{pmatrix} = \frac{a+d}{2} I_d + \frac{a-d}{2}\sigma_3 + b\sigma_1 \equiv \alpha I_d + \delta\sigma_3 + \beta\sigma_1 . \tag{13.21}$$

The product in this algebra is defined by

$$\mathcal{O}'' = A \odot A' = \frac{1}{2}\left(AA' + A'A\right), \tag{13.22}$$

which entails on the level of components α, δ, β the relations:

$$\alpha'' = \alpha\alpha' + \delta\delta' + \beta\beta', \; \delta'' = \alpha\delta' + \alpha'\delta, \; \beta'' = \alpha\beta' + \alpha'\beta . \tag{13.23}$$

The simplest upper symbols and the lower symbols of nontrivial basic elements are, respectively, given by

$$\cos 2\theta = \check{\sigma}_3(\theta) = \frac{1}{2}\hat{\sigma}_3(\theta), \; \sin 2\theta = \check{\sigma}_1(\theta) = \frac{1}{2}\hat{\sigma}_1(\theta) . \tag{13.24}$$

There follows for the symmetric matrix (13.21) the two symbols

$$\check{A}(\theta) = \frac{a+d}{2} + \frac{a-d}{2}\cos 2\theta + b\sin 2\theta = \alpha + \delta\cos 2\theta + \beta\sin 2\theta , \tag{13.25}$$

$$\hat{A}(\theta) = \alpha + 2\delta\cos 2\theta + 2\beta\sin 2\theta = 2\check{A}(\theta) - \tfrac{1}{2}\,\mathrm{Tr}\, A . \tag{13.26}$$

For instance, the lower symbol of the "quantum angle" (13.17) is equal to

$$\langle\theta|A_\theta|\theta\rangle = \pi - \tfrac{1}{2}\sin 2\theta , \tag{13.27}$$

a π-periodic function that smoothly varies between the two eigenvalues of A_θ, as shown in Figure 13.1.

One should notice that all these symbols belong to the subspace $\mathcal{V}_A$ of real Fourier series that is the closure of the linear span of the three functions 1, $\cos 2\theta$, and $\sin 2\theta$. Also note that $\hat{A}(\theta)$ is defined up to the addition of any function $\mathrm{N}(\theta)$ that makes (13.19) vanish. Such a function lives in the orthogonal complement of $\mathcal{V}_A$.

The Jordan multiplication law (13.22) is commutative but not associative and its counterpart on the level of symbols is the so-called $\star$-product. For instance, we have for the upper symbols

$$\widehat{A \odot A'}(\theta) \equiv \hat{A}(\theta) \star \hat{A}'(\theta) = \alpha\hat{A}'(\theta) + \alpha'\hat{A}(\theta) + \delta\delta' + \beta\beta' - \alpha\alpha' , \tag{13.28}$$

and the formula for lower symbols is the same.

We recall that the terminology of *lower/upper* is justified by two Berezin–Lieb inequalities, which follow from the symbol formalism. Let us make explicit this

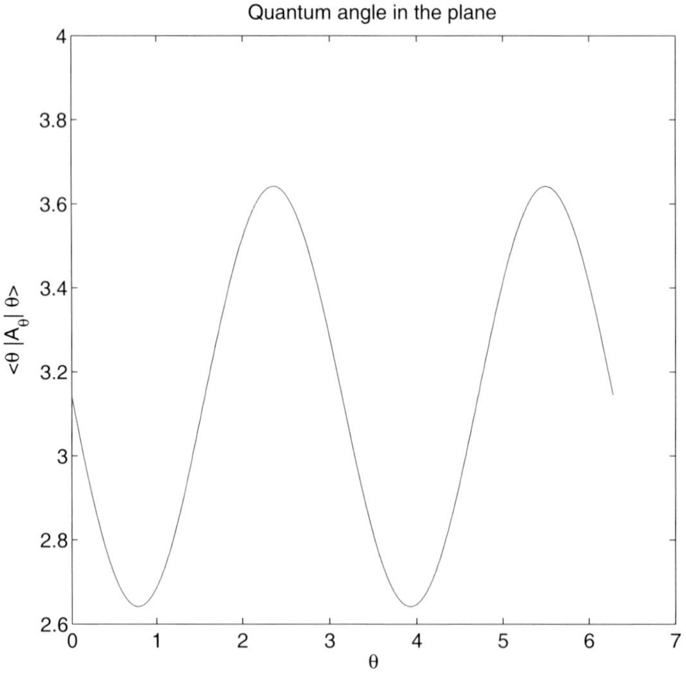

Fig. 13.1 The lower symbol of the angle operator in the two-dimensional quantization of the circle is the π-periodic function $\pi - \frac{1}{2}\sin 2\theta$ that varies between the two eigenvalues of A_θ.

inequality in the present context. Let g be a convex function. Denoting by $\lambda_\pm$ the eigenvalues of the symmetric matrix A, we have

$$\frac{1}{\pi}\int_0^{2\pi} g(\check{A}(\theta))\,d\theta \le \operatorname{Tr} g(A) = g(\lambda_+) + g(\lambda_-) \le \frac{1}{\pi}\int_0^{2\pi} g(\hat{A}(\theta))\,d\theta\,. \tag{13.29}$$

This double inequality is not trivial. Independently of the Euclidean context it reads as

$$\langle g(t + r\cos\theta)\rangle \le \tfrac{1}{2}\left[g(t+r) + g(t-r)\right] \le \langle g(t + 2r\cos\theta)\rangle\,, \tag{13.30}$$

where $t \in \mathbb{R}$, $r \ge 0$ and $\langle\cdot\rangle$ denotes the mean value on a period. If we apply (13.30) to the exponential function $g(t) = e^t$, we get an intertwining of inequalities involving Bessel functions of the second kind and the hyperbolic cosine:

$$\cdots \le I_0(x) \le \cosh x \le I_0(2x) \le \cosh 2x \le \cdots \quad \forall x \in \mathbb{R}\,. \tag{13.31}$$

13.5.2
Probabilistic Aspects

Behind the resolution of the identity (13.15) lies an interesting interpretation in terms of geometrical probability. Let us consider a Borel subset Δ of the interval

$[0, 2\pi)$ and the restriction to Δ of the integral (13.15):

$$a(\Delta) = \frac{1}{\pi} \int_{\Delta} d\theta |\theta\rangle\langle\theta| \,. \tag{13.32}$$

One easily verifies the following properties:

$$\begin{aligned} &a(\emptyset) = 0 \,, \quad a([0, 2\pi)) = I_d \,, \\ &a(\cup_{i \in J} \Delta_i) = \sum_{i \in J} a(\Delta_i) \,, \quad \text{if } \Delta_i \cap \Delta_j = \emptyset \quad \text{for all } i \neq j \,. \end{aligned} \tag{13.33}$$

The application $\Delta \mapsto a(\Delta)$ defines a normalized measure on the σ-algebra of the Borel sets in the interval $[0, 2\pi)$, assuming its values in the set of positive linear operators on the Euclidean plane. Denoting the measure density $(1/\pi)|\theta\rangle\langle\theta|\, d\theta$ by $a(d\theta)$, we shall also write

$$a(\Delta) = \int_{\Delta} a(d\theta) \,. \tag{13.34}$$

Let us now put into evidence the probabilistic nature of the measure $a(\Delta)$. Let $|\phi\rangle$ be a unit vector. The application

$$\Delta \mapsto \langle\phi|a(\Delta)|\phi\rangle = \frac{1}{\pi} \int_{\Delta} \cos^2(\theta - \phi)\, d\theta \tag{13.35}$$

is clearly a probability measure. It is positive, of total mass 1, and it inherits σ-additivity from $a(\Delta)$. Now, the quantity $\langle\phi|a(\Delta)|\phi\rangle$ means that direction $|\phi\rangle$ is examined from the point of view of the family of vectors $\{|\theta\rangle, \theta \in \Delta\}$. As a matter of fact, it has a *geometrical probability* interpretation in the plane [181]. With no loss of generality, let us choose $\phi = 0$. Recall here the canonical equation describing a straight line $D_{\theta,p}$ in the plane:

$$\langle\theta|u\rangle \equiv \cos\theta\, x + \sin\theta\, y = p \,, \tag{13.36}$$

where $|\theta\rangle$ is the direction normal to $D_{\theta,p}$ and the parameter p is equal to the distance of $D_{\theta,p}$ to the origin. It follows that $dp\, d\theta$ is the (nonnormalized) probability measure element on the set $\{D_{\theta,p}\}$ of the lines randomly chosen in the plane. Picking a certain θ, we consider the set $\{D_{\theta,p}\}$ of the lines normal to $|\theta\rangle$ that intersect the segment with origin O and length $|\cos\theta|$ equal to the projection of $|\theta\rangle$ onto $|0\rangle$ as shown in Figure 13.2

The measure of this set is equal to

$$\left(\int_0^{\cos^2\theta} dp \right) d\theta = \cos^2\theta\, d\theta \,. \tag{13.37}$$

Integrating (13.37) over all directions $|\theta\rangle$ gives the area of the unit circle. Hence, we can construe $\langle\phi|a(\Delta)|\phi\rangle$ as the probability of a straight line in the plane belonging to the set of secants of segments that are projections $\langle\phi|\theta\rangle$ of the unit vectors $|\theta\rangle$, $\theta \in \Delta$ onto the unit vector $|\phi\rangle$. One could think in terms of *polarizer* $\langle\theta|$ and *analyzer* $|\theta\rangle$ "sandwiching" the directional signal $|\phi\rangle$.

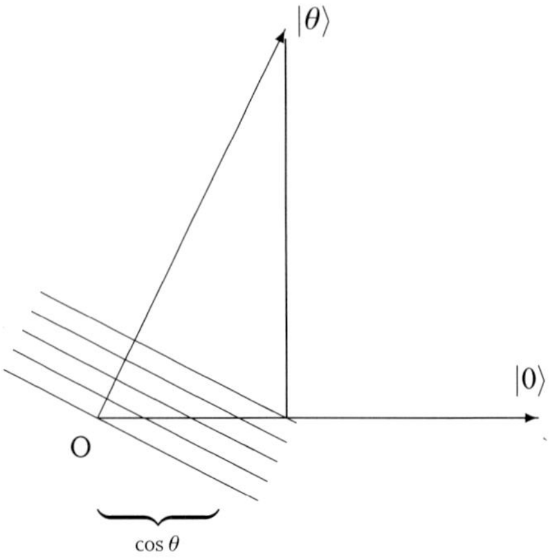

Fig. 13.2 Set $\{D_{\theta,p}\}$ of straight lines normal to $|\theta\rangle$ that intersect the segment with origin O and length $|\cos\theta|$ equal to the projection of $|\theta\rangle$ onto $|0\rangle$.

13.5.2.1
Remark: A Two-Dimensional Quantization of the Interval $[0,\pi)$ through a Continuous Frame for the Half-Plane

From a strictly quantal point of view, we should have made equivalent any vector $|\nu\rangle$ of the Euclidean plane with its symmetric $-|\nu\rangle$, which amounts to dealing with the half-plane viewed as the coset $\mathbb{R}^2/\mathbb{Z}_2$. Following the same procedure as above, we start from the Hilbert space $L^2\left([0,\pi), \frac{2}{\pi}\,d\theta\right)$ and we choose the *same* subset $\{\cos\theta, \sin\theta\}$, which is *still* orthonormal. Note that $\frac{2}{\pi}\cos^2\theta$ and $\frac{2}{\pi}\sin^2\theta$ are now exactly Wigner semicircle distributions. We then consider the continuous family (3.8) of coherent states $|\theta\rangle$. They are normalized and they solve the identity exactly like in (3.10) (just change the factor $\frac{1}{\pi}$ into $\frac{2}{\pi}$). The previous material can be repeated *in extenso* but with the restriction that $\theta \in [0,\pi)$.

13.6
Quantization with *k*-Fermionic Coherent States

As another illustration of the coherent state quantization, let us sketch an application to a more elaborate mathematical structure, namely, a para-Grassmann algebra for which coherent states can be constructed [182–184].

Let us first proceed with the construction of these coherent states by following the scheme described in Chapter 5. The observation set X is the para-Grassmann algebra[16] Σ_k. The latter is defined as the linear span of $\left\{1, \theta, \ldots, \theta^{k-1}\right\}$ and their

16) A Grassmann (or exterior) algebra of a given vector space V over a field is the algebra generated by the exterior (or wedge) product for which all elements are nilpotent, $\theta^2 = 0$. Para-Grassmann algebras are generalizations for which, given an integer $k > 2$, all elements obey $\theta^k = 0$.

respective conjugates $\bar{\theta}^i$; here, θ is a Grassmann variable satisfying $\theta^k = 0$. A measure on X is defined as

$$\mu(d\theta d\bar{\theta}) = d\theta\, w(\theta,\bar{\theta})\; d\bar{\theta}\,. \tag{13.38}$$

Here, the integral over $d\theta$ and $d\bar{\theta}$ should be understood in the sense of Berezin–Majid–Rodríguez-Plaza integrals [185]:

$$\int d\theta\, \theta^n = 0 = \int d\bar{\theta}\, \bar{\theta}^n\,, \quad \text{for } n = 0, 1, \ldots, k-2, \tag{13.39}$$

with

$$\int d\theta\, \theta^{k-1} = 1 = \int d\bar{\theta}\, \bar{\theta}^{k-1}\,. \tag{13.40}$$

The "weight" $w(\theta,\bar{\theta})$ is given by the q-deformed polynomial

$$w(\theta,\bar{\theta}) = \sum_{n=0}^{k-1} \left([n]_q![n]_{\bar{q}}!\right)^{\frac{1}{2}}\, \theta^{k-1-n}\bar{\theta}^{k-1-n}\,. \tag{13.41}$$

The q-deformed numbers are defined by

$$[x]_q := \frac{1-q^x}{1-q}\,, \quad [n]_q! = [1]_q[2]_q \cdots [n]_q,\,. \tag{13.42}$$

In the present case, $q = e^{\frac{2\pi i}{k}}$ is a kth root of unity and $\theta\bar{\theta} = q\bar{\theta}\theta$.

The (nonnormalized) *para-Grassmann* or *k-fermionic* coherent states should be understood as elements of $\mathbb{C}^k \otimes \Sigma_k$. They read as

$$|\theta\rangle = \sum_{n=0}^{k-1} \frac{\theta^n}{([n]_q!)^{\frac{1}{2}}}\, |n\rangle\,, \tag{13.43}$$

where $\{|n\rangle\}$ is an orthonormal basis of the Hermitian space $\mathbb{C}^k$.

The coherent state quantization of the "spinorial" para-Grassmann algebra rests upon the resolution of the unity I_k in $\mathbb{C}^k$:

$$\iint d\theta|\theta\rangle\, w(\theta,\bar{\theta})\, \langle\theta|\, d\bar{\theta} = I_k\,. \tag{13.44}$$

The quantization of a para-Grassmann-valued function $f(\theta,\bar{\theta})$ maps f to the linear operator A_f on $\mathbb{C}^k$:

$$A_f = \iint d\theta|\theta\rangle :\, f(\theta,\bar{\theta}) w(\theta,\bar{\theta}) :\, \langle\theta|\, d\bar{\theta}\,. \tag{13.45}$$

Here we actually recover the $k \times k$-matrix realization of the so-called *k-fermionic algebra* F_k [184]. For instance, we have for the simplest functions

$$A_\theta = \sum_{n=0}^{k-1} \left([n+1]_q\right)^{\frac{1}{2}} |n\rangle\langle n+1| \,, \quad A_{\bar\theta} = \sum_{n=0}^{k-1} \left([n+1]_{\bar q}\right)^{\frac{1}{2}} |n+1\rangle\langle n| = A_\theta^\dagger \,. \tag{13.46}$$

Their (anti-)commutator reads as

$$[A_\theta, A_{\bar\theta}] = \sum_{n=0}^{k-1} \frac{\cos\pi\frac{2n+1}{2k}}{\cos\frac{\pi}{2k}} |n\rangle\langle n| \,, \quad \{A_\theta, A_{\bar\theta}\} = \sum_{n=0}^{k-1} \frac{\sin\pi\frac{2n+1}{2k}}{\sin\frac{\pi}{2k}} |n\rangle\langle n| \,. \tag{13.47}$$

In the purely fermionic case, $k = 2$, we recover the canonical anticommutation rule $\{A_\theta, A_{\bar\theta}\} = I_2$.

13.7 Final Comments

Let us propose in this conclusion, on an elementary level, some hints for understanding better the probabilistic duality lying at the heart of the coherent state quantization procedure. Let X be an observation set equipped with a measure ν and let a real-valued function $X \ni x \mapsto a(x)$ have the status of "observable." This means that there exists an experimental device, Δ_a, giving access to a set or "spectrum" of numerical outcomes $\Sigma_a = \{a_j,\ j \in \mathcal{J}\} \subset \mathbb{R}$, commonly interpreted as the set of all possible measured values of $a(x)$. To the set Σ_a are attached two probability distributions defined by the set of functions $\Pi_a = \{p_j(x),\ j \in \mathcal{J}\}$:

1. A family of probability distributions on the set $\mathcal{J}$, $\mathcal{J} \ni j \mapsto p_j(x)$, $\sum_{j\in\mathcal{J}} p_j(x) = 1$, indexed by the observation set X. This probability encodes what is precisely expected for this pair (observable $a(x)$, device Δ_a).
2. A family of probability distributions on the measure space (X, ν), $X \ni x \mapsto p_j(x)$, $\int_X p_j(x)\,\nu(dx) = 1$, indexed by the set $\mathcal{J}$.

Now, the exclusive character of the possible outcomes a_j of the measurement of $a(x)$ implies the existence of a set of "conjugate" functions $X \ni x \mapsto \alpha_j(x)$, $j \in \mathcal{J}$, playing the role of phases, and making the set of complex-valued functions

$$\phi_j(x) \stackrel{\text{def}}{=} \sqrt{p_j(x)}\, e^{i\alpha_j(x)} \tag{13.48}$$

an orthonormal set in the Hilbert space $L^2_{\mathbb{K}}(X, \nu)$, $\int_X \overline{\phi_j(x)}\phi_{j'}(x)\,\nu(dx) = \delta_{jj'}$. There follows the existence of the family of coherent states or frame in the Hilbertian closure $\mathcal{K}$ of the linear span of the ϕ_j's:

$$X \ni x \mapsto |x\rangle = \sum_{j\in\mathcal{J}} \overline{\phi_j(x)}\, |\phi_j\rangle, \ \langle x|x\rangle = 1 \,, \quad \int_X |x\rangle\langle x|\,\nu(dx) = I_{\mathcal{K}} \,. \tag{13.49}$$

Consistency conditions have to be satisfied together with this material: they follow from the quantization scheme resulting from the frame (13.49).

Condition 13.2 (Frame consistency conditions)

(*i*) ***Spectral condition***

The frame quantization of the observable $a(x)$ *produces an essentially self-adjoint operator* A_a *in* $\mathcal{K}$ *that is diagonal in the basis* $\{\phi_j,\ j \in \mathcal{J}\}$, *with spectrum precisely equal to* $\sum_a$:

$$A_a \stackrel{\text{def}}{=} \int_X a(x)\, |x\rangle\langle x|\, \nu(dx) = \sum_{j\in\mathcal{J}} a_j |\phi_j\rangle\langle\phi_j|\,. \tag{13.50}$$

(*ii*) ***Classical limit condition***

The frame $\{|x\rangle\}$ *depends on a parameter* $\kappa \in [0,\infty)$, $|x\rangle = |x,\kappa\rangle$ *such that the limit*

$$\hat{A}_a(x,\kappa) \stackrel{\text{def}}{=} \langle x,\kappa|A_a(\kappa)|x,\kappa\rangle \underset{\kappa\to 0}{\longrightarrow} a(x) \tag{13.51}$$

holds for a certain topology $\mathcal{T}_{\text{cl}}$ *assigned to the set of classical observables.*

Once these conditions have been verified, one can start the frame quantization of other observables, for instance, the quantization of the conjugate observables $\alpha_j(x)$, and check whether the observational or experimental consequences or constraints due to this mathematical formalism are effectively in agreement with our "reality".

Many examples will be presented in the next chapters. For some of them, we have in view possible connections with objects of noncommutative geometry (such as fuzzy spheres or pseudospheres). These examples show the extreme freedom we have in analyzing a set X of *data* or *possibilities* just equipped with a measure, by just following our coherent state quantization procedure. The crucial step lies in the choice of a countable orthonormal subset in $L^2(X,\mu)$ obeying the finitude condition (13.8). A $\mathbb{C}^N$ (or l^2 if $N = \infty$) unitary transform of this original subset would actually lead to the same specific quantization, and the latter could also be obtained by using unitarily equivalent *continuous* orthonormal distributions defined within the framework of some Gel'fand triplet. Of course, further structure such as a symplectic manifold combined with spectral constraints imposed on some specific observables will considerably restrict that freedom and should lead, hopefully, to a unique solution, such as Weyl quantization, deformation quantization, and geometrical quantization are able to achieve in specific situations. Nevertheless, we believe that the generalization of the Berezin–Klauder–Toeplitz quantization that has been described here, and that goes far beyond the context of classical and quantum mechanics, not only sheds light on the specific nature of the latter, but will also help to solve in a simpler way some quantization problems.

14
Coherent State Quantization of Finite Set, Unit Interval, and Circle

14.1
Introduction

The examples that are presented in this chapter are, although elementary, rather unusual. In particular, we deal with observation sets X that are not phase space, and such sets are far from having any physical meaning in the common sense. We first explore the coherent state quantization of finite sets and the unit interval. Then, we return to the unit circle, already considered in the previous chapter: we will fully exploit the complex Fourier series in order to propose a satisfying solution to the quantum phase problem.

14.2
Coherent State Quantization of a Finite Set with Complex 2 × 2 Matrices

An elementary (but not trivial!) exercise for illustrating the quantization scheme introduced in the previous chapter concerns an arbitrary N-element set $X = \{x_i \in X\}$ viewed as an observation set. An arbitrary nondegenerate measure on it is given by a sum of Dirac measures:

$$\mu(dx) = \sum_{i=1}^{N} a_i \delta_{x_i}\,, \quad a_i > 0\,. \tag{14.1}$$

The Hilbert space $L^2_{\mathbb{C}}(X,\mu)$, denoted by $L^2(X,\mu)$ in the sequel, is simply isomorphic to the Hermitian space $\mathbb{C}^N$. An obvious orthonormal basis is given by

$$\left\{ \frac{1}{\sqrt{a_i}} \chi_{\{x_i\}}(x)\,, \quad i = 1, \ldots, N \right\}\,, \tag{14.2}$$

where $\chi_{\{a\}}$ is the characteristic function of the singleton $\{a\}$.

Let us consider the two-element orthonormal set $\{\phi_1 \equiv \phi_\alpha \equiv |\boldsymbol{\alpha}\rangle, \phi_2 \equiv \phi_\beta \equiv |\boldsymbol{\beta}\rangle\}$ defined in the most generic way by

$$\phi_\alpha(x) = \sum_{i=1}^{N} \overline{\alpha_i} \frac{1}{\sqrt{a_i}} \chi_{\{x_i\}}(x)\,, \quad \phi_\beta(x) = \sum_{i=1}^{N} \overline{\beta_i} \frac{1}{\sqrt{a_i}} \chi_{\{x_i\}}(x)\,, \tag{14.3}$$

Coherent States in Quantum Physics. Jean-Pierre Gazeau

ISBN: 978-3-527-40709-5

where the complex coefficients α_i and β_i obey

$$\sum_{i=1}^{N} |\alpha_i|^2 = 1 = \sum_{i=1}^{N} |\beta_i|^2 \,, \quad \sum_{i=1}^{N} \alpha_i \overline{\beta_i} = 0 \,. \tag{14.4}$$

In a Hermitian geometry language, our choice of $\{\phi_\alpha\,, \phi_\beta\}$ amounts to selecting in $\mathbb{C}^N$ the two orthonormal vectors $\boldsymbol{\alpha} = \{\alpha_i\}$, $\boldsymbol{\beta} = \{\beta_i\}$, and this justifies our notation for the indices. These vectors span a two-dimensional subspace $\cong \mathbb{C}^2$ in $\mathbb{C}^N$.

The expression for the coherent states is

$$|x\rangle = \frac{1}{\sqrt{\mathcal{N}(x)}} \left[\overline{\phi_\alpha(x)}|\boldsymbol{\alpha}\rangle + \overline{\phi_\beta(x)}|\boldsymbol{\beta}\rangle\right] , \tag{14.5}$$

in which $\mathcal{N}(x)$ is given by

$$\mathcal{N}(x) = \sum_{i=1}^{N} \frac{|\alpha_i|^2 + |\beta_i|^2}{a_i} \chi_{\{x_i\}}(x) \,. \tag{14.6}$$

The resulting resolution of unity reads as

$$I_d = \sum_{i=1}^{N} \left(|\alpha_j|^2 + |\beta_j|^2\right) |x_i\rangle\langle x_i| \,. \tag{14.7}$$

The overlap between two coherent states is given by the following kernel:

$$\langle x_i | x_j \rangle = \frac{\overline{\alpha_i}\alpha_j + \overline{\beta_i}\beta_j}{\sqrt{|\alpha_i|^2 + |\beta_i|^2}\sqrt{|\alpha_j|^2 + |\beta_j|^2}} \,. \tag{14.8}$$

To any real-valued function $f(x)$ on X, that is, to any vector $\boldsymbol{f} \equiv (f(x_i))$ in $\mathbb{R}^N$, there corresponds the following Hermitian operator A_f in $\mathbb{C}^2$, expressed in matrix form with respect to the orthonormal basis (14.3):

$$\begin{aligned} A_f &= \int_X \mu(dx)\, \mathcal{N}(x)\, f(x) |x\rangle\langle x| \\ &= \begin{pmatrix} \sum_{i=1}^N |\alpha_i|^2 f(x_i) & \sum_{i=1}^N \alpha_i \overline{\beta_i} f(x_i) \\ \sum_{i=1}^N \overline{\alpha_i}\beta_i f(x_i) & \sum_{i=1}^N |\beta_i|^2 f(x_i) \end{pmatrix} \equiv \begin{pmatrix} \langle \mathbf{F} \rangle_\alpha & \langle \boldsymbol{\beta}|\mathbf{F}|\boldsymbol{\alpha}\rangle \\ \langle \boldsymbol{\alpha}|\mathbf{F}|\boldsymbol{\beta}\rangle & \langle \mathbf{F} \rangle_\beta \end{pmatrix} , \end{aligned} \tag{14.9}$$

where $\boldsymbol{F}$ holds for the diagonal matrix $\{(f(x_i))\}$. It is clear that, for a generic choice of the complex α_i's and β_i's, all possible Hermitian 2×2 matrices can be obtained in this way if $N \geq 4$. By *generic* we mean that the following $4 \times N$ real matrix has rank equal to 4:

$$\mathcal{C} = \begin{pmatrix} |\alpha_1|^2 & |\alpha_2|^2 & \cdots & |\alpha_N|^2 \\ |\beta_1|^2 & |\beta_2|^2 & \cdots & |\beta_N|^2 \\ \Re(\alpha_1\overline{\beta_1}) & \Re(\alpha_2\overline{\beta_2}) & \cdots & \Re(\alpha_N\overline{\beta_N}) \\ \Im(\alpha_1\overline{\beta_1}) & \Im(\alpha_2\overline{\beta_2}) & \cdots & \Im(\alpha_N\overline{\beta_N}) \,. \end{pmatrix} \tag{14.10}$$

The case $N = 4$ with $\det C \neq 0$ is particularly interesting since then one has uniqueness of upper symbols of Pauli matrices $\sigma_1 = \begin{pmatrix} 0 & 1 \\ 1 & 0 \end{pmatrix}$, $\sigma_2 = \begin{pmatrix} 0 & -i \\ i & 0 \end{pmatrix}$, $\sigma_3 = \begin{pmatrix} 1 & 0 \\ 0 & -1 \end{pmatrix}$, $\sigma_0 = I_d$, which form a basis of the four-dimensional Lie algebra of complex Hermitian 2×2 matrices. As a matter of fact, the operator (14.9) decomposes with respect to this basis as

$$A_f = \langle f \rangle_+ \sigma_0 + \langle f \rangle_- \sigma_3 + \Re\left(\langle \boldsymbol{\beta} | \boldsymbol{F} | \boldsymbol{\alpha} \rangle\right) \sigma_1 - \Im\left(\langle \boldsymbol{\beta} | \boldsymbol{F} | \boldsymbol{\alpha} \rangle\right) \sigma_2 \,, \tag{14.11}$$

where the symbols $\langle f \rangle_\pm$ stand for the following averagings:

$$\langle f \rangle_\pm = \tfrac{1}{2} \sum_{i=1}^{N} \left(|\alpha_i|^2 \pm |\beta_i|^2 \right) f(x_i) = \tfrac{1}{2} \left(\langle \boldsymbol{F} \rangle_{\boldsymbol{\alpha}} \pm \langle \boldsymbol{F} \rangle_{\boldsymbol{\beta}} \right) . \tag{14.12}$$

Here the quantity $\langle f \rangle_+$ alone has a mean value status, precisely with respect to the probability distribution

$$p_i = \tfrac{1}{2} \left(|\alpha_i|^2 + |\beta_i|^2 \right) . \tag{14.13}$$

These average values also appear in the spectral values of the quantum observable A_f:

$$\mathrm{Sp}(f) = \left\{ \langle f \rangle_+ \pm \sqrt{\left(\langle f \rangle_-\right)^2 + |\langle \boldsymbol{\beta} | \boldsymbol{F} | \boldsymbol{\alpha} \rangle|^2} \right\} . \tag{14.14}$$

Note that if the vector $\boldsymbol{\alpha} = (1, 0, \ldots, 0)$ is part of the canonical basis and $\boldsymbol{\beta} = (0, \beta_2, \ldots, \beta_n)$ is a unit vector orthogonal to $\boldsymbol{\alpha}$, then A_f is diagonal and $\mathrm{Sp}(f)$ is trivially reduced to $\left(f(x_1), \langle \boldsymbol{F} \rangle_{\boldsymbol{\beta}} \right)$. The simplest upper symbols for Pauli matrices read in vector form as

$$\hat{\boldsymbol{\sigma}}_0 = \begin{pmatrix} 1 \\ 1 \\ 1 \\ 1 \end{pmatrix} , \quad \hat{\boldsymbol{\sigma}}_1 = C^{-1} \begin{pmatrix} 0 \\ 0 \\ 1 \\ 0 \end{pmatrix} , \quad \hat{\boldsymbol{\sigma}}_2 = C^{-1} \begin{pmatrix} 0 \\ 0 \\ 0 \\ -1 \end{pmatrix} , \quad \hat{\boldsymbol{\sigma}}_3 = C^{-1} \begin{pmatrix} 1 \\ -1 \\ 0 \\ 0 \end{pmatrix} . \tag{14.15}$$

On the other hand, and for any N, components of the lower symbol of A_f are given in terms of another probability distribution, for which the value of each one is twice the value of its counterpart in (14.13):

$$\langle x_l | A_f | x_l \rangle = \check{A}_f(x_l) = \sum_{i=1}^{N} \pi_{li} f(x_i) \,, \tag{14.16}$$

with

$$\pi_{ll} = |\alpha_l|^2 + |\beta_l|^2 = 2 p_l \,, \quad \pi_{li} = \frac{|\overline{\alpha_l} \alpha_i + \overline{\beta_l} \beta_i|^2}{|\alpha_l|^2 + |\beta_l|^2} \,, \; i \neq l \,. \tag{14.17}$$

Note that the matrix (π_{li}) is stochastic. As a matter of fact, components of lower symbols of Pauli matrices are given by

$$\check{\sigma}_0(x_l) = 1\,, \quad \check{\sigma}_1(x_l) = \frac{2\Re\,(\overline{\alpha_l}\beta_l)}{|\alpha_l|^2 + |\beta_l|^2}\,, \tag{14.18}$$

$$\check{\sigma}_2(x_l) = \frac{2\Im\,(\overline{\alpha_l}\beta_l)}{|\alpha_l|^2 + |\beta_l|^2}\,, \quad \check{\sigma}_3(x_l) = \frac{|\alpha_l|^2 - |\beta_l|^2}{|\alpha_l|^2 + |\beta_l|^2}\,. \tag{14.19}$$

Let us attempt to interpret these formal manipulations in terms of localization in the set X. Consider $X = \{x_i \in X\}$ as a set of N real numbers. One can then view the real-valued function f defined by $f(x_i) = x_i$ as the *position* observable. On the "quantum" level determined by the choice of $\boldsymbol{\alpha} = \{\alpha_i\}, \boldsymbol{\beta} = \{\beta_i\}$, the measurement of this position observable has the two possible outcomes given by (14.14). Moreover, the *position* x_l is privileged to a certain (quantitative) extent in the expression of the average value of the *position* operator when computed in state $|x_l\rangle$.

Before ending this section, let us examine the lower-dimensional cases $N = 2$ and $N = 3$. When $N = 2$ the basis change (14.3) reduces to a $U(2)$ transformation with $SU(2)$ parameters $\alpha = \alpha_1$, $\beta = -\overline{\beta}_1$, $|\alpha|^2 - |\beta|^2 = 1$, and some global phase factor. The operator (14.9) now reads as

$$A_f = f_+ I_d + f_- \begin{pmatrix} |\alpha|^2 - |\beta|^2 & -2\alpha\beta \\ -2\overline{\alpha\beta} & |\beta|^2 - |\alpha|^2 \end{pmatrix}\,, \tag{14.20}$$

with $f_\pm := (f(x_1) \pm f(x_2))/2$. We thus obtain a two-dimensional commutative algebra of "observables" A_f, as is expected since we consider the quantization of $\mathbb{C}^2$ yielded by an orthonormal basis of $\mathbb{C}^2$! This algebra is generated by the identity matrix $I_{2\times 2} = \sigma_0$ and the $SU(2)$ transform of σ_3: $\sigma_3 \to g\sigma_3 g^\dagger$ with $g = \begin{pmatrix} \alpha & \beta \\ -\bar{\beta} & \bar{\alpha} \end{pmatrix} \in SU(2)$. As is easily expected in this case, lower symbols reduce to components:

$$\langle x_l | A_f | x_l \rangle = \check{A}_f(x_l) = f(x_l)\,, \quad l = 1, 2\,. \tag{14.21}$$

Finally, it is interesting to consider the $N = 3$ case when all vector spaces considered are real. The basis change (14.3) involves four real independent parameters, say, α_1, α_2, β_1, and β_2, all with modulus < 1. The counterpart of (14.10) reads here as

$$C_3 = \begin{pmatrix} (\alpha_1)^2 & (\alpha_2)^2 & 1 - (\alpha_1)^2 - (\alpha_2)^2 \\ (\beta_1)^2 & (\beta_2)^2 & 1 - (\beta_1)^2 - (\beta_2)^2 \\ \alpha_1\beta_1 & \alpha_2\beta_2 & -\alpha_1\beta_1 - \alpha_2\beta_2 \end{pmatrix}\,. \tag{14.22}$$

If $\det C_3 = (\alpha_1\beta_2 - \alpha_2\beta_1)(\beta_1\beta_2 - \alpha_1\alpha_2) \neq 0$, then one has uniqueness of upper symbols of Pauli matrices σ_1, σ_3, and $\sigma_0 = I_{2\times 2}$, which form a basis of the three-dimensional Jordan algebra of real symmetric 2×2 matrices. These upper symbols read in vector form as

$$\hat{\boldsymbol{\sigma}}_0 = \begin{pmatrix} 1 \\ 1 \\ 1 \end{pmatrix}\,, \quad \hat{\boldsymbol{\sigma}}_1 = C_3^{-1} \begin{pmatrix} 0 \\ 0 \\ 1 \end{pmatrix}\,, \quad \hat{\boldsymbol{\sigma}}_3 = C_3^{-1} \begin{pmatrix} 1 \\ -1 \\ 0 \end{pmatrix}\,. \tag{14.23}$$

Finally, the extension of this quantization formalism to N'-dimensional subspaces of the original $L^2(X, \mu) \simeq \mathbb{C}^N$ appears to be straightforward on a technical if not interpretational level (see [186] for more examples).

14.3 Coherent State Quantization of the Unit Interval

We explore here two-dimensional (and higher-dimensional) quantizations of the unit segment.

14.3.1 Quantization with Finite Subfamilies of Haar Wavelets

Let us consider the unit interval $X = [0, 1]$ of the real line, equipped with the Lebesgue measure dx, and its associated Hilbert space denoted by $L^2[0, 1]$.

We start out the study by simply selecting the first two elements of the orthonormal Haar basis [56], namely, the characteristic function $\mathbf{1}(x)$ of the unit interval and the Haar wavelet:

$$\phi_1(x) = \mathbf{1}(x), \ \phi_2(x) = \mathbf{1}(2x) - \mathbf{1}(2x-1) \,. \tag{14.24}$$

Then we have

$$\mathcal{N}(x) = \sum_{n=1}^{2} |\phi_n(x)|^2 = 2 \quad \text{a.e.} \tag{14.25}$$

The corresponding coherent states read as

$$|x\rangle = \frac{1}{\sqrt{2}} \left[\phi_1(x)|1\rangle + \phi_2(x)|2\rangle\right] \,. \tag{14.26}$$

To any integrable function $f(x)$ on the interval there corresponds the linear operator A_f on $\mathbb{R}^2$ or $\mathbb{C}^2$:

$$\begin{aligned} A_f &= 2\int_0^1 dx\, f(x)|x\rangle\langle x| \\ &= \left[\int_0^1 dx\, f(x)\right] \left[|1\rangle\langle 1| + |2\rangle\langle 2|\right] + \left[\int_0^1 dx\, f(x)\phi_2(x)\right] \left[|1\rangle\langle 2| + |2\rangle\langle 1|\right] , \end{aligned} \tag{14.27}$$

or, in matrix form with respect to the orthonormal basis (14.24),

$$\begin{aligned} A_f &= \begin{pmatrix} \int_0^1 dx\, f(x) & \int_0^1 dx\, f(x)\phi_2(x) \\ \int_0^1 dx\, f(x)\phi_2(x) & \int_0^1 dx\, f(x) \end{pmatrix} \\ &= \begin{pmatrix} \langle f\rangle_{[0,1]} & 2\left(\langle f\rangle_{[0,1/2]} - \langle f\rangle_{[1/2,1]}\right) \\ 2\left(\langle f\rangle_{[0,1/2]} - \langle f\rangle_{[1/2,1]}\right) & \langle f\rangle_{[0,1]} \end{pmatrix} , \end{aligned} \tag{14.28}$$

where $\langle f\rangle_{\mathrm{I}} = (1/\operatorname{Vol} \mathrm{I}) \int_{\mathrm{I}} f(x)\, dx$ denotes the average of f on the set I. In particular, with the choice $f = \phi_1$ we recover the identity, whereas for $f = \phi_2$ we get $A_{\phi_2} =$

$\left(\begin{smallmatrix} 0 & 1 \\ 1 & 0 \end{smallmatrix}\right) = \sigma_1$, the first Pauli matrix. With the choice $f(x) = x^p, \Re e\, p > -1$,

$$A_{x^p} = \frac{1}{p+1} \begin{pmatrix} 1 & 2^{-p} - 1 \\ 2^{-p} - 1 & 1 \end{pmatrix} . \tag{14.29}$$

For an arbitrary coherent state $|x_0\rangle$, $x_0 \in [0, 1]$, it is interesting to evaluate the lower symbol of A_{x^p}. This gives

$$\langle x_0 | A_{x^p} | x_0 \rangle = \begin{cases} \frac{2^{-p}}{p+1} & 0 \le x_0 \le \frac{1}{2} , \\ \frac{2-2^{-p}}{p+1} & \frac{1}{2} \le x_0 \le 1 , \end{cases} \tag{14.30}$$

the two possible values being precisely the eigenvalues of the above matrix. Note the average values of the "position" operator: $\langle x_0|A_x|x_0\rangle = 1/4$ if $0 \le x_0 \le 1/2$ and $3/4$ if $1/2 \le x_0 \le 1$.

Clearly, like in the $N = 2$ case in the previous section, all operators A_f commute, since they are linear combinations of the identity matrix and the Pauli matrix σ_1. The procedure is easily generalized to higher dimensions. Let us add to the previous set $\{\phi_1, \phi_2\}$ other elements of the Haar basis, say, up to the "scale" J:

$$\begin{aligned} &\{\phi_1(x), \phi_2(x), \phi_3(x) = \sqrt{2}\phi_2(2x),\ \phi_4(x) = \sqrt{2}\phi_2(2x-1) , \\ &\quad \ldots, \phi_s(x) = 2^{j/2}\phi_2(2x-k),\ \phi_N(x) = 2^{J/2}\phi_2(2x - 2^J + 1)\} , \end{aligned} \tag{14.31}$$

where, at given $j = 1, 2, \ldots, J$, the integer k assumes its values in the range $0 \le k \le 2^j - 1$. The total number of elements of this orthonormal system is $N = 2^{J+1}$. The normalization function is equal to $\mathcal{N}(x) = \sum_{n=1}^{N} = |\phi_n(x)|^2 = 2^{J+1}$, and this clearly diverges at the limit $J \to \infty$. Then, it is remarkable if not expected that spectral values as well as average values of the "position" operator are given by $\langle x_0|A_x|x_0\rangle = (2k+1)/2^{J+1}$ for $k/2^J \le x_0 \le (k+1)/2^J$ where $0 \le k \le 2^J - 1$. Hence, our quantization scheme yields a dyadic discretization of the localization in the unit interval.

14.3.2
A Two-Dimensional Noncommutative Quantization of the Unit Interval

Now we choose another orthonormal system, namely, the first two elements of the trigonometric Fourier basis,

$$\phi_1(x) = \mathbf{1}(x),\ \phi_2(x) = \sqrt{2}\sin 2\pi x . \tag{14.32}$$

Then we have

$$\mathcal{N}(x) = \sum_{n=1}^{2} |\phi_n(x)|^2 = 1 + 2\sin^2 2\pi x . \tag{14.33}$$

The corresponding coherent states read as

$$|x\rangle = \frac{1}{\sqrt{1 + 2\sin^2 2\pi x}} \left[|1\rangle + \sqrt{2}\sin 2\pi x |2\rangle \right] . \tag{14.34}$$

To any integrable function $f(x)$ on the interval there corresponds the linear operator A_f on $\mathbb{R}^2$ or $\mathbb{C}^2$ (in its matrix form):

$$A_f = \begin{pmatrix} \int_0^1 dx\, f(x) & \sqrt{2}\int_0^1 dx\, f(x)\sin 2\pi x \\ \sqrt{2}\int_0^1 dx\, f(x)\sin 2\pi x & 2\int_0^1 dx\, f(x)\sin^2 2\pi x \end{pmatrix} . \tag{14.35}$$

Like in the previous case, with the choice $f = \phi_1$ we recover the identity, whereas for $f = \phi_2$ we get $A_{\phi_2} = \sigma_1$, the first Pauli matrix.

We now have to deal with a Jordan algebra of operators A_f, like in the $N = 3$ real case in the previous section. It is generated by the identity matrix and the two real Pauli matrices σ_1 and σ_3.

In this context, the *position* operator is given by

$$A_x = \begin{pmatrix} \frac{1}{2} & -\frac{1}{\sqrt{2\pi}} \\ -\frac{1}{\sqrt{2\pi}} & \frac{1}{2} \end{pmatrix} ,$$

with eigenvalues $\frac{1}{2} \pm \frac{1}{\sqrt{2\pi}}$. Note its average values as a function of the coherent state parameter $x_0 \in [0, 1]$:

$$\langle x_0 | A_x | x_0 \rangle = \frac{1}{2} - \frac{2}{\pi}\frac{\sin 2\pi x_0}{1 + 2\sin^2 2\pi x_0} .$$

In Figure 14.1 we give the curve of $\langle x_0 | A_x | x_0$ as a function of x_0. It is interesting to compare with the two-dimensional Haar quantization presented in the previous subsection.

14.4 Coherent State Quantization of the Unit Circle and the Quantum Phase Operator

We now come back to the unit circle. We gave an example of quantization in Section 13.5, resulting from the orthonormal system $\{\cos\theta, \sin\theta\}$ in $L^2(S^1, d\theta/\pi)$. Here, we explore the coherent state quantization based on an arbitrary number of elements of the Fourier exponential basis [187]. As an interesting byproduct of this "fuzzy circle," we give an expression for the phase or angle operator, and we discuss its relevance in comparison with various phase operators proposed by other authors.

14.4.1 A Retrospective of Various Approaches

In classical physics, angle and action are conjugated canonical variables. It was hoped to get the exact correspondence for their quantum counterpart after some quantization procedure. Taking for granted (true for the harmonic oscillator) the correspondence $I \to \hat{N}$, $\hat{N}$ being the number operator, and $\theta \to \hat{\theta}$, we should get

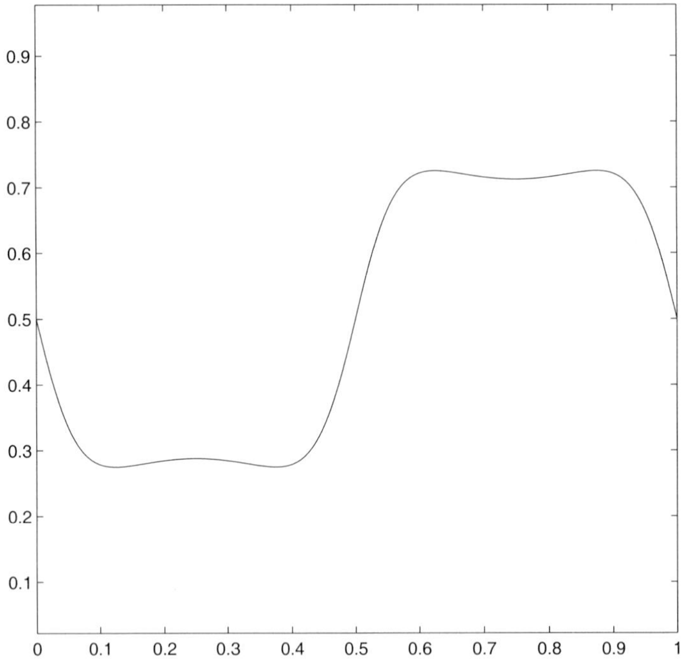

Fig. 14.1 Average value $\langle x_0 | A_x | x_0 \rangle$ of *position* operator A_x versus x_0 (cf. eigenvalues of A_x).

the canonical commutation relation

$$[\hat{N}, \hat{\theta}] = i I_d \,, \tag{14.36}$$

and from this the Heisenberg inequality $\Delta\hat{N}\,\Delta\hat{\theta} \geq \frac{1}{2}$.

Since the first attempt by Dirac in 1927 [163] various definitions of angle or phase operator have been proposed with more or less satisfactory success in terms of consistency with regard to the requirement that phase operators and number operators form a conjugate Heisenberg pair obeying (14.36).

To obtain this quantum-mechanical analog, the polar decomposition of raising and lowering operators

$$\hat{a} = \exp(i\hat{\theta})\hat{N}^{1/2} \,, \quad \hat{a}^\dagger = \hat{N}^{1/2}\exp(-i\hat{\theta}) \tag{14.37}$$

was originally proposed by Dirac, with the corresponding uncertainty relation

$$\Delta\hat{\theta}\,\Delta\hat{N} \geq \frac{1}{2} \,. \tag{14.38}$$

But the relation between operators (14.36) is misleading. The construction of a unitary operator is a delicate procedure and there are three main problems with it. First, we have that for a well-defined number state the uncertainty on the phase would be greater than 2π. This inconvenience, also present in the quantization of the angular momentum–angle pair, adds to the well-known contradiction, namely, "1 = 0", lying in the matrix elements of the commutator

$$-i\delta_{nn'} = \langle n'|[\hat{N}, \hat{\theta}]|n\rangle = (n - n')\langle n'|\hat{\theta}|n\rangle\,. \tag{14.39}$$

In the angular momentum case, this contradiction is avoided to a certain extent by introducing a proper periodical variable $\hat{\Phi}(\phi)$ [188]. If $\hat{\Phi}$ is just a sawtooth function, its discontinuities at $\phi = (2n + 1)\pi$, $n \in \mathbb{Z}$, give rise to Dirac peaks in the commutation relation:

$$[\hat{L}_z, \hat{\Phi}] = -i\left\{1 - 2\pi\sum_{n=-\infty}^{\infty}\delta(\phi - (2n + 1)\pi)\right\}. \tag{14.40}$$

The singularities in (14.40) can be excluded, as proposed by Louisell [189], taking sine and cosine functions of $\hat{\theta}$ to recover a valid uncertainty relation. Louisell introduced operators $\cos\hat{\theta}$ and $\sin\hat{\theta}$ and imposed the commutation rule $[\cos\hat{\theta}, \hat{N}] = i\sin\hat{\theta}$ and $[\sin\hat{\theta}, \hat{N}] = -i\cos\hat{\theta}$.

Nevertheless, the problem is harder in the number-phase case because, as shown by Susskind and Glogower [190], the decomposition (14.37) itself leads to the definition of nonunitary operators:

$$\exp(-i\hat{\theta}) = \sum_{n=0}^{\infty}|n\rangle\langle n + 1|\{+|\psi\rangle\langle 0|\},\ \text{and h.c.}\,, \tag{14.41}$$

and this nonunitarity explains the inconsistency revealed in (14.39). To overcome this handicap, a different polar decomposition was suggested in [190]:

$$\hat{a} = (\hat{N} + 1)^{\frac{1}{2}}\hat{E}_-, \quad \hat{a}^\dagger = \hat{E}_+(\hat{N} + 1)^{\frac{1}{2}}\,, \tag{14.42}$$

where the operators $\hat{E}_\pm$ are still nonunitary because of their action on the extremal state of the semibounded number basis [188]. Nevertheless, the addition of the restriction

$$\hat{E}_-|0\rangle = 0 \tag{14.43}$$

permits us to define Hermitian operators

$$\begin{aligned}\hat{C} &= \tfrac{1}{2}(\hat{E}_- + \hat{E}_+) = \hat{C}^\dagger\,,\\ \hat{S} &= \frac{1}{2i}(\hat{E}_- - \hat{E}_+) = \hat{S}^\dagger\,.\end{aligned} \tag{14.44}$$

These operators are named "cosine" and "sine" because they reproduce the same algebraic structure as the projections of the classical state in the phase space of the oscillator problem.

In an attempt to determine a Hermitian phase operator $\hat{\theta}$ fitting (14.36) in the classical limit, and for which constraints such as (14.43) would be avoided, Popov and Yarunin [191, 192] and, later, Barnett and Pegg [193] used an orthonormal set of eigenstates of $\hat{\theta}$ defined on the number state basis as

$$|\theta_m\rangle = \frac{1}{\sqrt{N}} \sum_{n=0}^{N-1} e^{in\theta_m} |n\rangle \,, \tag{14.45}$$

where, for a given N, the following equidistant subset of the angle parameter is selected:

$$\theta_m = \theta_0 + \frac{2\pi m}{N} \,, \quad m = 0, 1, \ldots, N-1 \,, \tag{14.46}$$

with θ_0 as a reference phase. Orthonormality stems from the well-known properties of the roots of the unity as happens with the discrete Fourier transform basis:

$$\sum_{n=0}^{N-1} e^{in(\theta_m - \theta_{m'})} = \sum_{n=0}^{N-1} e^{i2\pi(m-m')\frac{n}{N}} = N\delta_{mm'} \,. \tag{14.47}$$

Indeed, $z = e^{2\pi i \frac{q}{N}}$ for *any* $q \in \mathbb{Z}$ is a solution of

$$(1 - z^N) = (1 - z)(1 + z + \cdots + z^{N-1}) = 0 \,.$$

The inverse (unitary!) transform from the Pegg–Barnett orthonormal basis $\{|\theta_m\rangle\}$ to the number basis $\{|n\rangle\}$ is given by

$$|n\rangle = \frac{1}{\sqrt{N}} \sum_{m=0}^{N-1} e^{-in\,\theta_m} |\theta_m\rangle \,. \tag{14.48}$$

Hence, a system in a number state is equally likely to be found in any phase state, and the reciprocal relationship is also true.

The (Pegg–Barnett) phase operator on $\mathbb{C}^N$ is then defined by the spectral conditions

$$\hat{\theta}_{\text{bp}} |\theta_m\rangle = \theta_m |\theta_m\rangle \,, \tag{14.49}$$

or, equivalently, constructed through the spectral resolution

$$\hat{\theta}_{\text{bp}} = \sum_{m=0}^{N-1} \theta_m |\theta_m\rangle\langle\theta_m| \,. \tag{14.50}$$

Its expression in the number state basis reads as

$$\hat{\theta}_{\text{bp}} = \left(\theta_0 + \pi \frac{N-1}{N}\right) I_N + \frac{2\pi}{N} \sum_{n \neq n'}^{N-1} \frac{e^{i(n-n')\theta_0}}{e^{2\pi i \frac{n-n'}{N}} - 1} |n\rangle\langle n'| \,. \tag{14.51}$$

On the other hand, the expression of the number operator $\hat{N}$ in the phase basis reads as

$$\hat{N} = \sum_{n=0}^{N-1} |n\rangle\langle n| = \frac{N-1}{2} I_N + \sum_{m \neq m'}^{N-1} \frac{1}{e^{-2\pi i \frac{m'-m}{N}} - 1} |\theta_m\rangle\langle\theta_{m'}| \,. \tag{14.52}$$

Note the useful summation formula valid for any $z = e^{2\pi i \frac{q}{N}}$, $q \in \mathbb{Z}$, $q \neq 0$:

$$\sum_{n=0}^{N-1} n z^n = \frac{N}{z-1} .$$

The commutation rule with the number operator expressed in the number state basis results:

$$[\hat{N}, \hat{\theta}_{\text{bp}}] = \frac{2\pi}{N} \sum_{n \neq n'}^{N-1} (n-n') \frac{e^{i(n-n')\theta_0}}{e^{2\pi i \frac{n-n'}{N}} - 1} |n\rangle\langle n'| . \tag{14.53}$$

In the phase state basis it is

$$[\hat{N}, \hat{\theta}_{\text{bp}}] = \frac{2\pi}{N} \sum_{m \neq m'}^{N-1} (m-m') \frac{1}{1 - e^{-2\pi i \frac{m-m'}{N}}} |\theta_m\rangle\langle \theta_{m'}| . \tag{14.54}$$

Now, from the asymptotic behavior at large $N \gg m, m', n, n'$,

$$\begin{aligned} \langle \theta_m | [\hat{N}, \hat{\theta}_{\text{bp}}] | \theta_{m'} \rangle &\approx i(1 - \delta_{mm'}) , \\ \langle n | [\hat{N}, \hat{\theta}_{\text{bp}}] | n' \rangle &\approx -i(1 - \delta_{nn'}) e^{i(n-n')\theta_0} , \end{aligned} \tag{14.55}$$

it seems hopeless to get something like canonical commutation rules with this phase operator resulting from the unitary change of basis [number basis $\rightarrow$ phase basis].

Let us turn our attention to the unitary $e^{i\hat{\theta}_{\text{bp}}}$. Since $\hat{\theta}_{\text{bp}}$ is Hermitian, the operator $e^{i\hat{\theta}_{\text{bp}}}$ is unitary. Its expression in the phase state basis is immediate:

$$e^{i\hat{\theta}_{\text{bp}}} = \sum_{m=0}^{N-1} e^{i\theta_m} |\theta_m\rangle\langle\theta_m| = e^{i\theta_0} \sum_{m=0}^{N-1} e^{2\pi i \frac{m}{N}} |\theta_m\rangle\langle\theta_m| . \tag{14.56}$$

On the other hand, its expression in number state basis is off-diagonal and "circular":

$$e^{i\hat{\theta}_{\text{bp}}} = \sum_{n=0}^{N-2} |n\rangle\langle n+1| + e^{iN\theta_0} |N-1\rangle\langle 0| . \tag{14.57}$$

One can form from $e^{i\hat{\theta}_{\text{bp}}}$ Hermitian cosine and sine combinations with the expected relations:

$$\begin{aligned} &\cos^2 \hat{\theta}_{\text{bp}} + \sin^2 \hat{\theta}_{\text{bp}} = 1 , \quad [\cos \hat{\theta}_{\text{bp}}, \sin \hat{\theta}_{\text{bp}}] = 0 , \\ &\langle n | \cos^2 \hat{\theta}_{\text{bp}} | n \rangle = \tfrac{1}{2} = \langle n | \sin^2 \hat{\theta}_{\text{bp}} | n \rangle . \end{aligned}$$

Finally, annihilation and creation operators acting on $\mathbb{C}^N$ can be defined through the natural factorization

$$a = e^{i\hat{\theta}_{\text{bp}}} \hat{N}^{1/2} = \sum_{n=0}^{N-2} \sqrt{n+1} |n\rangle\langle n+1| , \quad a^\dagger = \sum_{n=1}^{N-1} \sqrt{n} |n\rangle\langle n-1| .$$

We then recover the finite-dimensional version of the canonical quantization:

$$[a, a^\dagger] = I_N - N|N-1\rangle\langle N-1| .$$

The Pegg–Barnett construction, which amounts to an adequate change of orthornormal basis in $\mathbb{C}^N$, gives for the ground number state $|0\rangle$ a random phase that avoids some of the drawbacks encountered in previous developments. Note that taking the limit $N \to \infty$ is questionable within a Hilbertian framework; this process must be understood in terms of mean values restricted to some suitable subspace and the limit has to be taken afterwards. In [193] the pertinence of the states (14.45) was proved by the expected value of the commutator with the number operator. The problem appears when the limit is taken since it leads to an approximate result.

More recently an interesting approach to the construction of a phase operator was used by Busch, Lahti, and their collaborators within the framework of measurement theory [194, 195, 197] (see also [196]). Phase observables are characterized as phase-shift-covariant positive operator measures, with the number operator playing the part of the shift generator.

14.4.2
Pegg–Barnett Phase Operator and Coherent State Quantization

Let us now reformulate the construction of the Pegg–Barnett phase operator along the lines of coherent state quantization. The observation set X is the set of N equidistant angles on the unit circle:

$$X = \left\{ \theta_m = \theta_0 + \frac{2\pi m}{N},\ m = 0, 1, \dots, N-1 \right\} . \tag{14.58}$$

We equip X with the discrete equally weighted normalized measure:

$$\int_X f(x)\,\mu(dx) \stackrel{\text{def}}{=} \frac{1}{N}\sum_{m=0}^{N-1} f(\theta_m) . \tag{14.59}$$

The Hilbert space $L^2(X, \mu)$, equipped with the scalar product

$$\langle f|g\rangle = \frac{1}{N}\sum_{m=0}^{N-1} \overline{f(\theta_m)}\, g(\theta_m) , \tag{14.60}$$

is thus isomorphic to $\mathbb{C}^N$. We now just choose as an orthonormal set the discrete Fourier basis:

$$\phi_n(\theta_m) = e^{-in\theta_m} , \quad \text{with} \quad \sum_{n=0}^{N-1} |\phi_n(\theta_m)|^2 = N . \tag{14.61}$$

The coherent states emerging from this orthonormal set are just the Pegg–Barnett phase states:

$$|\theta_m\rangle = \frac{1}{\sqrt{N}}\sum_{n=0}^{N-1} e^{in\theta_m} |\phi_n\rangle , \tag{14.62}$$

where the kets $|\phi_n\rangle$ can be identified with the number states $|n\rangle$. Of course, we trivially have normalization and resolution of the unity in $\mathbb{C}^N$:

$$\langle\theta_m|\theta_m\rangle = 1\,, \quad \int_X |\theta_m\rangle\langle\theta_m|\, N\mu(dx) = I_N\,. \tag{14.63}$$

The resolution of the unity immediately allows the construction of diagonal operators in $\mathbb{C}^N$:

$$f(x) \mapsto \int_X f(x)|\theta_m\rangle\langle\theta_m|\, N\mu(dx) = \sum_{m=0}^{N-1} f(\theta_m)|\theta_m\rangle\langle\theta_m| = A_f\,. \tag{14.64}$$

As expected from the discussion in Chapter 13, this quantization yields a commutative algebra of operators. The Pegg–Barnett phase operator is nothing other than the quantization of the function angle $f(x) = x \in X$.

14.4.3
A Phase Operator from Two Finite-Dimensional Vector Spaces

As suggested in [193], the commutation relation will approximate better the canonical one (14.36) if one enlarges enough the Hilbert space of states. Hence, we follow a construction of the phase operator based on a coherent state quantization scheme that involves *two* finite-dimensional vector spaces. This will produce a suitable commutation relation at the infinite-dimensional limit, still at the level of mean values.

The observation set X is now the set of M equidistant angles on the unit circle:

$$X = \left\{\theta_m = \theta_0 + \frac{2\pi m}{M}\,,\ m = 0, 1, \ldots, M-1\right\}\,. \tag{14.65}$$

Set X is equipped with the discrete equally weighted normalized measure:

$$\int_X f(x)\mu(dx) \stackrel{\text{def}}{=} \frac{1}{M}\sum_{m=0}^{M-1} f(\theta_m)\,. \tag{14.66}$$

The Hilbert space $L^2(X, \mu)$, with the scalar product

$$\langle f|g\rangle = \frac{1}{M}\sum_{m=0}^{M-1} \overline{f(\theta_m)}g(\theta_m)\,, \tag{14.67}$$

is isomorphic to $\mathbb{C}^M$. Let us choose as an orthonormal set the Fourier set with N elements, $N < M$:

$$\phi_n(\theta_m) = e^{-in\theta_m}\,, \quad \text{with} \ \sum_{n=0}^{N-1} |\phi_n(\theta_m)|^2 = N\,. \tag{14.68}$$

The coherent states now read

$$|\theta_m\rangle = \frac{1}{\sqrt{N}}\sum_{n=0}^{N-1} e^{in\theta_m}|\phi_n\rangle\,, \tag{14.69}$$

where the kets $|\phi_n\rangle$ can be identified with the number states $|n\rangle$. These states, different from (14.62), are normalized and resolve the unity *in* $\mathbb{C}^N$:

$$\langle\theta_m|\theta_m\rangle = 1, \quad \int_X |\theta_m\rangle\langle\theta_m|\, N\mu(dx) = I_N \,. \tag{14.70}$$

The important difference from the case $M = N$ is that we have lost orthonormality:

$$\begin{aligned}\langle\theta_m|\theta_{m'}\rangle &= \frac{1}{N}\sum_{n=0}^{N-1} e^{2\pi i n \frac{m'-m}{M}} \\ &= \begin{cases} 1 & \text{if } m = m'\,, \\ \dfrac{1}{N} e^{i\pi(m'-m)\frac{N-1}{M}} \dfrac{\sin\pi\frac{N(m'-m)}{M}}{\sin\pi\frac{m'-m}{M}} & \text{if } m \neq m'\,. \end{cases}\end{aligned} \tag{14.71}$$

The coherent state quantization reads as

$$f(x) \mapsto \int_X f(\theta_m)|\theta_m\rangle\langle\theta_m|\, N\mu(dx) = \sum_{m=0}^{M-1} f(\theta_m)|\theta_m\rangle\langle\theta_m| = A_f\,, \tag{14.72}$$

and provides a *noncommutative* algebra of operators *as soon as* $M > N$. The operators that emerge from this coherent state quantization are now different from the Pegg–Barnett operators. In particular, we obtain as a phase operator

$$\begin{aligned}A_\theta &= \frac{N}{M}\sum_{m=0}^{M-1} \theta_m|\theta_m\rangle\langle\theta_m| \\ &= \left(\theta_0 + \pi\frac{M-1}{M}\right) I_N + \frac{2\pi}{M}\sum_{n\neq n'}^{N-1} \frac{e^{i(n-n')\theta_0}}{e^{2\pi i\frac{n-n'}{M}} - 1}|\phi_n\rangle\langle\phi_{n'}|\,.\end{aligned} \tag{14.73}$$

The coherent states are not phase eigenstates any more, $A_\theta|\theta_m\rangle \neq \theta_m|\theta_m\rangle$, and the expression of the lower symbols is quite involved:

$$\langle\theta_m|A_\theta|\theta_m\rangle = \frac{N}{M}\theta_m + \frac{1}{M}\sum_{m'=0, m'\neq m}^{M-1} \theta_{m'}\left(\frac{\sin\pi\frac{N(m'-m)}{M}}{\sin\pi\frac{m'-m}{M}}\right)^2 . \tag{14.74}$$

Let us examine its commutation relation with the number operator expressed in the number state basis:

$$[\hat{N}, A_\theta] = \frac{2\pi}{M}\sum_{n\neq n'}^{N-1} (n-n')\frac{e^{i(n-n')\theta_0}}{e^{2\pi i\frac{n-n'}{M}} - 1}|\phi_n\rangle\langle\phi_{n'}|\,. \tag{14.75}$$

Its lower symbol reads as

$$\langle\theta_m|[\hat{N}, A_\theta]|\theta_m\rangle = \frac{2\pi}{NM}\sum_{n\neq n'}^{N-1} (n-n')\frac{e^{2\pi i(n-n')\frac{m}{M}}}{e^{2\pi i\frac{n-n'}{M}} - 1}\,. \tag{14.76}$$

Its asymptotic behavior at large $M \gg N$,

$$\langle \theta_m | [\hat{N}, A_\theta] | \theta_{m'} \rangle \approx -\frac{i}{N} \sum_{n \neq n'}^{N-1} e^{2\pi i (n-n') \frac{m}{M}}$$

$$= i - \frac{i}{N} \left(\frac{\sin \pi \frac{N(m)}{M}}{\sin \pi \frac{m}{M}} \right)^2 , \tag{14.77}$$

shows that we have, to some extent, improved the situation.

14.4.4
A Phase Operator from the Interplay Between Finite and Infinite Dimensions

We show now that there is no need to discretize the angle variable as in [193] to recover a suitable commutation relation. We adopt instead the Hilbert space of square-integrable functions on the circle as the natural framework for defining an appropriate phase operator in a finite-dimensional subspace.

So we take as an observation set X the unit circle S^1 provided with the measure $\mu(d\theta) = d\theta/2\pi$. The Hilbert space is $L^2(X,\mu) = L^2(S^1, d\theta/2\pi)$ and has the inner product

$$\langle f | g \rangle = \int_0^{2\pi} \overline{f(\theta)} g(\theta) \frac{d\theta}{2\pi} . \tag{14.78}$$

In this space we choose as an orthonormal set the N first Fourier exponentials with negative frequencies:

$$\phi_n(\theta) = e^{-in\theta} , \quad \text{with } \mathcal{N}(\theta) = \sum_{n=0}^{N-1} |\phi_n(\theta)|^2 = N . \tag{14.79}$$

The phase states are now defined as the corresponding "coherent states":

$$|\theta) = \frac{1}{\sqrt{N}} \sum_{n=0}^{N-1} e^{in\theta} |\phi_n\rangle , \tag{14.80}$$

where the kets $|\phi_n\rangle$ can be directly identified with the number states $|n\rangle$, and the round bracket denotes the continuous labeling of this family. We have, by construction, normalization and resolution of the unity in $\mathcal{H}_N \cong \mathbb{C}^N$:

$$(\theta|\theta) = 1, \quad \int_0^{2\pi} |\theta)(\theta| \, N\mu(d\theta) = I_N . \tag{14.81}$$

Unlike in (14.45), the states (14.80) are not orthogonal but overlap as

$$(\theta'|\theta) = \frac{e^{i\frac{N-1}{2}(\theta-\theta')}}{N} \frac{\sin \frac{N}{2}(\theta-\theta')}{\sin \frac{1}{2}(\theta-\theta')} . \tag{14.82}$$

Note that for N large enough these states contain all the Pegg–Barnett phase states and besides they form a continuous family labeled by the points of the circle. The coherent state quantization of a particular function $f(\theta)$ with respect to the continuous set (14.80) yields the operator A_f defined by

$$f(\theta) \mapsto \int_X f(\theta)|\theta)(\theta|\, N\mu(d\theta) \overset{\text{def}}{=} A_f \,. \tag{14.83}$$

An analogous procedure has been used within the framework of positive-operator-valued measures [194, 195]: the phase states are expanded over an infinite orthogonal basis with the known drawback of defining the convergence of the series $|\phi\rangle = \sum_n e^{in\theta}|n\rangle$ from the Hilbertian arena and the related questions concerning operator domains. When it is expressed in terms of the number states, the operator (14.83) takes the form

$$A_f = \sum_{n,n'=0}^{N-1} c_{n'-n}(f)|n\rangle\langle n'| \,, \tag{14.84}$$

where $c_n(f)$ is the Fourier coefficient of the function $f(\theta)$,

$$c_n(f) = \int_0^{2\pi} f(\theta) e^{-in\theta} \frac{d\theta}{2\pi} \,.$$

Therefore, the existence of the quantum version of f is ruled by the existence of its Fourier transform. Note that A_f is self-adjoint only when $f(\theta)$ is real-valued. In particular, a self-adjoint phase operator of the Toeplitz matrix type is readily obtained by choosing $f(\theta) = \theta$:

$$A_\theta = -i \sum_{\substack{n \neq n' \\ n,n'=0}}^{N-1} \frac{1}{n - n'} |n\rangle\langle n'| \,, \tag{14.85}$$

an expression that has to be compared with (12.13). Its lower symbol or expectation value in a coherent state is given by

$$(\theta|A_\theta|\theta) = \frac{i}{N} \sum_{\substack{n,n'=0 \\ n \neq n'}}^{N-1} \frac{e^{i(n-n')\theta}}{n' - n} \,. \tag{14.86}$$

Owing to the continuous nature of the set of $|\theta)$, all operators produced by this quantization are different from the Pegg–Barnett operators. As a matter of fact, the commutator $[\hat{N}, A_\theta]$ expressed in terms of the number basis reads as

$$[\hat{N}, A_\theta] = -i \sum_{\substack{n,n'=0 \\ n \neq n'}}^{N-1} |n\rangle\langle n'| = i I_N + (-i)\mathcal{I}_N \,, \tag{14.87}$$

and has all diagonal elements equal to 0. Here $\mathcal{I}_N = \sum_{n,n'=0}^{N-1} |n\rangle\langle n'|$ is the $N \times N$ matrix with all entries equal to 1. The spectrum of this matrix is reduced to the values 0 (degenerate $N-1$ times) and N, that is, $(1/N)\,\mathcal{I}_N$ is an orthogonal projector on one vector. More precisely, the normalized eigenvector corresponding to the eigenvalue N is along the diagonal in the first quadrant in $\mathbb{C}^N$:

$$|\nu_N\rangle = |\theta = 0) = \frac{1}{\sqrt{N}} \sum_{n=0}^{N-1} |n\rangle\,. \tag{14.88}$$

Other eigenvectors span the hyperplane orthogonal to $|\nu_N\rangle$. We can choose them as the set with $N-1$ elements:

$$\left\{ |\nu_n\rangle \overset{\text{def}}{=} \frac{1}{\sqrt{2}}(|n+1\rangle - |n\rangle)\,, \quad n = 0, 1, \ldots, N-2 \right\}. \tag{14.89}$$

The matrix $\mathcal{I}_N$ is just N times the projector $|\nu_N\rangle\langle\nu_N|$. Hence, the commutation rule reads as

$$[\hat{N}, \hat{A}_\theta] = -i \sum_{\substack{n \neq n' \\ n,n'=0}}^{N-1} |n\rangle\langle n'| = i\,(I_d - N|\nu_N\rangle\langle\nu_N|)\,. \tag{14.90}$$

A further analysis of this relation through its lower symbol yields, for the matrix $\mathcal{I}_N$, the function

$$(\theta|\mathcal{I}_N|\theta) = \frac{1}{N} \sum_{n,n'=0}^{N-1} e^{i(n-n')\theta} = \frac{1}{N} \frac{\sin^2 N\frac{\theta}{2}}{\sin^2 \frac{\theta}{2}}\,. \tag{14.91}$$

In the limit at large N this function is the Dirac comb (a well-known result in diffraction theory):

$$\lim_{N\to\infty} \frac{1}{N} \frac{\sin^2 N\frac{\theta}{2}}{\sin^2 \frac{\theta}{2}} = \sum_{k\in\mathbb{Z}} \delta(\theta - 2k\pi)\,. \tag{14.92}$$

Recombining this with (14.90) allows us to recover the canonical commutation rule through its lower symbol and with the addition of a Dirac comb:

$$(\theta|[\hat{N}, A_\theta]|\theta) \approx_{N\to\infty} i - i \sum_{k\in\mathbb{Z}} \delta(\theta - 2k\pi)\,. \tag{14.93}$$

This expression is the expected one for any periodical variable as was seen in (14.40).

Note that (14.93) is found through the expected value over phase coherent states and not in any physical state like in [193]. This shows that states (14.80), as standard coherent states, are the closest to classical behavior. Another main feature is that any of these states is equally weighted over the number basis that confirms a total indeterminacy on the eigenstates of the number operator. The opposite is also true:

the number state is equally weighted over all the family (14.80) and in particular this coincides with the results in [193].

The creation and annihilation operators are obtained using first the quantization (14.83) with $f(\theta) = e^{\pm i\theta}$,

$$A_{e^{\pm i\theta}} = \int_0^{2\pi} e^{\pm i\theta} N|\theta)(\theta|\frac{d\theta}{2\pi}, \tag{14.94}$$

and then including the number operator as $A_{e^{i\theta}} \hat{N}^{\frac{1}{2}} \equiv \hat{a}$ in a similar way to that in [193], where the authors used instead $e^{i\hat{\theta}_{PB}} \hat{N}^{\frac{1}{2}}$. The commutation relation between both operators is

$$[\hat{a}, \hat{a}^\dagger] = 1 - N|N-1\rangle\langle N-1|, \tag{14.95}$$

which converges to the common result only when the expectation value is taken on states where extremal state components vanish as N tends to infinity.

As the phase operator is not built from a spectral decomposition, it is clear that $A_{\theta^2} \neq A_\theta^2$ and the link with an uncertainty relation is not straightforward as in [193]. Instead, as suggested in [195], a different definition for the variance should be used.

The phase operator constructed here has most of the advantages of the Pegg–Barnett operator but allows more freedom within the Hilbertian framework. It is clear that a well-defined phase operator must be parameterized by all points in the circle to have convergence to the commutation relation consistent with the classical limit. It is also clear that the inconveniences due to the nonperiodicity of the phase pointed out in [190] are avoided from the very beginning by the choice of $X \equiv S^1$.

15
Coherent State Quantization of Motions on the Circle, in an Interval, and Others

15.1
Introduction

In most of the introductory textbooks devoted to quantum mechanics the quantum versions of two simple models are presented, namely, the motion of a particle on the circle and in an interval (the infinite square well potential). In this chapter, we revisit these examples in the light of coherent state quantization of the corresponding phase spaces, namely, the cylinder [198] and an infinite strip in the plane [199, 200]. We also apply our method to motion on the 1 + 1 de Sitter space-time, since the corresponding phase space is topologically a cylinder. We finally consider under the same angle a more exotic example, the motion on a discrete set of points as was presented in [201] under the name of "shadow" Schrödinger quantum mechanics.

15.2
Motion on the Circle

15.2.1
The Cylinder as an Observation Set

Quantization of the motion of a particle on the circle (like the quantization of polar coordinates in the plane) is an old question with so far mildly evasive answers. A large body of literature exists concerning the subject, more specifically devoted to the problem of angular localization and related Heisenberg inequalities; see, for instance, [202].

Let us apply our scheme of coherent state quantization to this particular problem. Let us first recall the material presented in the last section of Chapter 5. The observation set X is the phase space of a particle moving on the circle, precisely the cylinder $S^1 \times \mathbb{R} = \{x \equiv (\beta, J), |0 \leq \beta < 2\pi, \ J \in \mathbb{R}\}$, equipped with the measure $\mu(dx) = \frac{1}{2\pi}\, dJ\, d\beta$. The functions $\phi_n(x)$ forming the orthonormal system needed to

Coherent States in Quantum Physics. Jean-Pierre Gazeau

ISBN: 978-3-527-40709-5

construct coherent states are suitably weighted Fourier exponentials:

$$\phi_n(x) = \left(\frac{\varepsilon}{\pi}\right)^{1/4} e^{-\frac{\varepsilon}{2}(J-n)^2}\, e^{in\beta}\,, \quad n \in \mathbb{Z}\,, \tag{15.1}$$

where $\varepsilon > 0$ is a regularization parameter that can be arbitrarily small. The coherent states [62, 63, 67] read as

$$|x\rangle \equiv |J,\beta\rangle = \frac{1}{\sqrt{\mathcal{N}(J)}} \left(\frac{\varepsilon}{\pi}\right)^{1/4} \sum_{n\in\mathbb{Z}} e^{-\frac{\varepsilon}{2}(J-n)^2}\, e^{-in\beta} |e_n\rangle\,, \tag{15.2}$$

where the states $|e_n\rangle$'s, in one-to-one correspondence with the ϕ_n's, form an orthonormal basis of some separable Hilbert space $\mathcal{H}$. For instance, they can be considered as Fourier exponentials $e^{in\beta}$ forming the orthonormal basis of the Hilbert space $L^2(S^1, d\theta/2\pi) \cong \mathcal{H}$. They are the *spatial modes* in this representation. The normalization factor is a periodic train of normalized Gaussians and is proportional to an elliptic theta function,

$$\mathcal{N}(J) = \sqrt{\frac{\varepsilon}{\pi}} \sum_{n\in\mathbb{Z}} e^{-\varepsilon(J-n)^2} \underset{\text{Poisson}}{=} \sum_{n\in\mathbb{Z}} e^{2\pi i n J}\, e^{-\frac{\pi^2}{\varepsilon}n^2}\,, \tag{15.3}$$

and satisfies $\lim_{\varepsilon\to 0} \mathcal{N}(J) = 1$.

It should be noted that the quantization of the observation set follows the selection (or *polarization*) in the (modified) Hilbert space

$$L^2\left(S^1 \times \mathbb{R}, \sqrt{\frac{\varepsilon}{\pi}} \frac{1}{2\pi} e^{-\varepsilon J^2}\, dJ\, d\beta\right)$$

of all Laurent series in the complex variable $z = e^{\varepsilon J - i\beta}$.

15.2.2
Quantization of Classical Observables

By virtue of (13.12), the quantum operator (acting on $\mathcal{H}$) associated with the classical observable $f(x)$ is obtained through

$$A_f := \int_X f(x)|x\rangle\langle x|\mathcal{N}(x)\mu(dx) = \sum_{n,n'} \left(A_f\right)_{nn'} |e_n\rangle\langle e_{n'}|\,, \tag{15.4}$$

where

$$\left(A_f\right)_{nn'} = \sqrt{\frac{\varepsilon}{\pi}}\, e^{-\frac{\varepsilon}{8}(n-n')^2} \int_{-\infty}^{+\infty} dJ\, e^{-\frac{\varepsilon}{2}\left(J-\frac{n+n'}{2}\right)^2} \frac{1}{2\pi} \int_0^{2\pi} d\beta\, e^{-i(n-n')\beta} f(J,\beta)\,. \tag{15.5}$$

If f is J-dependent only, then A_f is diagonal with matrix elements that are Gaussian transforms of $f(J)$:

$$\left(A_{f(J)}\right)_{nn'} = \delta_{nn'} \sqrt{\frac{\varepsilon}{\pi}} \int_{-\infty}^{+\infty} dJ\, e^{-\frac{\varepsilon}{2}(J-n)^2}\, f(J)\,. \tag{15.6}$$

For the most basic one, associated with the classical observable J, this yields

$$A_J = \int_X \mu(dx)\mathcal{N}(J)\, J\,|J,\beta\rangle\langle J,\beta| = \sum_{n\in\mathbb{Z}} n\,|e_n\rangle\langle e_n|, \tag{15.7}$$

and this is nothing but the angular momentum operator, which reads in angular position representation (Fourier series) as $A_J = -i\partial/\partial\beta$.

If f is β-dependent only, $f(x) = f(\beta)$, we have

$$A_{f(\beta)} = \int_X \mu(dx)\mathcal{N}(J) f(\beta)\,|J,\beta\rangle\langle J,\beta| \tag{15.8}$$

$$= \sum_{n,n'\in\mathbb{Z}} e^{-\frac{\varepsilon}{4}(n-n')^2}\, c_{n-n'}(f)|e_n\rangle\langle e_{n'}|, \tag{15.9}$$

where $c_n(f)$ is the nth Fourier coefficient of f. In particular, we have for

- the operator "angle,"

$$A_\beta = \pi\mathbb{I}_{\mathcal{H}} + \sum_{n\neq n'} i\,\frac{e^{-\frac{\varepsilon}{4}(n-n')^2}}{n-n'}\,|e_n\rangle\langle e_{n'}|, \tag{15.10}$$

- the operator "Fourier fundamental harmonic,"

$$A_{e^{i\beta}} = e^{-\frac{\varepsilon}{4}} \sum_n |e_{n+1}\rangle\langle e_n|. \tag{15.11}$$

In the isomorphic realization of $\mathcal{H}$ in which the kets $|e_n\rangle$ are the Fourier exponentials $e^{i\,n\beta}$, $A_{e^{i\beta}}$ is a multiplication operator by $e^{i\beta}$ up to the factor $e^{-\frac{\varepsilon}{4}}$ (which can be made arbitrarily close to 1).

15.2.3
Did You Say *Canonical*?

The "canonical" commutation rule

$$[A_J, A_{e^{i\beta}}] = A_{e^{i\beta}} \tag{15.12}$$

is canonical in the sense that it is in exact correspondence with the classical Poisson bracket

$$\left\{J, e^{i\beta}\right\} = i e^{i\beta}. \tag{15.13}$$

For other nontrivial commutators having this exact correspondence, see [203]. There could be interpretational difficulties with commutators of the type

$$[A_J, A_{f(\beta)}] = \sum_{n,n'} (n-n')e^{-\frac{\varepsilon}{4}(n-n')^2}\, c_{n-n'}(f)\,|e_n\rangle\langle e_{n'}|, \tag{15.14}$$

and, in particular, for the angle operator itself,

$$[A_J, A_\beta] = i \sum_{n \neq n'} e^{-\frac{\varepsilon}{4}(n-n')^2} |e_n\rangle\langle e_{n'}| \,, \tag{15.15}$$

to be compared with the classical $\{J, \beta\} = 1$.

We have already encountered such difficulties in Chapters 12 and 14, and we guess they are only apparent. They are due to the discontinuity of the 2π-periodic saw function $B(\beta)$ which is equal to β on $[0, 2\pi)$. They are circumvented if we examine, like we did in those chapters, the behavior of the corresponding lower symbols at the limit $\varepsilon \to 0$. For the angle operator,

$$\begin{aligned}\langle J_0, \beta_0 | A_\beta | J_0, \beta_0 \rangle &= \pi + \frac{1}{2}\left(1 + \frac{\mathcal{N}(J_0 - \frac{1}{2})}{\mathcal{N}(J_0)}\right) \sum_{n \neq 0} i \, \frac{e^{-\frac{\varepsilon}{2} n^2 + i n \beta_0}}{n} \\ &\underset{\varepsilon \to 0}{\sim} \pi + \sum_{n \neq 0} i \, \frac{e^{i n \beta_0}}{n} \,,\end{aligned} \tag{15.16}$$

where we recognize at the limit the Fourier series of $B(\beta_0)$. For the commutator,

$$\begin{aligned}\langle J_0, \beta_0 | [A_J, A_\beta] | J_0, \beta_0 \rangle &= \frac{1}{2}\left(1 + \frac{\mathcal{N}(J_0 - \frac{1}{2})}{\mathcal{N}(J_0)}\right)\left(-i + \sum_{n \in \mathbb{Z}} i e^{-\frac{\varepsilon}{2} n^2 + i n \beta_0}\right) \\ &\underset{\varepsilon \to 0}{\sim} -i + i \sum_n \delta(\beta_0 - 2\pi n) \,.\end{aligned} \tag{15.17}$$

So we (almost) recover the canonical commutation rule except for the singularity at the origin mod 2π.

15.3
From the Motion of the Circle to the Motion on 1 + 1 de Sitter Space-Time

The material in the previous section is now used to describe the quantum motion of a massive particle on a 1 + 1 de Sitter background, which means a one-sheeted hyperboloid embedded in a 2 + 1 Minkowski space. Here, we just summarize the content of [198]. The phase space X is also a one-sheeted hyperboloid,

$$J_1^2 + J_2^2 - J_0^2 = \kappa^2 > 0 \,, \tag{15.18}$$

with (local) canonical coordinates (J, β), as for the motion on the circle. Phase space coordinates are now viewed as basic classical observables,

$$J_0 = J, \; J_1 = J\cos\beta - \kappa \sin\beta \,, \quad J_2 = J \sin\beta + \kappa\cos\beta \,, \tag{15.19}$$

and obey the Poisson bracket relations

$$\{J_0, J_1\} = -J_2 \,, \quad \{J_0, J_2\} = J_1 \,, \quad \{J_1, J_2\} = J_0 \,. \tag{15.20}$$

They are, as expected, the commutation relations of $\mathfrak{so}(1,2) \simeq \mathfrak{sl}(2,\mathbb{R})$, which is the kinematical symmetry algebra of the system. Applying the coherent state quantization (15.4) at $\varepsilon \neq 0$ produces the basic quantum observables:

$$A_{J_0} = \sum_n n|e_n\rangle\langle e_n| \,, \tag{15.21a}$$

$$A^{\varepsilon}_{J_1} = \frac{1}{2} e^{-\frac{\varepsilon}{4}} \sum_n \left(n + \frac{1}{2} + i\kappa \right) |e_{n+1}\rangle\langle e_n| + \text{c.c.} \,, \tag{15.21b}$$

$$A^{\varepsilon}_{J_2} = \frac{1}{2i} e^{-\frac{\varepsilon}{4}} \sum_n \left(n + \frac{1}{2} + i\kappa \right) |e_{n+1}\rangle\langle e_n| - \text{cc} \,. \tag{15.21c}$$

The quantization is asymptotically exact for these basic observables since

$$[A_{J_0}, A^{\varepsilon}_{J_1}] = iA^{\varepsilon}_{J_2} \,, \quad [A_{J_0}, A^{\varepsilon}_{J_2}] = -iA^{\varepsilon}_{J_1} \,, \quad [A^{\varepsilon}_{J_1}, A^{\varepsilon}_{J_2}] = -ie^{-\frac{\varepsilon}{2}} A_{J_0} \,. \tag{15.22}$$

Moreover, the quadratic operator

$$C^{\varepsilon} = (A^{\varepsilon}_{J_1})^2 + (A^{\varepsilon}_{J_2})^2 - e^{-\frac{\varepsilon}{2}} (A_{J_0})^2 \tag{15.23}$$

commutes with the Lie algebra generated by the operators A_{J_0}, $A^{\varepsilon}_{J_1}$, $A^{\varepsilon}_{J_1}$, that is, it is the Casimir operator for this algebra. In the representation given in (15.21a)–(15.21c), its value is fixed to $e^{-\frac{\varepsilon}{2}} (\kappa^2 + \frac{1}{4})\mathbb{I}$ and so admits the limit

$$C^{\varepsilon} \underset{\varepsilon \to 0}{\sim} \left(\kappa^2 + \frac{1}{4} \right) I_d \,.$$

Hence, we have produced a coherent state quantization that leads asymptotically to the principal series of representations of $SO_0(1,2)$.

15.4 Coherent State Quantization of the Motion in an Infinite-Well Potential

15.4.1 Introduction

Even though the quantum dynamics in an infinite square well potential represents a rather unphysical limit situation, it is a familiar textbook problem and a simple tractable model for the confinement of a quantum particle. On the other hand, this model has a serious drawback when it is analyzed in more detail. Namely, when one proceeds to a canonical standard quantization, the definition of a momentum operator with the usual form $-i\hbar d/dx$ has a doubtful meaning. This subject has been discussed in many places (see, e.g., [134]), and the attempts to circumvent this anomaly range from self-adjoint extensions [134–137] to $\mathcal{PT}$ symmetry approaches [204].

First of all, the canonical quantization assumes the existence of a momentum operator (essentially) self-adjoint in $L^2(\mathbb{R})$ that respects some boundary conditions at the boundaries of the well. These conditions cannot be fulfilled by the usual derivative form without the consequence of losing self-adjointness. Moreover, there exists an uncountable set of self-adjoint extensions of such a derivative operator that makes truly delicate the question of a precise choice based on physical requirements [135–137, 200].

When the classical particle is trapped in an infinite-well interval Δ, the Hilbert space of quantum states is $L^2(\Delta, dx)$ and the quantization problem becomes similar, to a certain extent, to the quantization of the motion on the circle S^1. Notwithstanding the fact that boundary conditions are not periodic but impose the condition instead that the wave functions in position representation vanish at the boundary, the momentum operator P for the motion in the infinite well should be the counterpart of the angular momentum operator J_{op} for the motion on the circle. Since the energy spectrum for the infinite square well is like $\{n^2, n \in \mathbb{N}^*\}$, we should expect that the spectrum of P should be $\mathbb{Z}^*$, like the one for J_{op} without the null eigenvalue. We profit from this similarity between the two problems by adapting the coherent states for the motion on the circle to the present situation. More precisely, we introduce two-component vector coherent states, in the spirit of [205], as infinite superpositions of spinors that are eigenvectors of P. We then examine the consequences of our choice after coherent state quantization of basic quantum observables, such as position, energy, and a quantum version of the problematic momentum. In particular we focus on their mean values in coherent states ("lower symbols") and quantum dispersions. As will be shown, the classical limit is recovered after choosing appropriate limit values for some parameters present in the expression of our coherent states.

15.4.2
The Standard Quantum Context

As already indicated in Chapter 9, the wave function $\Psi(q, t)$ of a particle of mass m trapped inside the interval $q \in [0, L]$ (here we use the notation $L = \pi a$) has to obey the conditions

$$\psi \in L^2([0, L], dq), \quad \Psi(0, t) = \Psi(L, t) = 0 \quad \forall t\,. \tag{15.24}$$

Its Hamiltonian reads as

$$H \equiv H_{\text{w}} = -\frac{\hbar^2}{2m}\frac{d^2}{dx^2}\,, \tag{15.25}$$

and is self-adjoint [206] on an appropriate dense domain in (15.24). Factorizing Ψ as $\Psi(q, t) = e^{-\frac{i}{\hbar}Ht}\Psi(q, 0)$, we have for $\Psi(q, 0) \equiv \psi(q)$ the eigenvalue equation $H\psi(q) = E\psi(q)$, together with the boundary conditions (15.24). Normalized eigenstates and corresponding eigenvalues are then given by

$$\begin{aligned} \psi_n(q) &= \sqrt{\frac{2}{L}}\sin\left(n\pi\frac{q}{L}\right), \quad 0 \leqslant q \leqslant L, \\ H\psi_n &= E_n\psi_n\,, \quad n = 1, 2, \ldots, \end{aligned} \tag{15.26}$$

with

$$E_n = \frac{\hbar^2\pi^2}{2mL^2}n^2 \equiv \hbar\omega n^2\,, \qquad \omega = \frac{\hbar\hat{E}\pi^2}{2mL^2} \equiv \frac{2\pi}{T_r}\,, \tag{15.27}$$

where T_r is the revival time to be compared with the purely classical round-trip time.

15.4.3 Two-Component Coherent States

The classical phase space of the motion of the particle is the infinite strip $X = [0, L] \times \mathbb{R} = \{x = (q, p) | q \in [0, L],\ p \in \mathbb{R}\}$ equipped with the measure $\mu(dx) = dq\,dp$. Typically, we have two phases in the periodic particle motion with a given energy: one corresponds to positive values of the momentum, $p = mv$, while the other one is for negative values, $p = -mv$. This observation naturally leads us to introduce the Hilbert space of two-component complex-valued functions (or spinors), square-integrable with respect to $\mu(dx)$:

$$\begin{aligned} L^2_{\mathbb{C}^2}(X, \mu(dx)) &\simeq \mathbb{C}^2 \otimes L^2_{\mathbb{C}}(X, \mu(dx)) \\ &= \left\{ \Phi(x) = \begin{pmatrix} \phi_+(x) \\ \phi_-(x) \end{pmatrix},\ \phi_\pm \in L^2_{\mathbb{C}}(X, \mu(dx)) \right\}. \end{aligned} \tag{15.28}$$

Inspired by the coherent states for the motion on the circle, we choose our orthonormal system as formed of the following vector-valued functions $\Phi_{n,\kappa}(x)$, $\kappa = \pm$,

$$\Phi_{n,+}(x) = \begin{pmatrix} \phi_{n,+}(x) \\ 0 \end{pmatrix}, \quad \Phi_{n,-}(x) = \begin{pmatrix} 0 \\ \phi_{n,-}(x) \end{pmatrix},$$
$$\phi_{n,\kappa}(x) = \sqrt{c}\,\exp\left(-\frac{1}{2\varrho^2}(p - \kappa p_n)^2\right) \sin\left(n\pi\frac{q}{L}\right), \quad \kappa = \pm,\ n = 1, 2, \ldots, \tag{15.29}$$

where

$$c = \frac{2}{\varrho L\sqrt{\pi}}\,, \qquad p_n = \sqrt{2mE_n} = \frac{\hbar\pi}{L}n\,. \tag{15.30}$$

The half-width $\varrho > 0$ in the Gaussians is a parameter which has the dimension of a momentum, say, $\varrho = \hbar\pi\vartheta/L$, with $\vartheta > 0$ a dimensionless parameter. This parameter can be arbitrarily small, like for the classical limit. It can be arbitrarily large, for instance, in the case of a very narrow well. The functions $\Phi_{n,\kappa}(x)$ are continuous, vanish at the boundaries $q = 0$ and $q = L$ of the phase space, and obey the finiteness condition:

$$
\begin{aligned}
0 < \mathcal{N}(x) \equiv \mathcal{N}(q, p) \equiv \mathcal{N}_+(x) + \mathcal{N}_-(x) &= \sum_{\kappa=\pm} \sum_{n=1}^{\infty} \Phi^{\dagger}_{n,\kappa}(x) \Phi_{n,\kappa}(x) \\
&= c \sum_{n=1}^{\infty} \left[\exp\left(-\frac{1}{\varrho^2}(p - p_n)^2 \right) + \exp\left(-\frac{1}{\varrho^2}(p + p_n)^2 \right) \right] \\
&\quad \times \sin^2\left(n\pi \frac{q}{L} \right) < \infty \,.
\end{aligned} \tag{15.31}
$$

The expression of $\mathcal{N}(q, p) \equiv c\ \mathcal{S}(q, p)$ can be simplified to

$$
\mathcal{S}(q, p) = \Re \left\{ \frac{1}{2} \sum_{n=-\infty}^{\infty} \left[1 - \exp\left(\mathrm{i}2\pi n \frac{q}{L} \right) \right] \exp\left(-\frac{1}{\varrho^2}(p - p_n)^2 \right) \right\} . \tag{15.32}
$$

Series $\mathcal{N}$ and $\mathcal{S}$ can be expressed in terms of elliptic theta functions. Function $\mathcal{S}$ has no physical dimension, whereas $\mathcal{N}$ has the same dimension as c, that is, the inverse of an action.

We are now in a position to define our vector coherent states along the lines of [205]. We set up a one-to-one correspondence between the functions $\Phi_{n,\kappa}$'s and two-component states

$$
|e_n, \pm\rangle \stackrel{\text{def}}{=} |\pm\rangle \otimes |e_n\rangle\,, \quad |+\rangle = \begin{pmatrix} 1 \\ 0 \end{pmatrix}, \quad |-\rangle = \begin{pmatrix} 0 \\ 1 \end{pmatrix}, \tag{15.33}
$$

forming an orthonormal basis of some separable Hilbert space of the form $\mathcal{K} = \mathbb{C}^2 \otimes \mathcal{H}$. The latter can be viewed also as the subspace of $L^2_{\mathbb{C}^2}(X, \mu(dx))$ equal to the closure of the linear span of the set of $\Phi_{n,\kappa}$'s. We choose the following set of 2×2 diagonal real matrices for our construction of vectorial coherent states:

$$
\mathrm{F}_n(x) = \begin{pmatrix} \phi_{n,+}(q, p) & 0 \\ 0 & \phi_{n,-}(q, p) \end{pmatrix} . \tag{15.34}
$$

Note that $\mathcal{N}(x) = \sum_{n=1}^{\infty} \mathrm{tr}(\mathrm{F}_n(x)^2)$. Vector coherent states, $|x, \chi\rangle \in \mathbb{C}^2 \otimes \mathcal{H} = \mathcal{K}$, are now defined for each $x \in X$ and $\chi \in \mathbb{C}^2$ by the relation

$$
|x, \chi\rangle = \frac{1}{\sqrt{\mathcal{N}(x)}} \sum_{n=1}^{\infty} \mathrm{F}_n(x)\, |\chi\rangle \otimes |e_n\rangle\,. \tag{15.35}
$$

In particular, we single out the two orthogonal coherent states

$$
|x, \kappa\rangle = \frac{1}{\sqrt{\mathcal{N}(x)}} \sum_{n=1}^{\infty} \mathrm{F}_n(x) |e_n, \kappa\rangle\,, \quad \kappa = \pm\,. \tag{15.36}
$$

By construction, these states also satisfy the infinite square well boundary conditions, namely, $|x,\kappa\rangle_{q=0} = |x,\kappa\rangle_{q=L} = 0$. Furthermore they fulfill the normalizations

$$\langle x,\kappa|x,\kappa\rangle = \frac{\mathcal{N}_\kappa(x)}{\mathcal{N}(x)}\,, \quad \sum_{\kappa=\pm}\langle x,\kappa|x,\kappa\rangle = 1 \tag{15.37}$$

and the resolution of the identity in $\mathcal{K}$:

$$\begin{aligned}
&\sum_{\kappa=\pm}\int_X |x,\kappa\rangle\langle x,\kappa|\mathcal{N}(x)\mu(dx)\\
&\quad= \sum_{\kappa=\pm}\sum_{n,n'=1}^{\infty}\int_{-\infty}^{\infty}\int_0^L \mathrm{F}_n(q,p)\mathrm{F}_{n'}(q,p)|e_n,\kappa\rangle\langle e_{n'},\kappa|dq\,dp\\
&\quad= \sum_{\kappa=\pm}\sum_{n=1}^{\infty}|e_n,\kappa\rangle\langle e_n,\kappa| = \sigma_0\otimes I_{\mathcal{H}} = I_{\mathcal{K}}\,,
\end{aligned} \tag{15.38}$$

where σ_0 denotes the 2×2 identity matrix consistently with the Pauli matrix notation σ_μ to be used in the sequel.

15.4.4 Quantization of Classical Observables

The quantization of a generic function $f(q,p)$ on the phase space is given by the expression (15.4), which is, for our particular choice of coherent state

$$\begin{aligned}
f\mapsto A_f &= \sum_{\kappa=\pm}\int_{-\infty}^{\infty}\int_0^L f(q,p)|x,\kappa\rangle\langle x,\kappa|\mathcal{N}(q,p)dq\,dp\\
&= \sum_{n,n'=1}^{\infty}|e_n\rangle\langle e_{n'}|\otimes\begin{pmatrix}(f_+)_{nn'} & 0\\ 0 & (f_-)_{nn'}\end{pmatrix}\,,
\end{aligned} \tag{15.39}$$

where

$$(f_\pm)_{nn'} = \int_{-\infty}^{\infty}dp\int_0^L dq\,\phi_{n,\pm}(q,p)\,f(q,p)\,\phi_{n',\pm}(q,p)\,. \tag{15.40}$$

For the particular case in which f is a function of p only, $f(p)$, the operator is given by

$$\begin{aligned}
A_f &= \sum_{\kappa=\pm}\int_{-\infty}^{\infty}\int_0^L f(p)|x,\kappa\rangle\langle x,\kappa|\mathcal{N}(q,p)dq\,dp\\
&= \frac{1}{\varrho\sqrt{\pi}}\sum_{n=1}^{\infty}|e_n\rangle\langle e_n|\otimes\begin{pmatrix}(f_+)_{nn'} & 0\\ 0 & (f_-)_{nn'}\end{pmatrix}\,,
\end{aligned} \tag{15.41}$$

with

$$(f_\pm)_{nn'} = \int_{-\infty}^{\infty}dp\,f(p)\exp\left(-\frac{1}{\varrho^2}(p\mp p_n)^2\right)\,. \tag{15.42}$$

Note that this operator is diagonal on the $|n,\kappa\rangle$ basis.

15.4.4.1
Momentum and Energy

In particular, using $f(p) = p$, one gets the operator

$$P \equiv A_p = \sum_{n=1}^{\infty} p_n \, \sigma_3 \otimes |e_n\rangle\langle e_n| \, , \tag{15.43}$$

where $\sigma_3 = \begin{pmatrix} 1 & 0 \\ 0 & -1 \end{pmatrix}$.

For $f(p) = p^2$, which is proportional to the Hamiltonian, the quantum counterpart reads as

$$A_{p^2} = \frac{\varrho^2}{2} \mathbb{I}_{\mathcal{K}} + \sum_{n=1}^{\infty} p_n^2 \, \sigma_0 \otimes |e_n\rangle\langle e_n| = \frac{\varrho^2}{2} \mathbb{I}_{\mathcal{K}} + (\widehat{p}\,)^2 \, . \tag{15.44}$$

Note that this implies that the operator for the square of the momentum does not coincide with the square of the momentum operator. Actually they coincide up to $O(\hbar^2)$.

15.4.4.2
Position

For a general function of the position $f(q)$ our quantization procedure yields the following operator:

$$A_q = \sum_{n,n'=1}^{\infty} \exp\left(-\frac{1}{4\varrho^2}(p_n - p_{n'})^2\right) \left[d_{n-n'}(f) - d_{n+n'}(f)\right] \sigma_0 \otimes |e_n\rangle\langle e_{n'}| \, , \tag{15.45}$$

where

$$d_m(f) \equiv \frac{1}{L} \int_0^L f(q) \cos\left(m\pi \frac{q}{L}\right) dq \, . \tag{15.46}$$

In particular, for $f(q) = q$ we get the "position" operator

$$Q \equiv A_q = \frac{L}{2} \mathbb{I}_{\mathcal{K}} - \frac{2L}{\pi^2} \sum_{\substack{n,n' \geq 1, \\ n+n'=2k+1}}^{\infty} \exp\left(-\frac{1}{4\varrho^2}(p_n - p_{n'})^2\right) \left[\frac{1}{(n-n')^2} - \frac{1}{(n+n')^2}\right] \times \sigma_0 \otimes |e_n\rangle\langle e_{n'}| \, , \tag{15.47}$$

with $k \in \mathbb{N}$. Note the appearance of the classical mean value for the position on the diagonal.

15.4.4.3
Commutation Rules

Now, to see to what extent these momentum and position operators differ from their classical (canonical) counterparts, let us consider their commutator:

$$[Q, P] = \frac{2\hbar}{\pi} \sum_{\substack{n \neq n' \\ n+n'=2k+1}}^{\infty} C_{n,n'}\, \sigma_3 \otimes |e_n\rangle\langle e_{n'}| \tag{15.48}$$

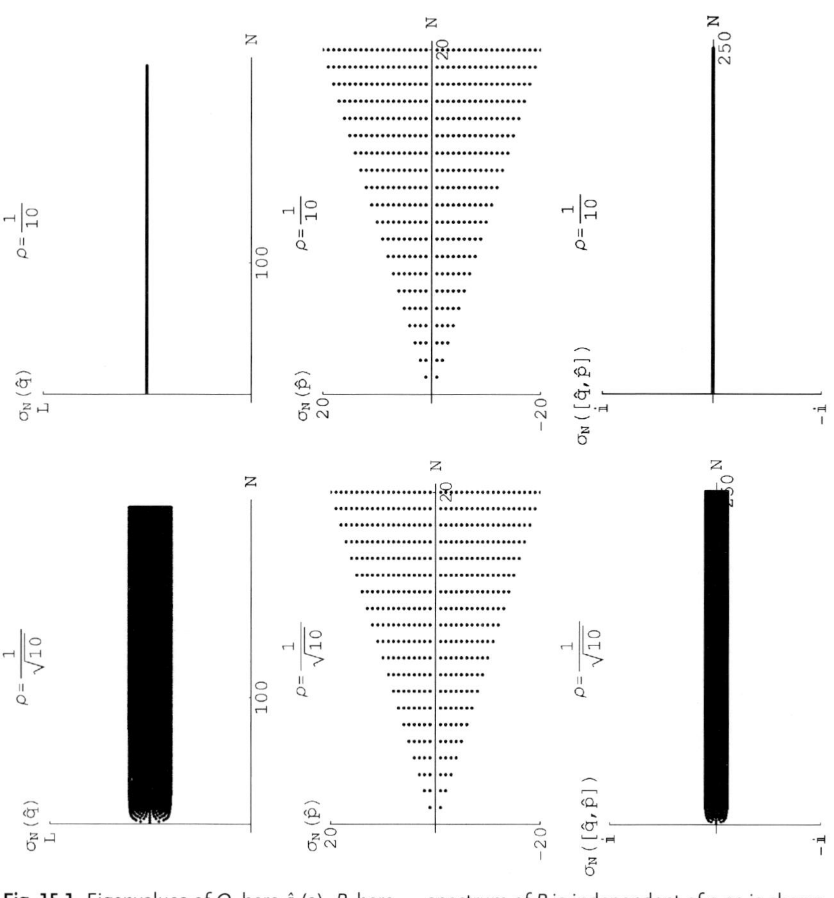

Fig. 15.1 Eigenvalues of Q, here $\hat{q}$ (a), P, here $\hat{p}$ (b), and $[Q, P]$ (c), for increasing values of the characteristic momentum $\varrho = \hbar\pi\vartheta/L$ of the system, and computed for $N \times N$ approximation matrices. Units were chosen such that $\hbar = 1$ and $L = \pi$ so that $\varrho = \vartheta$ and $p_n = n$. Note that for Q with ϱ small, the eigenvalues adjust to the classical mean value $L/2$. The spectrum of P is independent of ϱ as is shown in (15.43). For the commutator, the values are purely imaginary (reprinted from [Garcia de Leon, P., Gazeau, J.P., and Quéva, J.: The infinite well revisited: coherent states and quantization, Phys. Lett., 372, p. 3597, 2008] with permission from Elsevier).

$$C_{n,n'} = \exp\left(-\frac{1}{4\varrho^2}(p_n - p_{n'})^2\right)(n - n')\left[\frac{1}{(n-n')^2} - \frac{1}{(n+n')^2}\right]. \tag{15.49}$$

This is an infinite antisymmetric real matrix. The respective spectra of the finite matrix approximations of this operator and of position and momentum operators are compared in Figures 15.1 and 15.2 for various values of the regulator $\varrho = \hbar\pi\vartheta/L = \vartheta$ in units $\hbar = 1$ and $L = \pi$. When ϱ assumes large values, the eigenvalues of $[Q, P]$ accumulate around $\pm$i, that is, they become almost canonical. Conversely, when $\varrho \to 0$ all eigenvalues tend to 0, and this behavior corresponds to the classical limit.

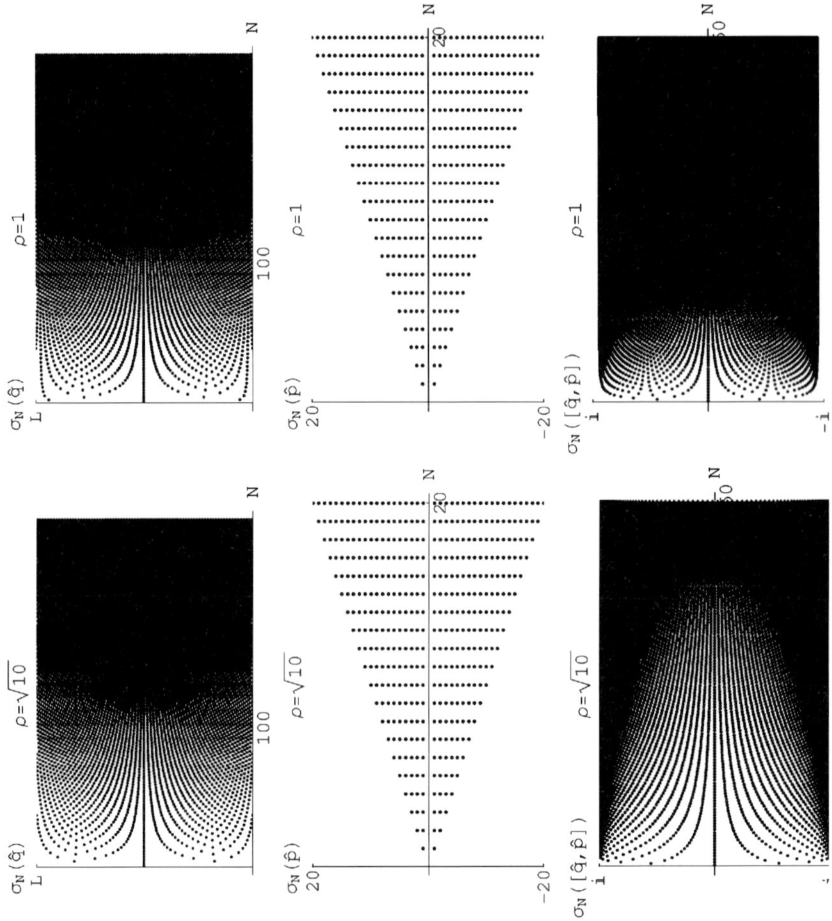

Fig. 15.2 Eigenvalues of Q, P, and $[Q, P]$ of $N \times N$ approximation matrices for increasingly larger values of $\varrho = \hbar\pi\vartheta/L = \vartheta$ in units $\hbar = 1$ and $L = \pi$. The spectrum of P is independent of ϱ as is shown in (15.43). For the commutator, the eigenvalues are purely imaginary and tend to accumulate around $i\hbar$ and $-i\hbar$ as ϱ increases (reprinted from [Garcia de Leon, P., Gazeau, J.P., and Quéva, J.: The infinite well revisited: coherent states and quantization, Phys. Lett., 372, p. 3597, 2008] with permission from Elsevier).

15.4.5
Quantum Behavior through Lower Symbols

Lower symbols are computed with normalized coherent states. The latter are denoted as follows:

$$|x\rangle = |x,+\rangle + |x,-\rangle\,. \tag{15.50}$$

Hence, the lower symbol of a quantum observable A should be computed as

$$\check{A}(x) = \langle x|A|x\rangle \equiv \check{A}_{++}(x) + \check{A}_{+-}(x) + \check{A}_{-+}(x) + \check{A}_{--}(x)\,.$$

This gives the following results for the observables previously considered.

15.4.5.1
Position

In the same way, the mean value of the position operator in a vector coherent state $|x\rangle$ is given by

$$\langle x|Q\,|x\rangle = \frac{L}{2} - \mathcal{Q}(q,p)\,, \tag{15.51}$$

where we can distinguish the classical mean value for the position corrected by the function

$$\begin{aligned}\mathcal{Q}(q,p) &= \frac{2L}{\pi^2}\frac{1}{\mathcal{S}}\sum_{\substack{n,n'=1,n\neq n'\\ n+n'=2k+1}}^{\infty} \exp\left(-\frac{1}{4\varrho^2}(p_n-p_{n'})^2\right)\left[\frac{1}{(n-n')^2}-\frac{1}{(n+n')^2}\right]\\ &\quad\times\left[\exp\left(-\frac{1}{2\varrho^2}[(p-p_n)^2+(p-p_{n'})^2]\right)+\right.\\ &\quad\left.+\exp\left(-\frac{1}{2\varrho^2}[(p+p_n)^2+(p+p_{n'})^2]\right)\right]\sin\left(n\pi\frac{q}{L}\right)\sin\left(n'\pi\frac{q}{L}\right)\,.\end{aligned} \tag{15.52}$$

15.4.5.2
Momentum

The mean value of the momentum operator in a vector coherent state $|x\rangle$ is given by the affine combination:

$$\begin{aligned}\langle x|P|x\rangle &= \frac{\mathcal{M}(x)}{\mathcal{N}(x)}\,,\\ \mathcal{M}(x) &= c\sum_{n=1}^{\infty} p_n\left[\exp\left(-\frac{1}{\varrho^2}(p-p_n)^2\right)-\exp\left(-\frac{1}{\varrho^2}(p+p_n)^2\right)\right]\\ &\quad\times\sin^2\left(n\pi\frac{q}{L}\right)\,.\end{aligned} \tag{15.53}$$

15.4.5.3
Position–Momentum Commutator

The mean value of the commutator in a normalized state $\Psi = \binom{\phi_+}{\phi_-}$ is the pure imaginary expression

$$\langle\Psi|[Q,P]|\Psi\rangle = \frac{2i\hbar}{\pi} \sum_{\substack{n\neq n' \\ n+n'=2k+1}}^{\infty} \exp\left(-\frac{1}{4\varrho^2}(p_n - p_{n'})^2\right)(n-n') \times \\ \times \left[\frac{1}{(n-n')^2} - \frac{1}{(n+n')^2}\right] \Im\left(\langle\phi_+|n\rangle\langle n'|\phi_+\rangle - \langle\phi_-|n\rangle\langle n'|\phi_-\rangle\right) . \tag{15.54}$$

Given the symmetry and the real-valuedness of states (15.36), the mean value of the commutator when Ψ is one of our coherent states vanishes, even if the operator does not. This result is due to the symmetry of the commutator spectrum with respect to 0. As is shown on right in Figures 15.1 and 15.2, the eigenvalues of the commutator tend to $\pm i\hbar$ as ϱ, or equivalently ϑ, increases. Still, there are some points with modulus less than $\hbar$. This leads to dispersions $\Delta Q \Delta P$ in coherent states $|x\rangle$ that are no longer bounded from below by $\hbar/2$. Actually, the lower bound of this product, for a region in the phase space as large as we wish, decreases as ϑ diminishes. A numerical approximation is shown in Figure 15.3.

15.4.6
Discussion

From the mean values of the operators obtained here, we verify that the coherent state quantization gives well-behaved momentum and position operators. The classical limit is reached once the appropriate limit for the parameter ϑ is found. If we consider the behavior of the observables as a function of the dimensionless quantity $\vartheta = \varrho L/\hbar\pi$, in the limit $\vartheta \to 0$ and when the Gaussian functions for the momentum become very narrow, the lower symbol of the position operator is $\mathcal{Q}(q,p) \sim L/2$. This corresponds to the classical average value position in the well. On the other hand, at the limit $\vartheta \to \infty$, for which the Gaussians involved spread to constant functions, the function $\check{Q}(q,p)$ converges numerically to the function q. In other words, the position operator obtained through this coherent state quantization yields a fair quantitative description for the quantum localization within the well. Clearly, if a classical behavior is sought, the values of ϑ have to be chosen near 0. This gives localized values for the observables. Consistent with these observations, the behavior of the product $\Delta Q \Delta P$ for low values of ϑ shows uncorrelated observables at any point in the phase space, whereas for large values of this parameter the product is constant and almost equal to the canonical quantum lowest limit $\hbar/2$. This is shown in Figure 15.3.

It is interesting to note that if we replace the Gaussian distribution, used here for the p variable in the construction of the coherent states, by any positive even probability distribution $\mathbb{R} \in p \mapsto \pi(p)$ such that $\sum_n \pi(p-n) < \infty$ the results are

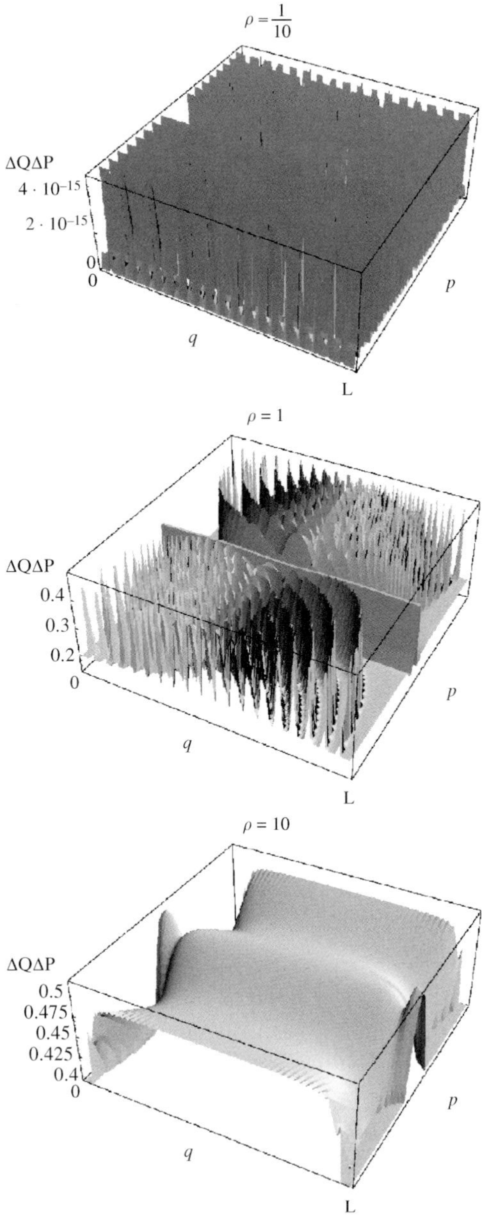

Fig. 15.3 Product $\Delta Q \Delta P$ for various values of $\varrho = \hbar\pi\vartheta/L = \vartheta$ in units $\hbar = 1$ and $L = \pi$. Note the modification of the vertical scale from one picture to another. Again, the position–momentum pair tends to decorrelate at low values of the parameter, as it should do in the classical limit. On the other hand it approaches the usual quantum-conjugate pair at high values of ϱ (reprinted from [Garcia de Leon, P., Gazeau, J.P., and Quéva, J.: The infinite well revisited: coherent states and quantization, Phys. Lett., 372, p. 3597, 2008] with permission from Elsevier).

not so different! The momentum spectrum is still $\mathbb{Z}$ and the energy spectrum has the form $\{n^2 + \text{constant}\}$.

15.5
Motion on a Discrete Set of Points

Now let us consider a problem inspired by modern quantum geometry, where geometrical entities are treated as quantum observables, as they have to be for them to be promoted to the status of objects and not to be simply considered as a substantial arena in which physical objects "live".

The content and terminology of this section were mainly inspired by [201], in which a toy model of quantum geometry is described. The game is to rebuild a "shadow" Schrödinger quantum mechanics on all possible discretizations of the real line. A simple way to do this is to adapt the previous frame quantization of the motion on the circle to an arbitrary discretization of the line. Actually, we deal with the same phase space as for the motion of the particle on the line, that is, the plane, but provided with a different measure, to get an arbitrarily discretized quantum position. Let us enumerate the successive steps of the coherent state quantization procedure applied to this specific situation.

- The observation set X is the plane $\mathbb{R}^2 = \{x \equiv (q, p)\}$.
- The measure (actually a functional) on X is partly of the "Bohr type":

$$\mu(f) = \int_{-\infty}^{+\infty} dq \lim_{T \to +\infty} \frac{1}{T} \int_{-\frac{T}{2}}^{\frac{T}{2}} dp\, f(q, p)\,. \tag{15.55}$$

- We choose as an orthonormal system of functions $\phi_n(x)$ suitably weighted Fourier exponentials associated with a discrete subset $\gamma = \{a_n\}$ of the real line:

$$\phi_n(x) = \left(\frac{\varepsilon}{\pi}\right)^{\frac{1}{4}} e^{-\frac{\varepsilon}{2}(q-a_n)^2}\, e^{-ia_n p}\,. \tag{15.56}$$

  Notice again that the continuous distribution $x \mapsto |\phi_n(x)|^2$ is the normal law centered at a_n for the position variable q.
- The "graph" (in the Ashtekar language) γ is supposed to be uniformly discrete (there exists a nonzero minimal distance between successive elements) in such a way that the aperiodic train of normalized Gaussians or "generalized" theta function

$$\mathcal{N}(x) \equiv \mathcal{N}(q) = \sqrt{\frac{\varepsilon}{\pi}} \sum_n e^{-\varepsilon(q-a_n)^2} \tag{15.57}$$

  converges. Poisson summation formulas can exist, depending on the structure of the graph γ [207].

- Accordingly, the coherent states read as

$$|x\rangle = |q, p\rangle = \frac{1}{\sqrt{\mathcal{N}(q)}} \sum_n \overline{\phi_n(x)} |\phi_n\rangle . \tag{15.58}$$

- Quantum operators acting on $\mathcal{H}$ are yielded by using

$$A_f = \int_X f(x)|x\rangle\langle x|\mathcal{N}(x)\mu(dx) . \tag{15.59}$$

What are the issues of such a quantization scheme in terms of elementary observables such as position and momentum? Concerning position, the algorithm that follows illustrates well the *auberge espagnole*[17] character of our quantization procedure (and actually of *any* quantization procedure), since the outcome is precisely that space itself is quantized, as expected from our choice (15.56). We indeed obtain for the position

$$\int_X \mu(dx)\mathcal{N}(q)\, q\, |q, p\rangle\langle q, p| = \sum_n a_n\, |\phi_n\rangle\langle\phi_n| . \tag{15.60}$$

Hence, the graph γ *is* the "quantization" of space when the latter is viewed from the point of view of coherent states precisely based on γ (...). On the other hand, this predetermination of the accessible space on the quantum level has dramatic consequences for the quantum momentum. Indeed, the latter experiences the following "discrete" catastrophe

$$\begin{aligned}
&\int_X \mu(dx)\mathcal{N}(q)\; p|q, p\rangle\langle q, p| \\
&\quad = \sum_{n,n'} \lim_{T\to+\infty} \left[\frac{\sin\left((a_n - a_{n'})\frac{T}{2}\right)}{(a_n - a_{n'})} - \frac{2}{T}\frac{\sin\left((a_n - a_{n'})\frac{T}{2}\right)}{(a_n - a_{n'})^2} \right] \\
&\quad\times e^{-\frac{\varepsilon}{4}(a_n - a_{n'})^2} |\phi_n\rangle\langle\phi_{n'}| ,
\end{aligned} \tag{15.61}$$

and the matrix elements do not exist for $a_n \neq a_{n'}$ and are ∞ for $a_n = a_{n'}$. Nevertheless, a means of circumventing the problem is to deal with Fourier exponentials $e^{i\lambda p}$ as classical observables, instead of the mere function p. This choice is naturally justified by our original option of a measure adapted to the almost-periodic structure of the space generated by the functions (15.56). For a single "frequency" λ we get

$$\int_X \mu(dx)\mathcal{N}(q)\, e^{i\lambda p}|q, p\rangle\langle q, p| = \sum_{n,n'} e^{-\frac{\varepsilon}{4}(a_n - a_{n'})^2}\, \delta_{\lambda, a_{n'} - a_n} |\phi_n\rangle\langle\phi_{n'}| . \tag{15.62}$$

17) An *auberge espagnole* (*Spanish inn*) is a French metaphor for a place where you get what you bring.

The generalization to superpositions of Fourier exponentials $e^{i\lambda p}$ is straightforward. In particular, one can define discretized versions of the momentum by considering coherent state quantized versions of finite differences of Fourier exponentials

$$\frac{1}{i}\frac{e^{i\lambda' p} - e^{i\lambda p}}{\lambda' - \lambda}, \tag{15.63}$$

in which "allowed" frequencies should belong to the set of "interpositions" $\gamma - \gamma'$ in the graph γ.

As a final comment, let us repeat here the remark we made at the end of the previous section: the Gaussian distribution in (15.56) is *not* the unique choice we have at our disposal. Any even probability distribution $q \mapsto \pi(q)$ such that $\sum_n \pi(q-a_n) < \infty$ will yield similar results.

16
Quantizations of the Motion on the Torus

16.1
Introduction

This chapter is devoted to the coherent states associated with the discrete Weyl–Heisenberg group and to their utilization for the quantization of the chaotic motion on the torus. The material presented here is part of a work by Bouzouina and De Bièvre [208] on the problem of the equipartition of the eigenfunctions of quantized ergodic maps on the torus. We describe two quantization procedures. The first one is based on those coherent states, along the lines of the book. The second one is an adaptation of the *Weyl quantization* to the underlying discrete symmetry. We then examine the classical limit of certain area-preserving ergodic maps on the two-dimensional torus $\mathbb{T}^2$, viewed as a phase space, with canonical coordinates $(q, p) \in [0, a) \times [0, b)$. We will see how the desired equipartition result can be easily proved for a large class of models.

The content of this chapter, although more mathematically involved, was selected to show the wide range of applications of coherent states in connection with chaotic dynamics on both quantum and classical levels.

16.2
The Torus as a Phase Space

The two-dimensional torus $\mathbb{T}^2$ is the Cartesian product of two circles, $\mathbb{T}^2 \cong S^1 \times S^1$. It is considered here as the Cartesian product of two cosets:

$$\mathbb{T}^2 = \mathbb{R}/a\mathbb{Z} \times \mathbb{R}/b\mathbb{Z} = \mathbb{R}^2/\Gamma\,, \quad a > 0\,,\ b > 0\,, \tag{16.1}$$

where Γ is the lattice $\Gamma = \{(ma, nb)\,,\ (m, n) \in \mathbb{Z}^2\}$. Its topology is that of the Cartesian product of two semiopen intervals: $\mathbb{T}^2 = [0, a) \times [0, b) = \{(q, p),\ 0 \leq q < a,\ 0 \leq p < b\}$. The variable q could be considered as the position (or angle), whereas p would be the momentum. In this sense, the torus is a phase space or symplectic manifold equipped with the 2-form $\omega = dq \wedge dp$ and the resulting normalized measure reads as $\mu = dq\, dp/(ab)$.

Coherent States in Quantum Physics. Jean-Pierre Gazeau

ISBN: 978-3-527-40709-5

Let us consider invertible maps $\Phi : \mathbb{T}^2 \mapsto \mathbb{T}^2$ preserving the measure in the sense that $\mu\left(\Phi^{-1}E\right) = \mu(E)$ for all measurable set $E \subset \mathbb{T}^2$. A first example of such transformations is provided by the irrational translations of the torus. Given two numbers, α_1 and α_2, that are incommensurable to the periods a and b, respectively, that is, $\alpha_1 \notin a\,\mathbb{Q}$ and $\alpha_2 \notin b\,\mathbb{Q}$, the transformation

$$\tau_{\boldsymbol{\alpha}} : \mathbb{T}^2 \ni (q, p) \mapsto (q + \alpha_1, p + \alpha_2) \in \mathbb{T}^2 , \quad \boldsymbol{\alpha} \equiv (\alpha_1, \alpha_2) \tag{16.2}$$

clearly leaves invariant the measure μ.

Another elementary transformation, which we call skew translation (or momentum affine) and denote by $\Phi_{\beta,k}$, is also measure-preserving:

$$\Phi_{\beta,k} = \tau_{(0,\beta)} \circ K , \quad \beta \notin b\mathbb{Q} , \quad K = \begin{pmatrix} 1 & k \\ 0 & 1 \end{pmatrix} , \quad k \in \mathbb{Z} . \tag{16.3}$$

The matrix K is an element of the group $SL(2, \mathbb{Z})$ and acts on the pairs (q, p) viewed as 2-vectors.

On a more general level, we consider the *hyperbolic automorphisms* of the torus. They are defined by matrices of the type

$$A = \begin{pmatrix} \alpha & \beta\frac{a}{b} \\ \gamma\frac{b}{a} & \delta \end{pmatrix} , \quad A\begin{pmatrix} q \\ p \end{pmatrix} = \begin{pmatrix} \alpha q + \beta\frac{a}{b}p \\ \gamma\frac{b}{a}q + \delta p \end{pmatrix} , \quad \alpha , \beta , \gamma , \delta \in \mathbb{Z} , \tag{16.4}$$

with $|\operatorname{Tr} A| = |\alpha + \delta| > 2$ and $\det A = \alpha\delta - \beta\gamma = 1$. As solutions of $\lambda^2 - \lambda \operatorname{Tr} A + 1 = 0$, the eigenvalues $\lambda_\pm = \frac{1}{2}\left(\operatorname{Tr} A \pm \sqrt{(\operatorname{Tr} A)^2 - 4}\right)$ are quadratic algebraic integers such that $|\lambda_+| > 1$ and $\lambda_- = 1/\lambda_+$. This action on the torus, known as *Anosov diffeomorphism*, stretches in the eigendirection corresponding to λ_+, whereas it contracts in the direction determined by λ_-. It gives rise to deformations of figures such as the famous *Arnold cat*. The set of such transformations forms the subgroup of $SL(2, \mathbb{R})$ leaving invariant the lattice Γ. This subgroup is isomorphic to $SL(2, \mathbb{Z})$. There exist periodic points in the torus for such hyperbolic actions. They are given by

$$\mathbf{x} = \left(\frac{q}{a}, \frac{p}{b}\right) \in \mathbb{Q}^2/\mathbb{Z}^2 .$$

We have shown three examples of invertible transformations Φ of the torus that are measure-preserving. The triplet $\sigma \stackrel{\text{def}}{=} (\mathbb{T}^2, \Phi, \mu)$ is a *dynamical system*. It is said to be *ergodic* if for all integrable functions on the torus, $f \in L^1(\mathbb{T}^2, \mu)$, its "temporal" average is equal to its "spatial" average:

$$\mu(f) \stackrel{\text{def}}{=} \underbrace{\int_{\mathbb{T}^2} f(\mathbf{x})\,\mu(d\mathbf{x})}_{\text{"spatial"}} = \underbrace{\lim_{T\to\infty} \frac{1}{T} \sum_{k=0}^{T-1} f \circ \Phi^k(\mathbf{x})}_{\text{"temporal"}} \quad \text{a.e.} \tag{16.5}$$

Now, there holds a general result for characterizing ergodicity of a dynamical system (X, Φ, μ), where X is a measure space with measure μ, here $X = \mathbb{T}^2$, and Φ is an invertible and measure-preserving map.

Theorem 16.1

The dynamical system $\sigma = (\mathbb{T}^2, \Phi, \mu)$ is ergodic if and only if at least one of the following conditions is fulfilled:

(i) Any measurable set $E \subset \mathbb{T}^2$, which is Φ-invariant (i.e., $\Phi^{-1}E = E$), is such that $\mu(E) = 0$ or $\mu(\mathbb{T}^2 \setminus E) = 0$.

(ii) If $f \in L^\infty(\mathbb{T}^2, \mu)$, that is, is bounded almost everywhere with respect to μ, is Φ-invariant, that is, $f \circ \Phi = f$, then it is constant almost everywhere with respect to μ.

If σ is ergodic, then the set of periodic orbits of the map Φ is μ-negligible. Within this ergodicity context, one notices that the hyperbolic automorphims (16.4) are *mixing*. Generally, a map Φ is mixing if for all $f, g \in L^2(\mathbb{T}^2, \mu)$ the following condition is fulfilled:

$$\lim_{k\to\infty} \int_{\mathbb{T}^2} f(\Phi^k(\mathbf{x}))\, g(\mathbf{x})\, \mu(d\mathbf{x}) = \mu(f)\,\mu(g)\,. \tag{16.6}$$

In particular, by choosing for f and g the characteristic functions $f = \chi_{E_1}$ and $g = \chi_{E_2}$ of two measurable sets $E_1,\ E_2 \subset \mathbb{T}^2$, a mixing map verifies

$$\lim_{k\to\infty} \mu\left(\left(\Phi^k E_1\right) \cap E_2\right) = \mu(E_1)\,\mu(E_2)\,. \tag{16.7}$$

In other words, the part of the set $\Phi^k E_1$ intersecting E_2 has a measure that is asymptotically proportional to the measure of E_2, which means that E_1 (if $\mu(E_1) > 0$) spreads uniformly in $\mathbb{T}^2$.

16.3 Quantum States on the Torus

To set up the quantum states on the torus, let us first recall two main features of quantum states on the plane viewed as a phase space:

(i) The (pure) quantum states of a particle having the line $\mathbb{R}$ as a configuration space and so the plane $\mathbb{R}^2$ as a phase space are most generally tempered distributions $\psi \in \mathcal{S}'(\mathbb{R})$, or simply $\mathcal{S}'$, for example, plane waves or Dirac distributions, and, when normalizable, are elements of the Hilbert space $L^2(\mathbb{R})$, the latter being viewed as a subspace of $\mathcal{S}'(\mathbb{R})$.

(ii) The translations acting on the classical phase space realize, on the quantum level, as transformations of $L^2(\mathbb{R})$ that belong to a unitary irreducible representation of the Weyl–Heisenberg group

$$\text{Weyl–Heisenberg group} \ni (\phi, q, p) \mapsto U(\phi, q, p) = e^{-\frac{i}{\hbar}\phi}\, \underbrace{e^{\frac{i}{\hbar}pQ - qP}}_{\equiv D(z)}\,, \tag{16.8}$$

with the notational modification $\phi \equiv -s$ with regard to Chapter 3. The composition law here reads as

$$U(\phi, q, p)\, U(\phi', q', p') = U\left(\phi + \phi' + \tfrac{1}{2}(q p' - p q'), q + q', p + p'\right) .$$

Such a representation is extendable to tempered distributions.

Let us adapt this Weyl–Heisenberg formalism to the torus. Since the phase space is now $\mathbb{T}^2 = \mathbb{R}^2 \setminus \Gamma$, with $\Gamma = \{(ma, nb),\ (m, n) \in \mathbb{Z}^2\}$, it is justified to impose the following periodic conditions on the quantum states ψ:

$$U(a, 0)\, \psi = e^{-i\kappa_1 a}\, \psi\,, \quad U(0, b)\, \psi = e^{i\kappa_2 b}\, \psi\,, \tag{16.9}$$

where, for simplification, $U(q, p)$ stands for $U(0, q, p)$. The pairs

$$\boldsymbol{\kappa} = (\kappa_1, \kappa_2) \in \left[0, \frac{2\pi}{a}\right) \times \left[0, \frac{2\pi}{b}\right) \tag{16.10}$$

are elements of what we naturally call the *reciprocal torus* T^2.

A crucial point of the construction is that, with the periodicity conditions (16.9), the quantization is not trivial if and only if the torus is "quantized" in the sense that there exists $N \in \mathbb{N}$ such that

$$a\, b = 2\pi\hbar N\,. \tag{16.11}$$

Then, the states ψ assume in "position" representation the following form:

$$\psi = \sqrt{\frac{a}{N}} \sum_{m \in \mathbb{Z}} c_m(\psi)\, \delta_{x_m}\,, \quad \text{with } x_m = \frac{ab}{2\pi N}\kappa_2 + m\frac{a}{N}\,, \tag{16.12}$$

where the expansion coefficients are subjected to the consistent periodicity condition $c_{m+N}(\psi) = e^{i\kappa_1 a} c_m(\psi)$.

The representation (16.12) *is unique.* Let us prove this statement.

Condition (16.9) means that ψ is a common eigenfunction of $U(a, 0)$ and $U(0, b))$. Therefore,

$$U(0, b)\, U(a, 0)\psi = U(a, 0)\, U(0, b)\psi = e^{-\frac{i}{\hbar}ab}\, U(0, b)\, U(a, 0)\psi\,,$$

the latter equality resulting from the Weyl–Heisenberg commutation rule on the group level. It implies the quantization of the torus: $\frac{ab}{2\pi\hbar} = N \in \mathbb{N}$.

Next, we have for $\psi \in \mathcal{S}'$

$$U(0, b)\psi(x) = e^{\frac{i}{\hbar}bQ}\psi(x) = e^{\frac{i}{\hbar}bx}\psi(x) = e^{i\kappa_2 b}\psi(x)\,,$$

and thus

$$\left(e^{\frac{ib}{\hbar}(x - \hbar\kappa_2)} - 1\right)\psi(x) = 0\,.$$

The solutions to this equation *in* the space of tempered distributions $\mathcal{S}'$ read as translated Dirac distributions:

$$\psi \propto \delta_{x_m}, \quad \text{with} \quad x_m = \hbar\kappa_2 + 2\pi m\frac{\hbar}{b} = \frac{ab}{2\pi N}\kappa_2 + m\frac{a}{N}.$$

Hence, by superposition,

$$\psi = \sqrt{\frac{a}{N}} \sum_{m\in\mathbb{Z}} c_m(\psi)\delta_{x_m},$$

where we have included the global normalization factor $\sqrt{\frac{a}{N}}$. Finally, from the eigenvalue equation $U(a,0)\phi = e^{-i\kappa_1 a}\psi$ and the action $U(a,0)\delta_{x_m}(x) = \delta_{x_m}(x+a) = \delta_{x_m-a}(x) = \delta_{x_{m-N}}(x)$, we easily derive the periodicity condition on the coefficients themselves:

$$c_{m+N}(\psi) = e^{i\kappa_1 a} c_m(\psi).$$

□

The linear span of these solutions is a vector space of dimension N, which we denote by $\mathcal{S}'(\boldsymbol{\kappa}, N)$. Let us say more about it. First, it can be obtained from the Schwartz space $\mathcal{S}(\mathbb{R})$ of rapidly decreasing smooth functions through the *symmetrization* operator $\mathrm{P}_{\boldsymbol{\kappa}}$ defined as follows:

$$\mathcal{S}(\mathbb{R}) \ni \psi \overset{\mathrm{P}_{\boldsymbol{\kappa}}}{\mapsto} \mathrm{P}_{\boldsymbol{\kappa}}\psi = \sum_{m,n\in\mathbb{Z}} (-1)^{Nmn} e^{i(\kappa_1 ma - \kappa_2 nb)}\, U(ma, nb)\psi \in \mathcal{S}'(\boldsymbol{\kappa}, N). \tag{16.13}$$

The proof of $\mathcal{S}'(\boldsymbol{\kappa}, N) = \mathrm{P}_{\boldsymbol{\kappa}}\,\mathcal{S}(\mathbb{R})$ is based on the Poisson summation formula.

As a Hilbert space, $\mathcal{S}'(\boldsymbol{\kappa}, N)$ is equipped with the scalar product

$$\langle\psi|\psi'\rangle_{\boldsymbol{\kappa},N} = \sum_{j=0}^{N-1} \overline{c_j}(\psi)c_j(\psi'). \tag{16.14}$$

This Hilbert space will be denoted by $\mathcal{H}_\hbar(\boldsymbol{\kappa})$. With respect to this inner product, the operators $U\left(n_1\frac{a}{N}, n_2\frac{b}{N}\right)$, $(n_1, n_2) \in \mathbb{Z}^2$ are unitary. They represent elements of the form $\left(2\pi\hbar\, s, n_1\frac{a}{N}, n_2\frac{b}{N}\right)$, with $s \in \mathbb{Z}$. Such elements form, thanks to (16.11), a discrete subgroup of the Weyl–Heisenberg group, sometimes called the *discrete* Weyl–Heisenberg group. Note that the first element of the triplet has no effective action since it is represented as a unit phase, and so can be omitted in the notation. Now, it is a general result of group representation theory that this unitary representation decomposes into a direct sum of unitary irreducible representation $U_{\boldsymbol{\kappa}}$ of the discrete Weyl–Heisenberg group:

$$U\left(n_1\frac{a}{N}, n_2\frac{b}{N}\right) = \int_0^{\frac{2\pi}{a}}\int_0^{\frac{2\pi}{b}} \nu(d\boldsymbol{\kappa})\, U_{\boldsymbol{\kappa}}\left(n_1\frac{a}{N}, n_2\frac{b}{N}\right), \tag{16.15}$$

where

$$\nu(d\boldsymbol{\kappa}) \stackrel{\text{def}}{=} \frac{ab}{(2\pi)^2}\, d^2\boldsymbol{\kappa}\,. \tag{16.16}$$

Consistently, we have the Hilbertian decomposition:

$$L^2(\mathbb{R}) \cong \int_0^{\frac{2\pi}{a}} \int_0^{\frac{2\pi}{b}} \nu(d\boldsymbol{\kappa})\, \mathcal{H}_\hbar(\boldsymbol{\kappa})\,, \tag{16.17}$$

where the Hilbert space $\mathcal{H}_\hbar(\boldsymbol{\kappa})$ carries the representation $U_{\boldsymbol{\kappa}}$. This decomposition actually reflects the well-known Bloch theorem for wave functions in a periodic potential. It is proven that the representations $U_{\boldsymbol{\kappa}}$ and $U_{\boldsymbol{\kappa}'}$ are not equivalent if $\boldsymbol{\kappa} \neq \boldsymbol{\kappa}'$. We thus obtain a continuous family of nonequivalent quantizations labeled by the pairs $\boldsymbol{\kappa} = (\kappa_1, \kappa_2)$.

Let us turn our attention to the momentum representation $\mathcal{F}_\hbar\mathcal{S}'(\boldsymbol{\kappa}, N)$, obtained by Fourier transform,

$$\hat{\psi}(p) = (\mathcal{F}_\hbar\psi)\,(p) = \frac{1}{\sqrt{2\pi\hbar}} \int_{-\infty}^{+\infty} e^{-\frac{i}{\hbar}px}\, \psi(x)\, dx\,,$$

from the above representation. Such a Fourier transform is just the restriction to $\mathcal{S}'(\boldsymbol{\kappa}, N)$ of the Fourier transform of tempered distributions, $\mathcal{F}_\hbar : \mathcal{S}' \mapsto \mathcal{S}'$. The elements of $\mathcal{F}_\hbar\mathcal{S}'(\boldsymbol{\kappa}, N)$ have the following form:

$$\hat{\psi}^{\boldsymbol{\kappa}} = \sqrt{\frac{b}{N}} \sum_{n\in\mathbb{Z}} d_n(\hat{\psi})\, \delta_{\xi_n} \quad \text{with} \quad \xi_n = \frac{ab}{2\pi N}\kappa_1 + n\frac{b}{N}\,. \tag{16.18}$$

Similarly to the position representation, the coefficients obey the periodicity condition

$$d_{n+N}(\hat{\psi}) = e^{-i\kappa_2 b}\, d_n(\hat{\psi})\,. \tag{16.19}$$

The relation between respective coefficients in position and momentum representations is given by

$$d_n(\hat{\psi}) = \frac{1}{\sqrt{N}}\, e^{-i\frac{\kappa_2 b}{N}\left(\frac{a\kappa_1}{2\pi}+n\right)} \sum_{m=0}^{N-1} c_m(\psi) e^{-im\left(\frac{2\pi}{N}n+\frac{a\kappa_1}{N}\right)}\,. \tag{16.20}$$

Since we are dealing with N-dimensional Hilbert spaces (i.e., Hermitian spaces) $\mathcal{H}_\hbar(\boldsymbol{\kappa})$, it is useful to establish the one-to-one correspondence between the above functional representations and the Hermitian space $\mathbb{C}^N$. This correspondence is defined through the following set of N "orthogonal" Dirac combs in $\mathcal{S}'$ put in one-to-one correspondence with the canonical basis of $\mathbb{C}^N$:

$$e_j^{\boldsymbol{\kappa}} = \sqrt{\frac{a}{N}} \sum_m e^{-ima\kappa_1} \delta_{x_m^j}\,, \quad 0 \le j \le N-1\,, \tag{16.21a}$$

$$f_k^{\boldsymbol{\kappa}} = \sqrt{\frac{b}{N}} \sum_n e^{-inb\kappa_2} \delta_{\xi_n^k}\,, \quad 0 \le k \le N-1\,, \tag{16.21b}$$

with $x_m^j = \frac{ab\kappa_2}{2\pi N} + j\frac{a}{N} - ma$ and $\xi_n^k = \frac{ab\kappa_1}{2\pi N} + k\frac{b}{N} - nb$.

16.4 Coherent States for the Torus

The coherent states for the torus, here denoted by $\psi^{\mathcal{Z}}_{(q,p)}$, are obtained by applying the operator P_κ to the standard coherent states. However, as a preliminary, we modify the definitions and notation of the latter to adapt them to the present context. First, we modify the group-theoretical construction à la Perelomov of the standard coherent states by choosing as a ground state in position representation a phase-modulated Gaussian function:

$$\psi^{\mathcal{Z}}_{(0,0)}(x) \stackrel{\text{def}}{=} \left(\frac{\Im\mathcal{Z}}{\pi\hbar}\right)^{\frac{1}{4}} e^{i\frac{\Re\mathcal{Z}}{2\hbar}x^2}\, e^{-\frac{\Im\mathcal{Z}}{2\hbar}x^2}, \tag{16.22}$$

where $\mathcal{Z}$ is a complex parameter with $\Im\mathcal{Z} > 0$. We then define coherent states as the unitary transport of this ground state by the displacement operator. In the position representation they read as

$$\psi^{\mathcal{Z}}_{(q,p)}(x) \stackrel{\text{def}}{=} e^{\frac{i}{\hbar}(pQ-qP)}\, \psi^{\mathcal{Z}}_{(0,0)}(x) = \left(\frac{\Im\mathcal{Z}}{\pi\hbar}\right)^{\frac{1}{4}} e^{-\frac{ipq}{2\hbar}}\, e^{\frac{i}{\hbar}px}\, e^{i\frac{\mathcal{Z}}{2\hbar}(x-q)^2}. \tag{16.23}$$

These functions clearly belong the Schwartz space $\mathcal{S}$. We recall the resolution of the unity in $L^2(\mathbb{R})$:

$$\int_{\mathbb{R}^2} \frac{dq\,dp}{2\pi\hbar} \left|\psi^{\mathcal{Z}}_{(q,p)}\right\rangle \left\langle\psi^{\mathcal{Z}}_{(q,p)}\right| = I_d\,. \tag{16.24}$$

The coherent states on the torus are then defined through the operator P_κ:

$$\psi^{\mathcal{Z},\kappa}_{(q,p)} \stackrel{\text{def}}{=} \mathrm{P}_\kappa\, \psi^{\mathcal{Z}}_{(q,p)} = \sum_{j=0}^{N-1} c_j(q,p)\, e^\kappa_j\,, \tag{16.25}$$

where the parameters (q,p) are now restricted to the torus, $(q,p) \in [0,a) \times [0,b)$, and the coefficients in the expansion formula are given by

$$c_j(q,p) = \left(\frac{\Im\mathcal{Z}}{\pi\hbar}\right)^{\frac{1}{4}} \sqrt{\frac{a}{N}}\, e^{-\frac{ipq}{2\hbar}} \sum_{m\in\mathbb{Z}} e^{i\kappa_1 ma}\, e^{-\frac{i}{\hbar}x^j_m}\, e^{-\frac{i\mathcal{Z}}{2\hbar}\left(x^j_m-q\right)^2},$$

$$\text{with}\quad x^j_m = j\frac{a}{N} + \frac{ab\kappa_2}{2\pi N} - ma\,. \tag{16.26}$$

The convergence of the series holds because of the condition $\Im\mathcal{Z} > 0$. These coefficients are suitably expressed in terms of the elliptic theta function of the third kind, ϑ_3 [209], which is defined as

$$\vartheta_3(x;\tau) = \sum_{n\in\mathbb{Z}} \tau^{n^2} e^{2inx}\,, \quad |\tau| < 1\,. \tag{16.27}$$

Therefore,

$$c_j(q,p) = \left(\frac{\Im\mathcal{Z}}{\pi\hbar}\right)^{\frac{1}{4}} \sqrt{\frac{a}{N}}\, e^{-\frac{ipq}{2\hbar}}\, e^{\frac{i}{\hbar}p\left(j\frac{a}{N}+\frac{ab}{2\pi N}\kappa_2-q\right)}\, \vartheta_3(X_j(q,p);\tau)\,, \tag{16.28}$$

$$\text{with} \quad X_j = \frac{a\kappa_1}{2} - \frac{a\,p}{2\hbar} - \frac{a\mathcal{Z}}{2\hbar}\left(j\frac{a}{N} + \frac{ab}{2\pi N}\kappa_2 - q\right), \qquad \text{and} \quad \tau = e^{i\frac{\mathcal{Z}a^2}{2\hbar}}. \tag{16.29}$$

The norm of the coherent state (16.25) is computed in accordance with (16.14):

$$\|\psi^{\mathcal{Z},\boldsymbol{\kappa}}_{(q,p)}\|^2_{\mathcal{H}_\hbar(\boldsymbol{\kappa})} = \left(\frac{\Im\mathcal{Z}}{\pi\hbar}\right)^{\frac{1}{2}} \frac{a}{N} \sum_{j=1}^{N-1} \left|\vartheta_3(X_j;\tau)\right|^2 \equiv \mathcal{N}(q,p). \tag{16.30}$$

In concordance with the general construction, illustrated in Chapter 3, of coherent states on a measure space, $\left(\mathbb{T}^2, \mu(d\mathbf{x} = dq\,dp/(2\pi\hbar))\right)$ in the present case, these coefficients form a finite orthogonal set in the Hilbert space $L^2\left(\mathbb{T}^2, dq\,dp/(2\pi\hbar)\right)$ and an orthogonal basis for the N-dimensional subspace isometric to $\mathcal{H}_\hbar(\boldsymbol{\kappa})$. The direct demonstration is rather cumbersome and we will omit it here. Furthermore, since one is working here with a representation of the states as tempered distributions, the question of normalization is not pertinent and it is preferable to appeal to the Schur lemma and discrete Weyl–Heisenberg unitary irreducible representation arguments to prove that the following resolution of the identity holds in $\mathcal{H}_\hbar(\boldsymbol{\kappa})$:

$$\int_{\mathbb{T}^2} \frac{dq\,dp}{2\pi\hbar} \left|\psi^{\mathcal{Z},\boldsymbol{\kappa}}_{(q,p)}\right\rangle \left\langle\psi^{\mathcal{Z},\boldsymbol{\kappa}}_{(q,p)}\right| = I_{\mathcal{H}_\hbar(\boldsymbol{\kappa})}. \tag{16.31}$$

The isometric map $W(\boldsymbol{\kappa},\mathcal{Z}) : \mathcal{H}_\hbar(\boldsymbol{\kappa}) \mapsto W(\boldsymbol{\kappa},\mathcal{Z})\,\mathcal{H}_\hbar(\boldsymbol{\kappa}) \subset L^2\left(\mathbb{T}^2, \frac{dq\,dp}{2\pi\hbar}\right)$ is precisely defined by

$$\mathcal{H}_\hbar(\boldsymbol{\kappa}) \ni \psi \mapsto W(\boldsymbol{\kappa},\mathcal{Z})\psi \stackrel{\text{def}}{=} \left\langle \psi^{\mathcal{Z},\boldsymbol{\kappa}}_{(q,p)} \mid \psi \right\rangle. \tag{16.32}$$

The range of $W(\boldsymbol{\kappa},\mathcal{Z})$ is a reproducing kernel Hilbert subspace of $L^2\left(\mathbb{T}^2, \frac{dq\,dp}{2\pi\hbar}\right)$.

Note two important properties of these coherent states, proved in [208], which will be used for deriving some estimates in the sequel.

An estimate The overlap of two coherent states obeys the following inequality:

$$\exists C > 0 \quad \text{such that} \quad \exists \alpha > 0 \left|\left\langle \psi^{\mathcal{Z},\boldsymbol{\kappa}}_{(q,p)} \middle| \psi^{\mathcal{Z}',\boldsymbol{\kappa}}_{(q',p')} \right\rangle\right| \le C\,e^{-\alpha N}, \tag{16.33}$$

where α is a function of $|q-q'|$ and $|p-p'|$.

A "classical" limit In the limit of large N and by fixing the product $N\hbar$, we get the following limit for the norm of coherent states:

$$\lim_{N\to\infty,\,N\hbar=\text{cst}} \|\psi^{\mathcal{Z},\boldsymbol{\kappa}}_{(q,p)}\|_{\mathcal{H}_\hbar(\boldsymbol{\kappa})} = 1. \tag{16.34}$$

16.5 Coherent States and Weyl Quantizations of the Torus

16.5.1 Coherent States (or Anti-Wick) Quantization of the Torus

Once the resolution of the identity (16.31) on the torus has been proved, it becomes possible to proceed to the coherent state (anti-Wick) quantization. With any essentially bounded observable $f(q,p)$, that is, $f(q,p) \in L^{\infty}(\mathbb{T}^2, \mu)$, we associate the operator

$$Op^{CS}_{\hbar,\boldsymbol{\kappa}}(f) \stackrel{\text{def}}{=} \int_{\mathbb{T}^2} \frac{dq\,dp}{2\pi\hbar} f(q,p) \left|\psi^{\mathcal{Z},\boldsymbol{\kappa}}_{(q,p)}\right\rangle \left\langle \psi^{\mathcal{Z},\boldsymbol{\kappa}}_{(q,p)}\right| . \tag{16.35}$$

This operator is defined for all $\boldsymbol{\kappa} \in \mathcal{T}^2$ and for all $\hbar > 0$. Let us be more precise by stating the following properties of the coherent state quantization in the present context:

(i) $Op^{CS\,\dagger}_{\hbar,\boldsymbol{\kappa}}(f) = Op^{CS}_{\hbar,\boldsymbol{\kappa}}(\overline{f})$,

(ii) If $f \in C^{\infty}(\mathbb{T}^2)$, then we can extend the quantization to the whole Hilbert space (16.17) as follows [208]:

$$Op^{CS}_{\hbar}(f) = \int_0^{\frac{2\pi}{a}} \int_0^{\frac{2\pi}{b}} \nu(d\boldsymbol{\kappa})\, Op^{CS}_{\hbar,\boldsymbol{\kappa}}(f),$$

where the operator $Op^{CS}_{\hbar}(f)$ results from the standard coherent state quantization,

$$Op^{CS}_{\hbar}(f) \stackrel{\text{def}}{=} \int_{\mathbb{R}^2} \frac{dq\,dp}{2\pi\hbar} f(q,p) \left|\psi^{\mathcal{Z}}_{(q,p)}\right\rangle \left\langle \psi^{\mathcal{Z}}_{(q,p)}\right| . \tag{16.36}$$

(iii) For all $f \in C^{\infty}(\mathbb{T}^2)$ and $\boldsymbol{\kappa} \in \mathcal{T}^2$, we have the estimate

$$\left\| Op^{CS}_{\hbar,\boldsymbol{\kappa}}(f) \right\|_{\mathcal{L}(\mathcal{H}_{\hbar}(\boldsymbol{\kappa}))} \leq \|f\|_{\infty} .$$

(iv) The following semiclassical behavior for the lower symbols holds:

$$\lim_{N\to\infty} \left\langle \psi^{\mathcal{Z},\boldsymbol{\kappa}}_{(q,p)} \middle| Op^{CS}_{\hbar,\boldsymbol{\kappa}}(f)\, \psi^{\mathcal{Z},\boldsymbol{\kappa}}_{(q,p)} \right\rangle = f(q,p) \quad \forall\,(q,p) \in \mathbb{T}^2 \text{ and } f \in C^{\infty}(\mathbb{T}^2). \tag{16.37}$$

16.5.2 Weyl Quantization of the Torus

The Weyl quantization for the motion of a particle on the line consists in mapping the space $\mathcal{S}^0$ of smooth functions that have bounded derivatives at any order,

$$\mathcal{S}^0 \stackrel{\text{def}}{=} \left\{ f \in \mathbb{C}^{\infty}(\mathbb{R}^2), \quad \text{all derivatives are bounded} \right\} , \tag{16.38}$$

into the space $\mathcal{L}(\mathcal{S}(\mathbb{R}))$ of bounded operators in the Schwartz space. It is defined as follows:

$$\mathcal{S}^0 \ni f \mapsto Op_\hbar^W(f) \in \mathcal{L}(\mathcal{S}(\mathbb{R}))\,,$$
$$\mathcal{S} \ni \psi \mapsto Op_\hbar^W(f)\,\psi \equiv \psi'\,,\quad \psi'(x) = \int_{\mathbb{R}} \frac{dy\,dp}{2\pi\hbar}\, e^{\frac{i}{\hbar}(x-y)p}\, f\left(\frac{x+y}{2}, p\right)\psi(y)\,. \tag{16.39}$$

Equivalently,

$$Op_\hbar^W(f) = \int_{\mathbb{R}^2} \frac{dq\,dp}{2\pi\hbar}\, \check{f}(q,p)\, e^{\frac{i}{\hbar}(pQ-qP)}\,, \tag{16.40}$$

where $\check{f}$ stands for the symplectic Fourier transform of f:

$$\check{f}(q,p) \stackrel{\text{def}}{=} \int_{\mathbb{R}^2} e^{\frac{i}{\hbar}(q\xi - px)}\, f(x,\xi)\,. \tag{16.41}$$

It has been proven [210] that the map $Op_\hbar^W$ has a unique extension to the tempered distributions $\mathcal{S}'(\mathbb{R})$ and yields bounded operators in $L^2(\mathbb{R})$.

The Weyl quantization on the torus $\mathbb{T}^2$ is just the particularization of (16.40) to the discrete Weyl–Heisenberg group:

$$\mathcal{S}'(\mathbb{R}^2) \supset \mathcal{S}^0(\mathbb{R}^2) \supset C^\infty(\mathbb{T}^2) \ni f \mapsto Op_\hbar^W(f)\,,$$
$$Op_\hbar^W(f) \stackrel{\text{def}}{=} \sum_{m,n} f_{m,n}\, e^{i\left(\frac{2\pi}{a}nQ - \frac{2\pi}{b}mP\right)} = \sum_{m,n} f_{m,n}\, U\left(\frac{ma}{N}, \frac{nb}{N}\right)\,, \tag{16.42}$$

with coefficients $f_{m,n}$ defined by $f(q,p) = \sum_{m,n} f_{m,n} e^{i\left(\frac{2\pi}{a}nq - \frac{2\pi}{b}mp\right)}$.

Like for the coherent state quantization, let us list the main properties of the Weyl quantization of the torus:

(i) The restriction of $Op_\hbar^W(f)$ to the Hilbert space $\mathcal{H}_\hbar(\boldsymbol{\kappa})$ is well defined from the fact that $Op_\hbar^W(f)\,\mathcal{H}_\hbar(\boldsymbol{\kappa}) \subset \mathcal{H}_\hbar(\boldsymbol{\kappa})$ for all $f \in \mathbb{C}^\infty(\mathbb{T}^2)$.

(ii) The following decomposition results:

$$Op_\hbar^W(f) = \int_0^{\frac{2\pi}{a}} \int_0^{\frac{2\pi}{b}} \nu(d\boldsymbol{\kappa})\, Op_{\hbar,\boldsymbol{\kappa}}^W(f)\,.$$

(iii) The following estimate for the difference between the Weyl quantization of the product of two functions and the product of the two respective Weyl quantizations of the functions holds:

For all $f, g \in C^\infty(\mathbb{T}^2)$ there exists $C > 0$ such that

$$\forall\, N \geq 1\,,\ \left\| Op_{\hbar,\boldsymbol{\kappa}}^W(f)\, Op_{\hbar,\boldsymbol{\kappa}}^{W\,\dagger}(g) - Op_{\hbar,\boldsymbol{\kappa}}^W(f\,\bar{g}) \right\|_{\mathcal{L}(\mathcal{H}_\hbar(\boldsymbol{\kappa}))} \leq \frac{C}{N}\,.$$

(iv) In the limit $N \to \infty$, we have for the trace

$$\lim_{N\to\infty} \frac{1}{N} \operatorname{tr}\left(Op^{W}_{\hbar,\kappa}(f)\right) = \int_{\mathbb{T}^2} \frac{dq\,dp}{ab} f(q,p)\,.$$

(v) There exists the following estimate for the difference between Weyl and coherent state quantizations:

$$\left\| Op^{W}_{\hbar,\kappa}(f) - Op^{CS}_{\hbar,\kappa}(f) \right\|_{\mathcal{L}(\mathcal{H}_\hbar(\kappa))} = O\left(N^{-1}\right)\,. \tag{16.43}$$

In consequence, it can be asserted that the Weyl quantization and the coherent state quantization are identical up to $\hbar$:

$$\left\| Op^{W}_{\hbar}(f) - Op^{CS}_{\hbar}(f) \right\| = O(\hbar)\,. \tag{16.44}$$

16.6 Quantization of Motions on the Torus

We are now equipped to proceed with the quantization of the symplectic transformations of the torus introduced in the first section of this chapter.

16.6.1 Quantization of Irrational and Skew Translations

These transformations are proved to be ergodic and *uniquely ergodic* in the sense that there exists one and only one Φ-invariant probability measure.

Any irrational translation $\tau_{\boldsymbol{\alpha}}$, $\boldsymbol{\alpha} = (\alpha_1, \alpha_2)$, $\alpha_1/\alpha_2 \notin \mathbb{Q}$, is unitarily represented by the operator $M_\kappa(\tau_{\boldsymbol{\alpha}})$ on $\mathcal{H}_\hbar(\kappa)$ defined as

$$M_\kappa(\tau_{\boldsymbol{\alpha}}) = e^{i\frac{\alpha_1\alpha_2}{2\hbar}} M_\kappa\left(\tau_{(\alpha_1,0)}\right) M_\kappa\left(\tau_{(0,\alpha_2)}\right)\,, \tag{16.45}$$

with

$$M_\kappa\left(\tau_{(0,\alpha_2)}\right) e^\kappa_j \stackrel{\text{def}}{=} e^{i\left(\kappa_2 + \frac{2\pi}{b} j\right)\alpha_1} e^\kappa_j\,,$$

and

$$M_\kappa\left(\tau_{(\alpha_1,0)}\right) f^\kappa_j \stackrel{\text{def}}{=} e^{-i\left(\kappa_1 + \frac{2\pi}{a} j\right)\alpha_1} f^\kappa_j\,.$$

The decomposition formula and the eigenequations are consistent with $U(\alpha_1,\alpha_2) = e^{i\frac{\alpha_1\alpha_2}{2\hbar}} U(\alpha_1,0)\, U(0,\alpha_2)$ and the fact that $U(\alpha_1,0)$ (resp. $U(0,\alpha_2)$) is diagonal in the momentum (or spatial) representation.

Concerning the skew translations, we have the representation

$$\Phi_{\beta,k} \mapsto M_\kappa\left(\Phi_{\beta,k}\right) = M_\kappa\left(\tau_{(0,\beta)}\right) \circ M_\kappa(K)\,, \tag{16.46}$$

where $M_\kappa(K)$, for

$$K = \begin{pmatrix} 1 & k \\ 0 & 1 \end{pmatrix}$$

realizes a quantization of the group $SL(2,\mathbb{Z})$ that is explained in the next paragraph.

16.6.2
Quantization of the Hyperbolic Automorphisms of the Torus

The task now is to build a unitary representative of the hyperbolic transformation of the torus:

$$A = \begin{pmatrix} \alpha & \beta\frac{a}{b} \\ \gamma\frac{b}{a} & \delta \end{pmatrix} \mapsto \text{unitary } M(A)\,.$$

It is proven that for all $\hbar > 0$ and for all $\boldsymbol{\kappa} \in \mathcal{T}^2$, there exists $\boldsymbol{\kappa}' \in \mathcal{T}^2$ such that $M(A)\,\mathcal{H}_\hbar(\boldsymbol{\kappa}) \subset \mathcal{H}_\hbar(\boldsymbol{\kappa}')$, where

$$\begin{pmatrix} \kappa_1' \\ \kappa_2' \end{pmatrix} = A \begin{pmatrix} \kappa_1 \\ \kappa_2 \end{pmatrix} + \pi N \begin{pmatrix} \frac{\alpha\beta}{b} \\ \frac{\gamma\delta}{a} \end{pmatrix} \mod \begin{pmatrix} \frac{2\pi}{b} \\ \frac{2\pi}{a} \end{pmatrix},$$

and where the map

$$A = \begin{pmatrix} a_1 & a_2 \\ a_3 & a_4 \end{pmatrix} \mapsto M(A)$$

is, in general, defined on any function $\psi \in \mathcal{S}(\mathbb{R})$ by

$$(M(A)\,\psi)(x) = \left(\frac{i}{2\pi\hbar}\right)^{1/2} \int_{-\infty}^{+\infty} e^{\frac{i}{\hbar}S(x,y)}\,\psi(y)\,dy\,,$$
$$S(x,y) \stackrel{\text{def}}{=} \frac{1}{2}\frac{a_4}{a_2}x^2 - \frac{1}{a_2}xy + \frac{1}{2}\frac{a_1}{a_2}y^2\,. \tag{16.47}$$

The operator $M(A)$ is the *quantum propagator* associated with the classical discrete dynamics defined by the matrix A. It is extended by duality to the space of tempered distributions $\mathcal{S}'(\mathbb{R})$. It fulfills the important intertwining property

$$M(A)\,U(q,p)\,(M(A))^\dagger = U\left(A\begin{pmatrix} q \\ p \end{pmatrix}\right). \tag{16.48}$$

Let us now restrict the transformations A to be the element of the group $SL(2,\mathbb{Z})$ and such that $|\,\mathrm{Tr}\,A| > 2$. The following was proved in [208]:

Proposition 16.1

For all $\hbar > 0$, there exists $\boldsymbol{\kappa} \in \mathcal{T}^2$ such that the unitary representatives $M(A)$ stabilize the Hilbert space $\mathcal{H}_\hbar(\boldsymbol{\kappa})$ of quantum states on the torus,

$$M(A)\,\mathcal{H}_\hbar(\boldsymbol{\kappa}) \subset \mathcal{H}_\hbar(\boldsymbol{\kappa})\,,$$

if and only if $A \in SL(2,\mathbb{Z})$ has the form

$$A = \begin{pmatrix} \text{even} & \text{odd} \times \frac{a}{b} \\ \text{odd} \times \frac{b}{a} & \text{even} \end{pmatrix} \quad \text{or} \quad A = \begin{pmatrix} \text{odd} & \text{even} \times \frac{a}{b} \\ \text{even} \times \frac{b}{a} & \text{odd} \end{pmatrix}.$$

Note that the 2-component momentum $\boldsymbol{\kappa}$ can be chosen independently of $\hbar$.

16.6.3 Main Results

The applications of the previous results concern particularly the control of the perturbations of the hyperbolic automorphisms of the torus $\mathbb{T}^2$, the demonstration of a semiclassical Egorov theorem on the relation between quantization and temporal evolution, the propagation of coherent states on the torus, and the semiclassical behavior of the spectra of evolution operators of the type $M_{\kappa}(\Phi) \in \mathcal{U}\left(\mathcal{H}_{\hbar}(\kappa)\right)$.

Let us say more about this semiclassical Egorov theorem. It means that quantization and temporal evolution commute up to $\hbar$. More precisely,

Theorem 16.2

Let H be a smooth observable on the torus, that is, $H \in C^{\infty}(\mathbb{T}^2)$. For all $f \in C^{\infty}(\mathbb{T}^2)$, for all $t \in \mathbb{R}$, and for all $\kappa \in \mathcal{T}^2$, we have the following estimate:

$$\left\| e^{-i\frac{t}{\hbar}Op_{\hbar,\kappa}(H)}\, Op_{\hbar,\kappa}(f)\, e^{-i\frac{t}{\hbar}Op_{\hbar,\kappa}(H)} - Op_{\hbar,\kappa}\left(f \circ \psi_t^H\right) \right\|_{\mathcal{L}(\mathcal{H}_{\hbar}(\kappa))} = O(N^{-1})\,, \tag{16.49}$$

where ψ_t^H is the classical flow associated with the observable H, and the notation $Op_{\hbar,\kappa}$ stands for both types of quantization (coherent state or Weyl).

This theorem is exact in the case of hyperbolic automorphisms of the torus and the Weyl quantization:

$$Op^W_{\hbar,\kappa}(f \circ A) = (M(A))^{\dagger}\, Op^W_{\hbar,\kappa}(f) M(A)\,. \tag{16.50}$$

Concerning the spectral properties of the evolution operator, one gets precise information on the semiclassical behavior of the eigenvalues $e^{i\theta_j^N}$, $1 \le j \le N_1$, and the corresponding eigenfunctions $\phi^{i\theta_j^N}$, $1 \le j \le N_1$, of the unitary operators $M_{\kappa}(\Phi)$ representative of automorphisms $\Phi : \mathbb{T}^2 \mapsto \mathbb{T}^2$.

Proposition 16.2

[[211]] Let Φ a hyperbolic automorphism of the torus. Then the eigenvalues of the unitary representative $M_{\kappa}(\Phi)$ are Lebesgue uniformly distributed on the unit circle at the semiclassical limit:

$$\lim_{N \to \infty} \frac{\#\{j,\ \theta_j^N \in [\theta_0, \theta_0 + \eta]\}}{N} = \lambda[\theta_0, \theta_0 + \eta]\,, \tag{16.51}$$

where λ is the Lebesgue measure.

This result also holds for the Hamiltonian perturbations of Φ.

Finally, let us end this chapter by formulating the essential equipartition result in regard to the initial purpose to examine ergodicity on the torus on a semiclassical level. This amounts to examining the possible limits at $N \to \infty$ of the matrix elements (and particularly the lower symbols)

$$\langle \phi_k^N | Op_{\hbar,\kappa}(f) | \phi_j^N \rangle$$

of the quantized versions of observables for both types of quantizations.

Theorem 16.3

Suppose that for all smooth observable f on the torus, that is, $f \in C^\infty(\mathbb{T}^2)$, for all $k \in \mathbb{N}$, there exists $C > 0$ so that for all N

$$\left\| M_\kappa(\Phi)^{-k} \, Op_{\hbar,\kappa}(f) \, M_\kappa(\Phi)^k - Op_{\hbar,\kappa}\left(f \circ \Phi^k\right) \right\|_{\mathcal{L}(\mathcal{H}_\hbar(\kappa))} \leq \frac{C}{N} . \tag{16.52}$$

Here Φ is an area-preserving map on the torus such that $\forall N \exists \boldsymbol{\kappa} \in \left[[0, \frac{2\pi}{a}\right) \times \left[[0, \frac{2\pi}{b}\right)$ and the unitary representative $M_\kappa(\Phi)$ satisfies (16.52).

Write

$$M_\kappa(\Phi) \, \phi_j^N = \lambda_j^N \, \phi_j^N$$

for the eigenvalues and eigenfunctions of $M_\kappa(\Phi)$.

Then there exists a set of indices $E(N) \subset [1, N]$ satisfying $\lim_{N\to\infty} \frac{\# E(N)}{N} = 1$ such that, for all functions $f \in C^\infty(\mathbb{T}^2)$ and for all maps $\mathcal{J} : N \in \mathbb{N}^* \mapsto \mathcal{J}(N) \in E(N)$, we have the ergodicity property of the lower symbols:

$$\lim_{N\to\infty} \langle \phi_{\mathcal{J}(N)}^N | \, Op_{\hbar,\kappa}(f) \, | \phi_{\mathcal{J}(N)}^N \rangle = \int_{\mathbb{T}^2} f(q, p) \, \frac{dq \, dp}{ab} , \tag{16.53}$$

uniformly with respect to the map $\mathcal{J}$.

17
Fuzzy Geometries: Sphere and Hyperboloid

17.1
Introduction

This is an extension to the sphere S^2 and to the one-sheeted hyperboloid in $\mathbb{R}^3$, the "1 + 1 de Sitter space-time," of the quantization of the unit circle as was described in Chapter 14. We show explicitly how the coherent state quantization of these manifolds leads to fuzzy or noncommutative geometries. The quantization of the sphere rests upon the unitary irreducible representations $\mathcal{D}^j$ of $SO(3)$ or $SU(2)$, $j \in \mathbb{N}/2$. The quantization of the hyperboloid is carried out with unitary irreducible representations of $SO_0(1,2)$ or $SU(1,1)$ in the principal series, namely, $\mathcal{U}^{-\frac{1}{2}+i\varrho}$, $\varrho \in \mathbb{R}$. The limit at infinite values of the representation parameters j and ϱ, restores commutativity for the geometries considered.

It should be stressed that we proceed to a "noncommutative" reading of a given geometry and not of a given dynamical system. This means that we do not consider in our approach any time parameter and related evolution.

17.2
Quantizations of the 2-Sphere

In this section, we proceed to the fuzzy quantization [212] of the sphere S^2 by using the coherent states built in Chapter 6 from orthonormal families of spin spherical harmonics $\left({}_\sigma Y_{jm}\right)_{-j\leqslant m\leqslant j}$. We recall that for a given σ such that $2\sigma \in \mathbb{Z}$ and j such that $2|\sigma| \leqslant 2j \in \mathbb{N}$ there corresponds the continuous family of coherent states (6.9) living in a $(2j+1)$-dimensional Hermitian space. For a given j, we thus get $2j+1$ realizations, corresponding to the possible values of σ. These realizations yield a family of $2j+1$ nonequivalent quantizations of the sphere. We show in particular that the case $\sigma = 0$ is singular in the sense that it maps the Cartesian coordinates of the 2-sphere to null operators.

We then establish the link between this coherent state quantization approach to the 2-sphere and the original Madore construction [178] of the fuzzy sphere and we examine the question of equivalence between the two procedures. Note that a construction of the fuzzy sphere based on Gilmore–Perelomov–Radcliffe coher-

Coherent States in Quantum Physics. Jean-Pierre Gazeau

ISBN: 978-3-527-40709-5

ent states (in the case $\sigma = j$) was also carried out by Grosse and Prešnajder [213]. They proceeded with a covariant symbol calculus à la Berezin with its corresponding $\star$-product. However, their approach is different from the coherent state quantization illustrated here.

Appendix C contains a set of formulas concerning the group $SU(2)$, its unitary representations, and the spin spherical harmonics, specially needed for a complete description of our coherent state approach to the 2-sphere.

17.2.1
The 2-Sphere

We now apply our coherent state quantization to the unit sphere S^2 viewed as an observation set. Our approach should not be confused with the quantization of the phase space for the motion on the 2-sphere (i.e., quantum mechanics on the 2-sphere; see, e.g., [214, 216]). A point of X is denoted by its spherical coordinates, $x \equiv \hat{r} = (\theta, \phi)$. Through the usual embedding in $\mathbb{R}^3$, we may view x as a vector $\hat{r} = (x^i) \in \mathbb{R}^3$, the "ambient" space, obeying $\sum_{i=1}^{3}(x^i)^2 = 1$. We equip S^2 with the usual (nonnormalized) $SO(3)$-invariant measure $\mu(dx) = \sin\theta d\theta d\phi$.

17.2.2
The Hilbert Space and the Coherent States

At the basis of the coherent state quantization procedure is the choice of the Hermitian space $\mathcal{H} = \mathcal{H}^{\sigma j}$, which is a subspace of $L^2(S^2)$, and which carries a $(2j+1)$-dimensional unitary irreducible representation of the group $SU(2)$. We recall that $\mathcal{H} = \mathcal{H}^{\sigma j}$ is the vector space spanned by the spin spherical harmonics ${}_\sigma Y_{j\mu} \in L^2(S^2)$. With the usual inner product of $L^2(S^2)$, the spin spherical harmonics provide an orthonormal basis of $\mathcal{H}^{\sigma j}$. We also recall that the special case $\sigma = 0$ corresponds to the ordinary spherical harmonics ${}_0 Y_{jm} = Y_{jm}$.

The spin spherical harmonic basis allows us to identify $\mathcal{H}^{\sigma j}$ with $\mathbb{C}^{2j+1}$:

$$
{}_\sigma Y_{j\mu} \rightsquigarrow |\mu\rangle \hookrightarrow (0, \dots, 0, 1, 0, \dots, 0)^t\,, \quad \text{with} \quad \mu = -j, -j+1, \dots, j\,, \tag{17.1}
$$

where the 1 is at position μ and the superscript t denotes the transpose.

The normalized sigma-spin coherent states associated with the spin spherical harmonics are given by

$$
|x\rangle \equiv |\hat{r};\sigma\rangle = |\theta, \phi; \sigma\rangle = \frac{1}{\sqrt{\mathcal{N}(\hat{r})}} \sum_{\mu=-j}^{j} \overline{{}_\sigma Y_{j\mu}(\hat{r})} |\sigma j \mu\rangle\,, \quad |\hat{r};\sigma\rangle \in \mathcal{H}_{\sigma j}\,, \tag{17.2}
$$

with

$$
\mathcal{N}(\hat{r}) = \sum_{\mu=-j}^{j} |{}_\sigma Y_{j\mu}(\hat{r})|^2 = \frac{2j+1}{4\pi}\,.
$$

In the sequel, we will keep the shortcup notation $|x\rangle$ for $|\hat{r};\sigma\rangle$ as far as possible.

17.2.3
Operators

Let us denote by $\mathcal{O}^{\sigma j} \equiv \mathrm{End}(\mathcal{H}^{\sigma j})$ the space of linear operators (endomorphisms) acting on $\mathcal{H}^{\sigma j}$. This is a complex vector space of dimension $(2j+1)^2$ and an algebra for the natural composition of endomorphisms. The spin spherical harmonic basis allows us to write a linear endomorphism of $\mathcal{H}^{\sigma j}$ (i. e. an element of $\mathcal{O}^{\sigma j}$) in a matrix form. This provides the algebra isomorphism

$$\mathcal{O}^{\sigma j} \cong M(2j+1, \mathbb{C}) \equiv M_{2j+1} ,$$

the algebra of complex matrices of order $2j+1$, equipped with the matrix product.

The projector $|x\rangle\langle x|$ is a particular linear endomorphism of $\mathcal{H}^{\sigma j}$, that is, an element of $\mathcal{O}^{\sigma j}$. Being Hermitian by construction, it may be seen as a Hermitian matrix of order $2j+1$, that is, an element of $\mathrm{Herm}_{2j+1} \subset M_{2j+1}$. Note that Herm_{2j+1} and M_{2j+1} have respective (complex) dimensions $(j+1)(2j+1)$ and $(2j+1)^2$.

17.2.4
Quantization of Observables

According to our now-familiar prescription, the coherent state quantization associates with the *classical* observable $f : S^2 \mapsto \mathbb{C}$ the *quantum* observable

$$\begin{aligned} f \mapsto A_f &= \int_{S^2} \mu(dx) f(x) \mathcal{N}(x) |x\rangle\langle x| \\ &= \sum_{\mu,\nu=-j}^{j} \int_{S^2} \mu(dx) f(x) \overline{{}_\sigma Y_{j\mu}(x)}\, {}_\sigma Y_{j\nu}(x) |\mu\rangle\langle\nu| . \end{aligned} \tag{17.3}$$

This operator is an element of $\mathcal{O}^{\sigma j}$. Of course its existence is subjected to the convergence of (17.3) in the weak sense as an operator integral. The expression above gives directly its expression as a matrix in the spin spherical harmonic basis, with matrix elements $(A_f)_{\mu\nu}$:

$$A_f = \sum_{\mu,\nu=-j}^{j} (A_f)_{\mu\nu} |\mu\rangle\langle\nu|$$

$$\text{with} \quad (A_f)_{\mu\nu} = \int_{S^2} \mu(dx) f(x) \overline{{}_\sigma Y_{j\mu}(x)}\, {}_\sigma Y_{j\nu}(x) . \tag{17.4}$$

When f is real-valued, the corresponding matrix belongs to $\mathrm{Herm}_{(2j+1)}$. Also, we have $A_{\bar{f}} = (A_f)^\dagger$ (matrix transconjugate), where we have used the same notation for the operator and the associated matrix.

Note the interesting expression of the lower symbol of A_f, easily obtained from the expression (6.81) of the overlap of two sigma-spin coherent states, and viewed as a sort of Jacobi transform on the sphere of the function f:

$$\begin{aligned}\langle x|A_f|x\rangle &= \int_{S^2} \mu(dx')f(x')\mathcal{N}(x')|\langle x|x'\rangle|^2 \\ &= \frac{2j+1}{4\pi}\int_{S^2} \sin\theta\, d\theta\, d\phi \left(\frac{1+\hat{r}\cdot\hat{r}'}{2}\right)^{2\sigma} \left|P_{j-\sigma}^{(0,2\sigma)}(\hat{r}\cdot\hat{r}')\right|^2 f(\hat{r}')\,.\end{aligned} \tag{17.5}$$

Also note that the map (17.3) can be extended to a class of distributions on the sphere, in the spirit of Chapter 12.

17.2.5
Spin Coherent State Quantization of Spin Spherical Harmonics

The quantization of an arbitrary spin harmonic ${}_\nu Y_{kn}$ yields an operator in $\mathcal{H}^{\sigma j}$ whose $(2j+1)\times(2j+1)$ matrix elements are given by the following integral resulting from (17.3):

$$\begin{aligned}\left[A_{{}_\nu Y_{kn}}\right]_{\mu\mu'} &= \int_X {}_\sigma Y^*_{j\mu}(x)\,{}_\sigma Y_{j\mu'}(x)\,{}_\nu Y_{kn}(x)\mu(dx) \\ &= \int_X (-1)^{\sigma-\mu}{}_{-\sigma}Y_{j-\mu}(x)\,{}_\sigma Y_{j\mu'}(x)\,{}_\nu Y_{kn}(x)\,\mu(dx)\,.\end{aligned} \tag{17.6}$$

17.2.6
The Usual Spherical Harmonics as Classical Observables

As asserted in Appendix C (Section C.7), it is only when $\nu-\sigma+\sigma=0$, that is, when $\nu=0$, that the integral (17.6) is given in terms of a product of two $3j$ symbols. Therefore, the matrix elements of $A_{Y_{\ell m}}$ in the spin spherical harmonic basis are given in terms of the $3j$ symbols by

$$\left(A_{Y_{\ell m}}\right)_{\mu\nu} = (-1)^{\sigma-\mu}(2j+1)\sqrt{\frac{(2\ell+1)}{4\pi}}\begin{pmatrix} j & j & \ell \\ -\mu & \nu & m\end{pmatrix}\begin{pmatrix} j & j & \ell \\ -\sigma & \sigma & 0\end{pmatrix}. \tag{17.7}$$

This generalizes formula (2.7) of [217]. This expression is a real quantity.

17.2.7
Quantization in the Simplest Case: $j = 1$

In the simplest case $j=1$, we find for the matrix elements (17.7)

$$\left[A_{Y_{10}}\right]_{mn} = \sigma\sqrt{\frac{3}{4\pi}}\frac{1}{j(j+1)}m\delta_{mn}\,, \tag{17.8}$$

$$\left[A_{Y_{11}}\right]_{mn} = -\sigma\sqrt{\frac{3}{4\pi}}\frac{1}{j(j+1)}\sqrt{\frac{(j-n)(j+n+1)}{2}}\delta_{mn+1}\,, \tag{17.9}$$

$$\left[A_{Y_{1-1}}\right]_{mn} = \sigma\sqrt{\frac{3}{4\pi}}\frac{1}{j(j+1)}\sqrt{\frac{(j+n)(j-n+1)}{2}}\delta_{mn-1}\,. \tag{17.10}$$

From a comparison with the actions (6.75a–6.75c) of the spin angular momentum on the spin-σ spherical harmonics, we have the identification:

$$A_{Y_{10}} = \sigma\sqrt{\frac{3}{4\pi}}\frac{1}{j(j+1)}\Lambda_3^{\sigma j}\,, \tag{17.11}$$

$$A_{Y_{11}} = -\sigma\sqrt{\frac{3}{8\pi}}\frac{1}{j(j+1)}\Lambda_+^{\sigma j}\,, \tag{17.12}$$

$$A_{Y_{1-1}} = \sigma\sqrt{\frac{3}{8\pi}}\frac{1}{j(j+1)}\Lambda_-^{\sigma j}\,. \tag{17.13}$$

The remarkable identification of the quantized versions of the components of $\hat{r} = (x^i)$ pointing to S^1 with the components of the spin angular momentum operator results:

$$A_{x^a} = K\Lambda_a^{\sigma j}\,, \quad \text{with} \quad K \equiv \frac{\sigma}{j(j+1)}\,. \tag{17.14}$$

17.2.8
Quantization of Functions

Any function f on the 2-sphere with reasonable properties (continuity, integrability, and so on) may be expanded in spherical harmonics as

$$f = \sum_{\ell=0}^{\infty}\sum_{m=-\ell}^{\ell} f_{\ell m} Y_{\ell m}\,, \tag{17.15}$$

from which follows the corresponding expansion of A_f. However, the $3j$ symbols are nonzero only when a triangular inequality is satisfied. This implies that the expansion is truncated at a finite value, giving

$$A_f = \sum_{\ell=0}^{2j}\sum_{m=-\ell}^{\ell} f_{\ell m} A_{Y_{\ell m}}\,. \tag{17.16}$$

This relation means that the $(2j+1)^2$ observables $(A_{Y_{\ell m}})$, $\ell \leqslant 2j$, $-\ell \leqslant m \leqslant \ell$ provide a second basis of $\mathcal{O}^{\sigma j}$.

The $f_{\ell m}$ are the components of the matrix $A_f \in \mathcal{O}^{\sigma j}$ in this basis.

17.2.9
The Spin Angular Momentum Operators

17.2.9.1
Action on Functions

The Hermitian space $\mathcal{H}^{\sigma j}$ carries a unitary irreducible representation of the group $SU(2)$ with generators $\Lambda_a^{\sigma j}$ defined in (6.71–6.73). The latter belong to $\mathcal{O}^{\sigma j}$. Their action is given in (6.75a–6.75c). and the above calculations have led to the crucial relations (17.14). We see here the peculiarity of the ordinary spherical harmonics ($\sigma = 0$) as an orthonormal system for the quantization procedure: they would lead to a trivial result for the quantized version of the Cartesian coordinates! On the other hand, the quantization based on the Gilmore–Radcliffe spin coherent states, $\sigma = j$, yields the maximal value: $K = 1/(j+1)$. Hereafter we assume $\sigma \neq 0$.

17.2.9.2
Action on Operators

The $SU(2)$ action on $\mathcal{H}^{\sigma j}$ induces the following canonical (infinitesimal) action on $\mathcal{O}^{\sigma j} = \mathrm{End}(\mathcal{H}^{\sigma j})$

$$\mathcal{L}_a^{\sigma j} :\mapsto \mathcal{L}_a^{\sigma j} A \equiv [\Lambda_a^{\sigma j}, A] \quad \text{(the commutator)}\,, \tag{17.17}$$

here expressed through the generators.

We prove in Appendix D, (D4), that

$$\mathcal{L}_a^{\sigma j} A_{Y_{\ell m}} = A_{J_a Y_{\ell m}}\,, \tag{17.18}$$

from which we get

$$\mathcal{L}_3^{\sigma j} A_{Y_{\ell m}} = m A_{Y_{\ell m}} \quad \text{and} \quad (\mathcal{L}^{\sigma j})^2 A_{Y_{\ell m}} = \ell(\ell+1) A_{Y_{\ell m}}\,. \tag{17.19}$$

We recall that the $(A_{Y_{\ell m}})_{\ell \leqslant 2j}$ form a basis of $\mathcal{O}^{\sigma j}$. The relations above make $A_{Y_{\ell m}}$ appear as the unique (up to a constant) element of $\mathcal{O}^{\sigma j}$ that is common eigenvector to $\mathcal{L}_3^{\sigma j}$ and $(\mathcal{L}^{\sigma j})^2$, with eigenvalues m and $\ell(\ell+1)$ respectively. This implies by linearity that, for all f such that A_f makes sense,

$$\mathcal{L}_a^{\sigma j} A_f = A_{J_a f} \quad \text{and} \quad (\mathcal{L}^{\sigma j})^2 A_f = A_{J^2 f}\,. \tag{17.20}$$

17.3
Link with the Madore Fuzzy Sphere

17.3.1
The Construction of the Fuzzy Sphere à la Madore

Let us first recall the Madore construction of the fuzzy sphere as it was originally presented in [178] (p. 148), which we slightly modify to make the correspondence with the coherent state quantization. It starts from the expansion of any smooth function $f \in C^\infty(S^2)$ in terms of spherical harmonics,

$$f = \sum_{\ell=0}^{\infty} \sum_{m=-\ell}^{\ell} f_{\ell m} Y_{\ell m}\,. \tag{17.21}$$

Let us denote by V^ℓ the $(2\ell+1)$-dimensional vector space generated by the $Y_{\ell m}$, at fixed ℓ.

Through the embedding of S^2 into $\mathbb{R}^3$ any function in S^2 can be considered as the restriction of a function on $\mathbb{R}^3$ (which we write with the same notation), and, under some mild conditions, such functions are generated by the homogeneous polynomials in $\mathbb{R}^3$. This allows us to express (17.21) in a polynomial form in $\mathbb{R}^3$:

$$f(x) = f_{(0)} + \sum_{(i_1)} f_{(i)} x^i + \ldots + \sum_{(i_1 i_2 \ldots i_\ell)} f_{(i_1 i_2 \ldots i_\ell)} x^{i_1} x^{i_2} \ldots x^{i_\ell} + \ldots , \qquad (17.22)$$

where each subsum is restricted to a V^ℓ and involves all symmetric combinations of the i_k indices, each one varying from 1 to 3. This gives, for each fixed value of ℓ, $2\ell + 1$ coefficients $f_{(i_1 i_2 \ldots i_\ell)}$ (ℓ fixed), which are those of a symmetric traceless $3 \times 3 \times \ldots \times 3$ (ℓ times) tensor.

The fuzzy sphere with $2j+1$ cells is usually written as $\left(S_{\text{fuzzy}}, j\right)$, with j an integer or semi-integer. We here extend the Madore procedure and this leads to a σ-indexed family of fuzzy spheres $\left\{\left({}_\sigma S_{\text{fuzzy}}, j\right)\right\}$. Let us list the steps of this construction:

1. We consider a $(2j + 1)$-dimensional irreducible unitary representation of $SU(2)$. The standard construction considers the vector space V^j of dimension $2j + 1$, on which the three generators of $SU(2)$ are expressed as the usual $(2j+1) \times (2j+1)$ Hermitian matrices J_a. We make instead a different choice, namely, the three $\Lambda_a^{\sigma j}$, which correspond to the choice of the representation space $\mathcal{H}^{\sigma j}$ (instead of V^j in the usual construction). Since they obey the commutation relations of $\mathfrak{su}(2)$,

$$[\Lambda_a^{\sigma j}, \Lambda_b^{\sigma j}] = i\varepsilon_{abc}\Lambda_c^{\sigma j}, \qquad (17.23)$$

   the usual procedure may be applied. As we have seen, $\mathcal{H}^{\sigma j}$ can be realized as the Hilbert space spanned by the orthonormal basis of spin spherical harmonics $\{{}_\sigma Y_{j\mu}\}_{\mu=-j \ldots j}$, with the usual inner product.

   Since the standard derivation of all properties of the fuzzy sphere rests only upon the abstract commutation rules (17.23), nothing but the representation space changes if we adopt the representation space $\mathcal{H}$ instead of V.
2. The operators $\Lambda_a^{\sigma j}$ belong to $\mathcal{O}^{\sigma j}$ and have a Lie algebra structure through the skew products defined by the commutators. But the *symmetrized* products of operators provide a second algebra structure, which we write as $\mathcal{O}^{\sigma j}$, at the basis of the construction of the fuzzy sphere: these symmetrized products of the $\Lambda_a^{\sigma j}$, up to power $2j$, generate the algebra $\mathcal{O}^{\sigma j}$ (of dimension $(2j+1)^2$) of all linear endomorphisms of $\mathcal{H}^{\sigma j}$, exactly like the ordinary J_a's do in the original Madore construction. This is the analog of the standard construction of the fuzzy sphere, with the J_a and V^j replaced by $\Lambda_a^{\sigma j}$ and $\mathcal{H}^{\sigma j}$.
3. The construction of the fuzzy sphere of radius R is defined by associating an operator $\hat{f}$ in $\mathcal{O}^{\sigma j}$ with any function f. Explicitly, this is done by first replacing each coordinate x^i by the operator

$$\hat{x}^a \equiv \kappa \Lambda_a^{\sigma j}, \quad \text{with} \quad \kappa = \frac{R}{\sqrt{j(j+1)}} \qquad (17.24)$$

in the above expansion (17.22) of f (in the usual construction, this would be J_a instead of $\Lambda_a^{\sigma j}$). We immediately note that the three operators (17.24) obey the constraint

$$(\hat{x}^a)^2 + (\hat{x}^a)^2 + (\hat{x}^a)^2 = R^2 \tag{17.25}$$

and the commutation rules

$$\left[\hat{x}^a, \hat{x}^b\right] = i\varepsilon^{ab}{}_c\, \hat{x}^c\,. \tag{17.26}$$

This is the "noncommutative sphere of radius R".

4. Next, we replace in (17.22) the usual product by the symmetrized product of operators, and we truncate the sum at index $\ell = 2j$. This associates with any function f an operator $\hat{f} \in \mathcal{O}^{\sigma j}$.
5. The vector space M_{2j+1} of $(2j+1) \times (2j+1)$ matrices is linearly generated by a number $(2j+1)^2$ of independent matrices. According to the above construction, a basis of M_{2j+1} can be selected as formed by all the products of the $\Lambda_a^{\sigma j}$ up to power $2j+1$ (which is necessary and sufficient to close the algebra).
6. The commutative algebra limit is restored by letting j go to infinity while parameter κ goes to zero and κj is fixed to $\kappa j = R$.

The geometry of the fuzzy sphere $\left(S_{\text{fuzzy}}, j\right)$ is thus constructed after making the choice of the algebra of the matrices of the representation, with their matrix product. It is taken as the algebra of operators, which generalize the functions. The rank $(2j+1)$ of the matrices allows us to view them as endomorphisms in a Hermitian space of dimension $(2j+1)$. This is exactly what allows the coherent state quantization introduced in the previous section.

17.3.2
Operators

We have defined the action on $\mathcal{O}^{\sigma j}$:

$$\mathcal{L}_a^{\sigma j} A \equiv [\Lambda_a^{\sigma j}, A]\,.$$

The formula (17.22) expresses any function f of V^ℓ as the reduction to S^2 of a homogeneous polynomial of order ℓ:

$$f = \sum_{\alpha,\beta,\gamma} f_{\alpha,\beta,\gamma}(x^1)^\alpha(x^2)^\beta(x^3)^\gamma;\, \alpha+\beta+\gamma=\ell\,.$$

The action of the ordinary momentum operators J_3 and J^2 is straightforward. Namely,

$$J_3 f = \sum_{\alpha,\beta,\gamma} f_{\alpha,\beta,\gamma}(-i)\left[\beta(x^1)^{\alpha+1}(x^2)^{\beta-1}(x^3)^\gamma - \alpha(x^1)^{\alpha-1}(x^2)^{\beta+1}(x^3)^\gamma\right]\,, \tag{17.27}$$

and similarly for J_1 and J_2.

On the other hand, we have by definition

$$\hat{f} = \sum_{\alpha,\beta,\gamma} f_{\alpha,\beta,\gamma} \mathcal{S}\left((\hat{x^1})^\alpha (\hat{x^2})^\beta (\hat{x^3})^\gamma \right), \tag{17.28}$$

where $\mathcal{S}(\cdot)$ means symmetrization. Recalling $\hat{x}^a = \kappa \Lambda_a^{\sigma j}$, and using (17.23), we apply the operator $\mathcal{L}_3^{\sigma j}$ to this expression:

$$\mathcal{L}_3^{\sigma j} \hat{f} \equiv \left[\Lambda_3^{\sigma j}, \hat{f} \right] = \sum_{\alpha,\beta,\gamma} f_{\alpha,\beta,\gamma} \left[\Lambda_3^{\sigma j}, S\left(\hat{x^1}^\alpha \hat{x^2}^\beta \hat{x^3}^\gamma \right) \right]. \tag{17.29}$$

We prove in Appendix E that the commutator of the symmetrized term is the symmetrized commutator. Then, using the identity

$$[J, AB \cdots M] = [J, A]B \cdots M + A[J, B] \cdots M + \cdots + AB \cdots [J, M],$$

which results easily (by induction) from $[J, AB] = [J, A]B + A[J, B]$, it follows that

$$\mathcal{L}_3^{\sigma j} \hat{f} \equiv \left[\Lambda_3^{\sigma j}, \hat{f} \right] = \sum_{\alpha,\beta,\gamma} f_{\alpha,\beta,\gamma} \left(i\alpha \hat{x^1}^{\alpha-1} \hat{x^2}^{\beta+1} \hat{x^3}^\gamma - i\beta \hat{x^1}^{\alpha+1} \hat{x^2}^{\beta-1} \hat{x^3}^\gamma \right). \tag{17.30}$$

We thus have proven

$$\mathcal{L}_3^{\sigma j} \hat{f} = \widehat{J_3 f}. \tag{17.31}$$

Similar identities hold for $\mathcal{L}_1^{\sigma j}$, $\mathcal{L}_2^{\sigma j}$, and thus for $(\mathcal{L}^{\sigma j})^2$.

It follows that $\widehat{Y_{\ell m}}$ appears as an element of $\mathcal{O}^{\sigma j}$ that is a common eigenvector of $\mathcal{L}_3^{\sigma j}$, with eigenvalue m, and of $(\mathcal{L}^{\sigma j})^2$, with eigenvalue $\ell(\ell+1)$. Since we proved above that such an element is unique (up to a constant), it results that each $\widehat{Y_{\ell m}} \propto A_{Y_{\ell m}}$. Thus, the $\widehat{Y_{\ell m}}$'s, for $\ell \le j$, $-j \le m \le j$, form a basis of $\mathcal{O}^{\sigma j}$.

Then, the Wigner–Eckart theorem (see D) implies that $A_{Y_{\ell m}} = C(\ell)\widehat{Y_{\ell m}}$, where the proportionality constant $C(\ell)$ does not depend on m (this can actually be checked directly). These coefficients can be calculated directly, after noting that

$$\widehat{Y_{\ell\ell}} \propto (\Lambda_+)^\ell \propto \left(\hat{x^1} + i\hat{x^2} \right)^\ell.$$

In fact,

$$\widehat{Y_{\ell\ell}} = a(\ell) \left(\hat{x}^1 + i\hat{x}^2 \right)^\ell; \quad a(\ell) = \frac{\sqrt{(2\ell+1)!}}{2^{\ell+1}\sqrt{\pi}\ell!}.$$

We obtain

$$C(\ell) = 2^\ell \frac{(-1)^{j+\sigma-2\ell}(2j+1)}{\kappa^\ell} \sqrt{\frac{(2j-\ell)!}{(2j+\ell+1)!}} \begin{pmatrix} j & j & \ell \\ -\sigma & \sigma & 0 \end{pmatrix}. \tag{17.32}$$

17.4 Summary

Two families of quantization of the sphere have been presented in the previous sections, namely,

(i) the usual construction of the fuzzy sphere, which depends on the parameter j. This parameter defines the "size" of the discrete cell.

(ii) the coherent state quantization approach, which depends on two parameters, j and $\sigma \neq 0$.

These two quantizations may be formulated as involving the same algebra of operators (quantum observables) $\mathcal{O}$, acting on the same Hermitian space $\mathcal{H}$. In Table 17.1 we compare these two types of "fuzzyfication" of the sphere. Note that $\mathcal{H}$ and $\mathcal{O}$ are not the Hermitian space and algebra appearing in the usual construction of the fuzzy sphere (when we consider them as embedded in the space of functions on the sphere, and of operators acting on them), but they are isomorphic to them, and nothing is changed.

The difference lies in the fact that the quantum counterparts A_f and $\hat{f}$ of a given classical observable f differ in both approaches. Thus, the coherent state quantization really differs from the usual fuzzy sphere quantization. It follows from the calculations above that all properties of the usual fuzzy sphere are shared by the coherent state quantized version. The only point to be checked is whether it gives the sphere manifold in some classical limit. The answer is positive as far as the classical limit is correctly defined. Simple calculations show that it is obtained as the limit $j \mapsto \infty, \sigma \mapsto \infty$, provided that the ratio σ/j tends to a finite value. Thus, one may consider that the coherent state quantization leads to a one (discrete)-

Table 17.1 Coherent state quantization of the sphere is compared with the Madore construction of the fuzzy sphere through correspondence formulas.

	Coherent states fuzzy sphere	Madore-like fuzzy sphere
Hilbert space	$\mathcal{H} = \mathcal{H}^{\sigma j} = \mathrm{span}({}_\sigma Y_{j\mu}) \subset L^2(S^2)$	
Endomorphisms	$\mathcal{O} = \mathcal{O}^{\sigma j} = \mathrm{End}\,\mathcal{H}^{\sigma j}$	
Spin angular momentum operators	$\Lambda_a^{\sigma j} \in \mathcal{O}$	
Observables	$A_f \in \mathcal{O}^{\sigma j};\ A_{x^a} = K\Lambda_a^{\sigma j}$	$\hat{f} \in \mathcal{O}^{\sigma j};\ \hat{x}^a = \kappa\Lambda_a^{\sigma j}$
Action of angular momentum	$\mathcal{L}_a^{\sigma j} A_f \equiv \left[\Lambda_a^{\sigma j}, A_f\right] = A_{J_a f}$	$\mathcal{L}_a^{\sigma j} \hat{f} \equiv \left[\Lambda_a^{\sigma j}, \hat{f}\right] = \widehat{J_a f}$
Correspondence	$A_{Y_{\ell m}} = C(\ell)\widehat{Y_{\ell m}}$	

parameter family of fuzzy spheres if we impose relations of the type $\sigma = j - \sigma_0$, for fixed $\sigma_0 > 0$, for instance.

17.5 The Fuzzy Hyperboloid

In a continuation of the previous sections, we now turn our attention to the construction of a fuzzy version of the two-dimensional "de Sitter" hyperboloid by using a coherent state quantization [218]. Two-dimensional de Sitter space-time can be viewed as a one-sheeted hyperboloid embedded in a three-dimensional Minkowski space:

$$M_H = \{x \in \mathbb{R}^3 \ : \ x^2 = \eta_{\alpha\beta}\, x^\alpha x^\beta = (x^0)^2 - (x^1)^2 - (x^2)^2 = -H^{-2}\} . \qquad (17.33)$$

In Figure 17.1 we show this model of space-time in its four-dimensional version. We recall (see, e.g., [219] and references therein) that the de Sitter space-time is the unique maximally symmetric solution of the vacuum Einstein equations with positive cosmological constant Λ. This constant is linked to the constant Ricci curvature 4Λ of this space-time. There exists a fundamental length $H^{-1} := \sqrt{3/(c\Lambda)}$. The isometry group of the de Sitter manifold is, in the four-dimensional case, the 10-parameter de Sitter group $SO_0(1,4)$. The latter is a deformation of the proper orthochronous Poincaré group $\mathcal{P}_+^\uparrow$.

In the case of our two-dimensional toy model, the isometry group is $SO_0(1,2)$ or its double covering $SU(1,1) \simeq SL(2,\mathbb{R})$, as already mentioned in Section 14.3. Its Lie algebra is spanned by the three Killing vectors $K_{\alpha\beta} = x_\alpha \partial_\beta - x_\beta \partial_\alpha$ (K_{12}: com-

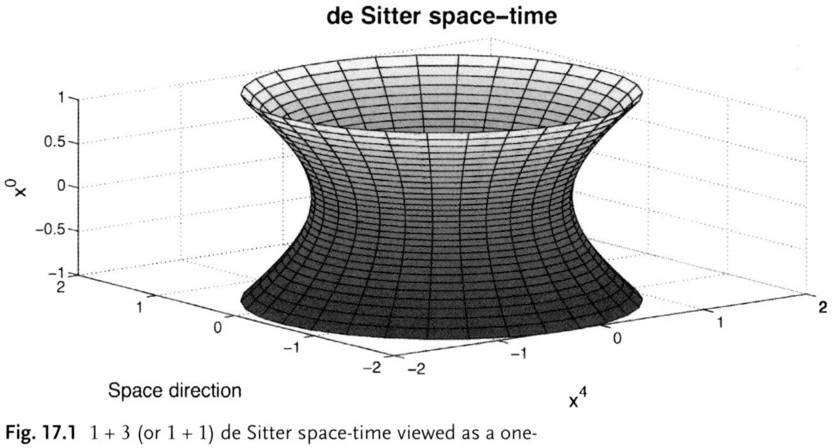

Fig. 17.1 1 + 3 (or 1 + 1) de Sitter space-time viewed as a one-sheeted hyperboloid embedded in a five (or three)-dimensional Minkowski space (the bulk). Coordinate x^0 plays, to some extent, the role of de Sitter time, and x^4 (or x^2) is the extra dimension needed to embed the de Sitter hyperboloid into the ambient 1 + 4 (or 1 + 2) Minkowski space.

pact, for "space translations"; K_{02}: noncompact, for "time translations"; K_{01}: noncompact, for Lorentz boosts). These Killing vectors are represented as (essentially) self-adjoint operators in a Hilbert space of functions on M_H, square-integrable with respect to some invariant inner (Klein–Gordon type) product.

The quadratic Casimir operator has eigenvalues that determine the unitary irreducible representations:

$$Q = -\tfrac{1}{2} M_{\alpha\beta} M^{\alpha\beta} = -j(j+1)\, I_d = (\varrho^2 + \tfrac{1}{4})\, I_d\,, \tag{17.34}$$

where $j = -\frac{1}{2} + i\varrho$, $\varrho \in \mathbb{R}^+$ for the principal series.

Comparing the geometrical constraint (17.33) with the group-theoretical one (17.34) (in the principal series) suggests the "fuzzy" correspondence [218]:

$$x^\alpha \mapsto \hat{x}^\alpha = \frac{r}{2} \varepsilon^{\alpha\beta\gamma} M_{\beta\gamma}\,, \quad \text{i.e.,} \quad \hat{x}^0 = r M_{21}\,, \quad \hat{x}^1 = r M_{02}\,, \quad \hat{x}^2 = r M_{10}\,. \tag{17.35}$$

r being a constant with length dimension. The following commutation rules are expected:

$$[\hat{x}^0, \hat{x}^1] = i r \hat{x}^2\,, \ [\hat{x}^0, \hat{x}^2] = -i r \hat{x}^1\,, \ [\hat{x}^1, \hat{x}^2] = i r \hat{x}^0\,, \tag{17.36}$$

with $\eta_{\alpha\beta} \hat{x}^\alpha \hat{x}^\beta = -r^2 (\varrho^2 + \frac{1}{4})\, I_d$, and its "commutative classical limit", $r \to 0$, $\varrho \to \infty$, $r\varrho = H^{-1}$.

Let us now proceed to the coherent state quantization of the two-dimensional de Sitter hyperboloid. The "observation" set X is the hyperboloid M_H. Convenient intrinsic coordinates are those of the topologically equivalent cylindrical structure, (τ, θ), $\tau \in \mathbb{R}$, $0 \le \theta < 2\pi$, through the parameterization $x^0 = r\tau$, $x^1 = r\tau \cos\theta - H^{-1} \sin\theta$, $x^2 = r\tau \sin\theta + H^{-1} \cos\theta$, with the invariant measure $\mu(dx) = d\tau\, d\theta/(2\pi)$. The functions $\phi_m(x)$ forming the orthonormal system needed to construct coherent states are like those already chosen in Chapter 15 for the quantization of the motion on the circle:

$$\phi_m(x) = \left(\frac{\varepsilon}{\pi}\right)^{1/4} e^{-\frac{\varepsilon}{2}(\tau - m)^2}\, e^{i m \theta}\,, \ m \in \mathbb{Z}\,, \tag{17.37}$$

where the parameter $\varepsilon > 0$ can be arbitrarily small and represents a necessary regularization. Through the usual construction the coherent states read as

$$|\tau, \theta\rangle = \frac{1}{\sqrt{\mathcal{N}(\tau)}} \left(\frac{\varepsilon}{\pi}\right)^{1/4} \sum_{m \in \mathbb{Z}} e^{-\frac{\varepsilon}{2}(\tau - m)^2}\, e^{-i m \theta} |m\rangle\,, \tag{17.38}$$

where $|\phi_m\rangle \equiv |m\rangle$. We recall that the normalization factor

$$\mathcal{N}(\tau) = \sqrt{\frac{\varepsilon}{\pi}} \sum_{m \in \mathbb{Z}} e^{-\varepsilon(\tau - m)^2} < \infty$$

is a periodic train of normalized Gaussians and is proportional to an elliptic theta function.

The coherent state quantization scheme yields the quantum operator A_f, acting on $\mathcal{H}$ and associated with the classical observable $f(x)$. For the most basic one, associated with the coordinate τ, one gets

$$A_\tau = \int_X \tau\, |\tau,\theta\rangle\langle\tau,\theta| \mathcal{N}(\tau)\, \mu(dx) = \sum_{m\in\mathbb{Z}} m|m\rangle\langle m|\,. \tag{17.39}$$

This operator reads in angular position representation (Fourier series) as $A_\tau = -i\partial/\partial\theta$, and is easily identified as the compact representative M_{12} of the Killing vector K_{12} in the principal series of $SO_0(1,2)$ unitary irreducible representation. Thus, the "time" component x^0 is naturally quantized, with spectrum $r\mathbb{Z}$, through $x^0 \mapsto \hat{x}^0 = -rM_{12}$. For the two other ambient coordinates one gets

$$\hat{x}^1 = \frac{re^{-\frac{\varepsilon}{4}}}{2}\sum_{m\in\mathbb{Z}}\{p_m|m+1\rangle\langle m| + h.c\}\,,\ \hat{x}^2 = \frac{re^{-\frac{\varepsilon}{4}}}{2i}\sum_{m\in\mathbb{Z}}\{p_m|m+1\rangle\langle m| - h.c\}\,, \tag{17.40}$$

with $p_m = (m+\frac{1}{2}+i\varrho)$. The commutation rules are those of $so(1,2)$, which are those of (17.36) with a local modification to $[\hat{x}^1, \hat{x}^2] = -ire^{-\frac{\varepsilon}{2}}\hat{x}^0$. The commutative limit at $r \to 0$ is apparent. It is proved that the same holds for higher-degree polynomials in the ambient space coordinates.

18
Conclusion and Outlook

What can we retain after such a journey through a (little) part of the coherent state landscape, beyond the physical context? At the risk of imposing upon the reader a quite partial view of this wide subject, let us propose a few insights into the deep significance of coherent states, once this notion has been cleared of its quantum physics dressing. We have shown that at the heart of the existence of coherent states there is a set X of parameters or data potentially measurable through some experimental or observational protocol, say, $\mathcal{E}$. The set X is equipped with a measure so that we deal with a measure space $(X, \mu_{\mathcal{E}})$, a minimal structure that is also determined by $\mathcal{E}$, along with a mixing of symmetry or conservation principles and degree of confidence (see Section A.12).

The term (classical) "observable" designates a function $f(x)$ on X, susceptible to being measured within the framework imposed by $\mathcal{E}$. Ideally, once a function has been assigned as an observable, all the values it assumes are accessible to measurement. For instance, if X represents the phase space of the motion of a particle on the line, that is, the set $\{(q, p)\}$ of possible initial positions and velocities, all the values assumed by a function $f(q, p)$ acknowledged as an observable can be measured, and they constitute the *spectrum* of the observable f. The word "spectrum" is usually employed to designate an image[18] or a distribution, of components of physical quantities such as light, sound, and particles arranged according to characteristics such as wavelength, frequency, charge, and energy. An observable f can be viewed as a diagonal continuous or discrete, infinite or finite matrix, with elements the values found in its spectrum. A possible interpretation is to consider f as the multiplication operator M_f in the Hilbert space $L^2(X, \mu_{\mathcal{E}})$ provided by the measure space:

$$M_f : \phi \mapsto \phi' = M_f \phi \quad (M_f \phi)(x) = f(x)\phi(x)\,. \tag{18.1}$$

It is clear that the set of such observables forms a commutative algebra, in abstraction from operator domain restrictions.

Changing the experimental/observational protocol, substituting $\mathcal{E}$ with a new one, say, $\mathcal{E}'$, can lead to discrepancies in this ideal scheme, in the sense that certain

18) The reader is invited to scrutinize the image on the front cover of this book!

Coherent States in Quantum Physics. Jean-Pierre Gazeau

ISBN: 978-3-527-40709-5

functions $f(x)$ (think of the combination $(p^2 + q^2)/2$ in the example of one-dimensional motion) lose their status of observable in the previous sense: the spectrum is not the same as the set of all the values assumed by the function. It is then necessary to build a formalism able to account for this new evidence: the customary way is to view $f(x)$ as a nondiagonal matrix, a more general operator, acting on some linear space, with spectrum the set of observed values in conformity with $\mathcal{E}'$.

The question becomes finding the right mathematical framework in which $f(x)$ is realized as an operator. Clearly, the adoption of the new protocol $\mathcal{E}'$, possibly together with the new measure space $(X, \mu_{\mathcal{E}'})$, results in the emergence of a generically noncommutative algebra of operators. We have observed that this noncommutativity occurs along with a reduction of the Hilbert space $L^2(X, \mu_{\mathcal{E}'})$ to a closed subspace, say, $\mathcal{K}$. This reduction should be thought of as the action of a projector $\mathbb{P}_{\mathcal{K}}$ on that Hilbert space:

$$\mathcal{K} = \mathbb{P}_{\mathcal{K}} L^2(X, \mu_{\mathcal{E}'}) . \tag{18.2}$$

Note that the restriction of $\mathbb{P}_{\mathcal{K}}$ to the subspace $\mathcal{K}$ is the identity operator $I_{\mathcal{K}}$ on the latter. It is precisely at this point that the existence of a family of normalized vectors $|x\rangle$ resolving, as elements of $L^2(X, \mu_{\mathcal{E}'})$, the projector $\mathbb{P}_{\mathcal{K}}$, or, as elements of $\mathcal{K}$, resolving the unity $I_{\mathcal{K}}$,

$$\int_X |x\rangle\langle x| \, \nu_{\mathcal{E}'}(dx) = \mathbb{P}_{\mathcal{K}} \ \text{or} \ I_{\mathcal{K}} , \tag{18.3}$$

allows us to implement the construction of the operator corresponding to an observable *f*. Indeed, such a function, which was viewed as a multiplication operator M_f according to the former protocol $\mathcal{E}$, becomes, under the projection $\mathbb{P}_{\mathcal{K}}$, the operator

$$A_f = \mathbb{P}_{\mathcal{K}} M_f \mathbb{P}_{\mathcal{K}} = \int_X f(x) |x\rangle\langle x| \, \nu_{\mathcal{E}'}(dx) , \tag{18.4}$$

the quantized version of *f* under the quantization provided by the family $\{|x\rangle, x \in X\}$. Actually, we have followed a mathematical procedure known in specific situations as the Toeplitz quantization of the set *X*.

Of course, we could be faced with ambiguities or conflicts of the type: within the same protocol, two different observables could lead to different projections, and so to incompatible physics or to different interpretations of the original set *X*. A change of protocol is then necessary. These possibilities raise deep questions, beyond the scope of this book.

Appendix A
The Basic Formalism of Probability Theory

Many excellent textbooks exist on the subject. I recommend one of them, which is based on a course by Sinai [220].

A.1
Sigma-Algebra

Let X be a set. A family $\mathcal{F}$ of subsets of X is a σ-algebra if and only if it has the following properties:

(i) The empty set $\emptyset$ is in $\mathcal{F}$.

(ii) If A is in $\mathcal{F}$, then so is the complement of A.

(iii) If $A_1, A_2, A_3, \dots$ is a sequence in $\mathcal{F}$, then their (countable) union is also in $\mathcal{F}$.

From (i) and (ii) it follows that X is in $\mathcal{F}$; from (ii) and (iii) it follows that the σ-algebra is also closed under countable intersections. σ-algebras are mainly used to define measures on X, as are defined in the next section. An ordered pair $(X, \mathcal{F})$, where X is a set and $\mathcal{F}$ is a σ-algebra over X, is called a measurable space.

A.1.1
Examples

1. The family consisting only of the empty set $\emptyset$ and X is a σ-algebra over X, the so-called trivial σ-algebra. Another σ-algebra over X is given by the *power set* of X, that is, the set $\mathcal{P}(X)$ of all subsets of X.
2. If $\{\mathcal{F}_a\}$ is a family of σ-algebras over X, then the intersection of all $\mathcal{F}_a$ is also a σ-algebra over X.
3. If $\mathcal{U}$ is an arbitrary family of subsets of X, then we can form a special σ-algebra from $\mathcal{U}$, called the σ-algebra generated by $\mathcal{U}$. We denote it by $\sigma(\mathcal{U})$ and define it as follows. First note that there is a σ-algebra over X that contains $\mathcal{U}$, namely, the power set of X. Let Φ be the family of all σ-algebras over X that contain $\mathcal{U}$ (i.e., a σ-algebra $\mathcal{F}$ over X is in Φ if and only if $\mathcal{U}$ is a subset

Coherent States in Quantum Physics. Jean-Pierre Gazeau

ISBN: 978-3-527-40709-5

of $\mathcal{F}$.) Then we define $\sigma(\mathcal{U})$ to be the intersection of all σ-algebras in Φ. $\sigma(\mathcal{U})$ is then the smallest σ-algebra over X that contains $\mathcal{U}$.

4. This leads to the most important example: the *Borel algebra* over any topological space is the σ-algebra generated by the open sets (or, equivalently, by the closed sets). Note that this σ-algebra is not, in general, the whole power set.
5. On the Euclidean space $\mathbb{R}^n$, another σ-algebra is of importance: that of all Lebesgue measurable sets. This σ-algebra contains more sets than the Borel algebra on $\mathbb{R}^n$ and is preferred in integration theory.

A.2
Measure

A measure μ is a function defined on a σ-algebra $\mathcal{F}$ over a set X and taking values in $[0, \infty)$ such that the following properties are satisfied:

1. The empty set has measure zero:

$$\mu(\emptyset) = 0 . \tag{A1}$$

2. Countable additivity or σ-additivity: if E_1, E_2, E_3, is a countable sequence of pairwise disjoint sets in $\mathcal{F}$, the measure of the union of all the E_i is equal to the sum of the measures of each E_i:

$$\mu\left(\bigcup_{i=1}^{\infty} E_i\right) = \sum_{i=1}^{\infty} \mu(E_i) . \tag{A2}$$

The triple $(X, \mathcal{F}, \mu)$ is then called a measure space, and the members of $\mathcal{F}$ are called measurable sets.

A.3
Measurable Function

If $\mathcal{F}$ is a σ-algebra over X and $\mathcal{G}$ is a σ-algebra over Y, then a function $f : X \mapsto Y$ is measurable if the preimage of every set in $\mathcal{G}$ is in $\mathcal{F}$.

By convention, if Y is some topological space, such as the space of real numbers $\mathbb{R}$ or the complex numbers $\mathbb{C}$, then the Borel σ-algebra generated by the open sets on Y is used, unless otherwise specified.

If a function from one topological space to another is measurable with respect to the Borel σ-algebras on the two spaces, the function is also known as a *Borel function*. Continuous functions are Borel; however, not all Borel functions are continuous.

Given two measurable spaces $(X_1, \mathcal{F}_1)$ and $(X_2, \mathcal{F}_2)$, a measurable function $f : X_1 \mapsto X_2$ and a measure $\mu : \mathcal{F}_1 \mapsto [0, +\infty]$, the *pushforward* of μ is defined to be the measure $f * \mu : \mathcal{F}_2 \mapsto [0, +\infty]$ given by

$$f * \mu(B) \stackrel{\text{def}}{=} \mu\left(f^{-1}(B)\right) \quad \text{for} \quad B \in \mathcal{F}_2 \,. \tag{A3}$$

A.4 Probability Space

A probability space is a set Ω, together with a σ-algebra $\mathcal{F}$ on Ω and a measure P on that σ-algebra, such that $P(\Omega) = 1$. The set Ω is called the *sample space* and the elements of $\mathcal{F}$ are called the *events*. The measure P is called the *probability measure,* and $P(E)$ is the probability of the event $E \in \mathcal{F}$.

Every set A with nonzero probability defines another probability

$$P(B|A) = \frac{P(A \cap B)}{P(A)} \tag{A4}$$

on the space. This is usually read as "probability of B given A." If this *conditional* probability of B given A is the same as the probability of B, then B and A are said to be independent.

In the case that the sample space is finite or countably infinite, a probability function can also be defined by its values on the elementary events $\{\omega_1\}, \{\omega_2\}, \ldots$, where $\Omega = \{\omega_1, \omega_2, \ldots\}$.

A.5 Probability Axioms

In a context proper to probability theory, the properties of a probability measure $P : \mathcal{F} \mapsto [0, 1]$ are imposed under the form of axioms known as the *Kolmogorov axioms*:

K$_1$ For any set E,

$$0 \le P(E) \le 1 \,. \tag{A5}$$

That is, the probability of an event set is represented by a real number between 0 and 1.

K$_2$

$$P(\Omega) = 1 \,. \tag{A6}$$

That is, the probability that some elementary event in the entire sample set will occur is 1. More specifically, there are no elementary events outside the sample set. This is often overlooked in some mistaken probability calculations; if one cannot precisely define the whole sample set, then the probability of any subset cannot be defined either.

K3 Any countable sequence of mutually disjoint events $E_1, E_2, \ldots$ satisfies

$$P(E_1 \cup E_2 \cup \cdots) = P(E_1) + P(E_2) + \cdots . \tag{A7}$$

That is, the probability of an event set which is the union of other disjoint subsets is the sum of the probabilities of those subsets. This is called σ-additivity. If there is any overlap among the subsets, this relation does not hold.

A probability distribution assigns to every interval of the real numbers a probability, so that the probability axioms are satisfied. In technical terms, a probability distribution is a probability measure whose domain is the Borel algebra on the reals.

A.6
Lemmas in Probability

(i) From the Kolmogorov axioms one deduces other useful rules for calculating probabilities:

$$P(A \cup B) = P(A) + P(B) - P(A \cap B) . \tag{A8}$$

That is, the probability that A or B will happen is the sum of the probabilities that A will happen and that B will happen, minus the probability that A and B will happen. This can be extended to the inclusion–exclusion principle.

(ii)

$$P(\Omega - E) = 1 - P(E) . \tag{A9}$$

That is, the probability that any event will not happen is 1 minus the probability that it will.

(iii) Using conditional probability as defined above, it also follows immediately that

$$P(A \cap B) = P(A)\, P(B|A) . \tag{A10}$$

That is, the probability that A and B will happen is the probability that A will happen, times the probability that B will happen given that A happened; this relationship gives Bayes's theorem (see next). It then follows that A and B are independent if and only if

$$P(A \cap B) = P(A)\, P(B) . \tag{A11}$$

A.7
Bayes's Theorem

Bayes's theorem relates the conditional and marginal probabilities of events A and B, where B has a nonvanishing probability:

$$P(A|B) = \frac{P(B|A)\, P(A)}{P(B)} . \tag{A12}$$

More generally, let $\Omega = \bigcup_i A_i$, $A_i \cap A_j = \emptyset$ for $i \neq j$, be a partition of the event space. Then we have

$$P(A_i|B) = \frac{P(B|A_i)\,P(A_i)}{\sum_j P(B|A_j)\,P(A_j)}\,, \tag{A13}$$

for any A_i in the partition.

A.8
Random Variable

Let $(\Omega, \mathcal{F}, P)$ be a probability space and (Y, Σ) be a measurable space. Then a random variable ξ is formally defined as a measurable function $\xi : \Omega \mapsto Y$. In other words, the preimage of the "well-behaved" subsets of Y (the elements of Σ) are events, and hence are assigned a probability by P.

When the measurable space is the measurable space over the real numbers, one speaks of real-valued random variables. Then, the function ξ is a real-valued random variable if

$$\{\omega \in \Omega \mid \xi(\omega) \leq r\} \in \mathcal{F}\,, \quad \forall r \in \mathbb{R}\,. \tag{A14}$$

A.9
Probability Distribution

The probability distribution of a real-valued variable ξ can be uniquely described by its *cumulative* distribution function $F_\xi(x)$ (also called a distribution function), which is defined by

$$F_\xi(x) = P(\xi \leq x) \tag{A15}$$

for any x in $\mathbb{R}$.

More generally, given a random variable $\xi : \Omega \mapsto Y$ between a probability space $(\Omega, \mathcal{F}, P)$, the sample space, and a measurable space (Y, Σ), called the state space, a probability distribution on (Y, Σ) is a probability measure $\xi * P : \Sigma \mapsto [0, 1]$ on the state space, where $\xi * P$ is the push-forward measure of P.

A distribution is called discrete if its cumulative distribution function consists of a sequence of finite jumps, which means that it belongs to a discrete random variable ξ, a variable which can only attain values from a certain finite or countable set. Discrete distributions are characterized by a *probability mass function* p such that $P(\xi = x) = p(x)$.

A distribution is called continuous if its cumulative distribution function is continuous. A random variable ξ is called continuous if its distribution function is continuous. In that case $P(\xi = x) = 0$ for all $x \in \mathbb{R}$. Note also that there are probability distribution functions which are neither discrete nor continuous.

The so-called absolutely continuous distributions (frequently and loosely named continuous distributions) can be expressed by a *probability density function*: a non-negative Lebesgue integrable function f defined on the reals such that

$$F(x) = P(\xi \leq x) = \int_{-\infty}^{x} f(u)\, du \,. \tag{A16}$$

Discrete distributions do not admit such a density. There are continuous distributions (e.g., the devil's staircase) that also do not admit a density; they are called singular continuous distributions.

A.10
Expected Value

If ξ is a random variable defined on a probability space $(\Omega, \mathcal{F}, P)$, then the expected value of ξ, denoted $E(\xi)$, or sometimes $\langle \xi \rangle$, is defined as

$$E(\xi) = \int_{\Omega} \xi\, dP \,, \tag{A17}$$

where the notation stands for the Lebesgue integral resulting from the measure P. Note that not all random variables have an expected value, since the integral may not exist (e.g., Cauchy distribution).

If ξ is a discrete random variable with probability mass function $p(x)$, then the expected value becomes

$$E(\xi) = \sum_{i} x_i \; p(x_i) \,. \tag{A18}$$

If the probability distribution of ξ admits a probability density function $f(x)$, then the expected value can be computed as

$$E(\xi) = \int_{-\infty}^{+\infty} x\, f(x)\, dx \,. \tag{A19}$$

The expected value of an arbitrary function of ξ, $g(\xi)$, with respect to the probability density function $f(x)$ is given by

$$E(g(\xi)) = \int_{-\infty}^{+\infty} g(x)\, f(x)\, dx \,. \tag{A20}$$

A.11
Conditional Probability Densities

There is also a version of Bayes's theorem for continuous distributions. It is somewhat harder to derive, since probability densities, strictly speaking, are not probabilities, so Bayes's theorem has to be established by a limit process. Bayes's theorem for probability densities is formally similar to the theorem for probabilities:

$$f_\xi(x|\eta = y) = \frac{f_{\xi,\eta}(x, y)}{f_\eta(y)} = \frac{f_\eta(y|\xi = x)\, f_\xi(x)}{f_\eta(y)}$$
$$= \frac{f_\eta(y|\xi = x)\, f_\xi(x)}{\int_{-\infty}^{+\infty} f_\eta(y|\xi = u)\, f_\xi(u)\, du}\,, \qquad \text{(A21)}$$

where the continuous analog of the law of total probability is used in the denominator,

$$f_\eta(y) = \int_{-\infty}^{+\infty} f_\eta(y|\xi = u)\, f_\xi(u)\, du\,. \qquad \text{(A22)}$$

In (A21),

- $f_{\xi,\eta}(x, y)$ is the joint distribution of ξ and η,
- $f_\xi(x|\eta = y)$ is the conditional density function for random variable ξ,
- and $f_\xi(x)$ and $f_\eta(y)$ are the marginal density functions for the random variables ξ and η, respectively.

A.12
Bayesian Statistical Inference

In cases of incomplete knowledge of the preparation, a probabilistic experiment may be modeled by a *family* of probability distributions, for example, $\{p_\theta(y),\, \theta \in \Theta\}$, where p_θ is a probability function in the discrete case and a probability density function in the continuous case for random variable η. Here we consider the case where the parameter θ which indexes the family is a continuous variable over the set Θ. For one (unknown) value of θ, the model is assumed to be predictive. For simplicity, we consider the case of a single parameter, but one may extend the reasoning to a multiparameter model.

After the experiment has been performed, and pointer value y_0, say, has been obtained, interest might center upon an inferred probability distribution over the parameter space Θ. In that case, we take Θ to be a measure space such as $(\Theta, \mu(d\theta))$. Suppose that $\mu(d\theta)$ can be written in terms of a density function $\mu(d\theta) = \Pi(\theta)\, d\theta$. Then Bayesian statistical inference consists of using Bayes's formula for conditional distributions in the following way:

$$f(\theta|y_0) = \frac{p_\theta(y_0)\, \Pi(\theta)}{\int_\Theta p_{\theta'}(y_0)\, \Pi(\theta')\, d\theta'}\,. \qquad \text{(A23)}$$

In statistical parlance,

- $f(\theta|y_0)$ is referred to as a posterior probability density function on the parameter space.
- $\mu(d\theta)$ is called an a priori measure on the parameter space.

- $\theta \mapsto p_\theta(y_0)$ for given y_o, as a function of θ, is referred to as a likelihood function.

Note that, in theory, we have a collection of possible likelihood functions in that there is one for each possible value of y.

A.13 Some Important Distributions

A.13.1 Degenerate Distribution

This is the probability distribution of a random variable which always has the same value. It is localized at a point x_0 on the real line.

The cumulative distribution function of the degenerate distribution is then the Heaviside step function $\theta(x-x_0)$, which is equal to 1 for $x \geq x_0$ and to 0 if $x < x_0$. As a discrete distribution, the degenerate distribution does not have a proper density. Actually, within the framework of distribution theory, Dirac's delta function can serve this purpose.

A.13.2 Uniform Distribution

The discrete uniform distribution is a discrete probability distribution that can be characterized by saying that all values of a finite set of possible values are equally probable. If the values of a random variable with a discrete uniform distribution are real, it is possible to express the cumulative distribution function in terms of the degenerate distribution:

$$F(x; n) = \frac{1}{n} \sum_{i=1}^{n} \theta(x - x_i) . \tag{A24}$$

The continuous uniform distribution is a family of probability distributions such that for each member of the family all intervals of the same length on the distribution's support are equally probable. The probability density function of the continuous uniform distribution is

$$f(x) = \begin{cases} \dfrac{1}{b-a} & \text{for } a \leq x \leq b , \\ 0 & \text{for } x < a \text{ or } x > b \end{cases} = \frac{\theta(x-a) - \theta(x-b)}{b-a} . \tag{A25}$$

A.13.2.1 Bernoulli Distribution

This is a discrete family of probability distributions indexed by parameter p, $0 \leq p \leq 1$. Put $q = 1 - p$. The random variable ξ has only two possible values, say, 0 and 1. Then the probability mass function is given by

$$P(\xi|x) = p^x (1-p)^{1-x} = \begin{cases} p & \text{if } x = 1\,, \\ q & \text{if } x = 0\,, \\ 0 & \text{otherwise}\,. \end{cases} \tag{A26}$$

A.13.2.2

Binomial Distribution

The family is indexed by parameters p and n, where $0 \le p \le 1$ and n is a positive integer. Put $q = 1 - p$. The binomial distribution is given by the terms in the expansion of $(q + p)^n$.

$$P(\xi = k) = \binom{n}{k} p^k q^{n-k} \quad \text{for} \quad k = 0, 1, \ldots, n\,. \tag{A27}$$

A model leading to this distribution may be given by the following. If n independent trials are performed and in each there is a probability p that an outcome ω occurs, then the number of trials in which ω occurs can be represented by a random variable ξ having the binomial distribution with parameters n and p.

When $n = 1$, the distribution is the Bernoulli distribution.

A.13.2.3

Hypergeometrical Distribution

This is a discrete probability distribution that describes the number of successes in a sequence of n draws from a finite population without replacement. For a random variable ξ following the hypergeometrical distribution with parameters N, D, and n, the probability of getting exactly k successes is given by

$$P(\xi = k) = \frac{\binom{D}{k}\binom{N-D}{n-k}}{\binom{N}{n}}\,. \tag{A28}$$

A.13.2.4

Negative Binomial Distribution

The family is indexed by parameters $P > 0$ and positive real number r. Put $Q = 1 + P$. The negative binomial distribution is given by the terms in the expansion of $(Q - P)^{-r}$. Thus,

$$P(\xi = k) = \binom{r + k - 1}{r - 1} \left(\frac{P}{Q}\right)^r \left(1 - \frac{P}{Q}\right)^r, \tag{A29}$$

for $k = 0, 1, 2, \ldots$. Unlike the binomial distribution, r need not be an integer. When r is an integer, the distribution is sometimes called the Pascal distribution.

When $r = 1$, we have $P(\xi = k) = (P/Q)^k Q^{-1}$, which is called the geometrical or Bose–Einstein distribution.

A model leading to this distribution when r is a positive integer may be the following. Put $P = (1 - p)/p$ and $r = m$. Let random variable ξ represent the number of independent trials necessary to obtain m occurrences of an event ω which has constant probability p of occurring at each trial. Then we have

$$P(\xi = m + k) = \binom{m+k-1}{m-1} p^m (1-p)^k \tag{A30}$$

for $k = 1, 2, 3, \ldots$.

A.13.2.5
Poisson Distribution

The family is indexed by parameter $\lambda > 0$. Then random variable ξ has the Poisson distribution when the probability distribution is given by

$$P(\xi = k) = \frac{e^{-\lambda}\lambda^k}{k!}, \tag{A31}$$

for $k = 1, 2, 3, \ldots$.

The Poisson distribution is the limit of a sequence of binomial distributions in which n tends to infinity and p tends to zero such that $n\,p$ (binomial mean) remains equal to λ (Poisson mean) [221].

A.13.2.6
Poisson Process [222]

A stochastic process $\{N(t), t \geq 0\}$ is said to be a counting process if $N(t)$ represents the total number of events ω that have occurred up to time t and satisfies the following conditions:

(i) $N(t) \geq 0$.
(ii) $N(t)$ is integer-valued.
(iii) If $s < t$, then $N(s) \leq N(t)$.
(iv) For $s < t$, $N(t) - N(s)$ equals the number of events that have occurred in the interval (s, t).

A counting process possesses independent increments if the number of events which occur in disjoint time intervals are independently distributed.

A counting process is a Poisson process with rate λ, $\lambda > 0$, if it satisfies the following conditions:

(i) $N(0) = 0$.
(ii) The process has independent increments.
(iii) The number of events in any interval of length t has the Poisson distribution with mean $\lambda\,t$.

A.13.2.7
Mixtures of Discrete Probability Distributions [223]

Consider a family of discrete distributions indexed by a parameter $\lambda \in \Lambda$. For random variable ξ, write $P(\xi = x) = p_x(\lambda)$. Let $G(d\lambda)$ represent a probability measure on Λ. Then the probability distribution

$$Pr(\xi = x) = \int_{\Lambda} p_x(\lambda)\, G(d\lambda)$$

is described as $p_x(\lambda)$ mixed on λ by $G(d\lambda)$.

For example, if $p_x(\lambda)$ is the Poisson distribution and $G(d\lambda)$ is the exponential distribution, then we have the result that ξ has the geometrical distribution. More generally, a Poisson variate mixed on λ by the gamma distribution (see below) has the negative binomial distribution.

A.13.2.8
Beta Distribution

The beta distribution is a continuous probability distribution with the probability density function defined on the interval [0, 1]:

$$f(x) = \frac{1}{B(a,b)}\, x^{a-1}\, (1-x)^{b-1}\,, \tag{A32}$$

where a and b are parameters that must be greater than zero, and $B(a,b) = \Gamma(a)\Gamma(b)/\Gamma(a+b)$ is the beta function expressed in terms of the gamma functions.

A.13.2.9
Wigner Semicircle Distribution

The Wigner semicircle distribution is a continuous probability distribution with the interval $[-R, R]$ as support. The graph of its probability density function f is a semicircle of radius R centered at the origin (actually a semiellipse with suitable normalization):

$$f(x) = \frac{2}{\pi R^2}\, \sqrt{1-x^2}\,, \tag{A33}$$

for $R < x < R$, and $f(x) = 0$ if $x > R$ or $x < -R$.

This distribution arises as the limiting distribution of eigenvalues of many random symmetric matrices as the size of the matrix approaches infinity.

A.13.2.10
Gamma Distribution

The gamma distribution is a continuous probability distribution. Its probability density function can be expressed in terms of the gamma function:

$$f(x;k,\theta) = \frac{e^{-x/\theta}}{\theta^k}\, \frac{x^{k-1}}{\Gamma(k)}\,, \tag{A34}$$

where $k > 0$ is the shape parameter and $\theta > 0$ is the scale parameter.

A.13.2.11
Normal Distribution

The normal distribution is certainly the most popular among continuous probability distributions. It is also called the Gaussian distribution. It is actually a family of distributions of the same general form, differing only in their location and scale parameters: the mean and standard deviation. Its probability density function with mean μ and standard deviation σ (equivalently, variance σ^2) is given by

$$f(x;\mu,\sigma) = \frac{1}{\sigma\sqrt{2\pi}} e^{(x-\mu)^2/2\sigma^2} . \tag{A35}$$

The standard normal distribution is the normal distribution with a mean of zero and a standard deviation of one. Because the graph of its probability density resembles a bell, it is often called a bell curve.

Approximately normal distributions occur in many situations, as a result of the central limit theorem. When there is reason to suspect the presence of a large number of small effects acting additively and independently, it is reasonable to assume that observations will be normal. There are statistical methods to empirically test that assumption.

Two examples of the occurrence of the normal law in physics:

Photon counts Light intensity from a single source varies with time, and is usually assumed to be normally distributed. However, quantum mechanics interprets measurements of light intensity as photon counting. Ordinary light sources that produce light by thermal emission should follow a Poisson distribution or Bose–Einstein distribution on very short time scales. On longer time scales (longer than the coherence time), the addition of independent variables yields an approximately normal distribution. The intensity of laser light, which is a quantum phenomenon, has an exactly normal distribution.

Measurement errors Repeated measurements of the same quantity are expected to yield results which are clustered around a particular value. If all major sources of errors have been taken into account, it is assumed that the remaining error must be the result of a large number of very small additive effects, and hence normal. Deviations from normality are interpreted as indications of systematic errors which have not been taken into account. Note that this is the central assumption of the mathematical theory of errors.

A.13.2.12
Cauchy Distribution

The Cauchy distribution is an example of a continuous distribution that does not have an expected value or a variance. In physics it is usually called a Lorentzian, and it is the distribution of the energy of an unstable state in quantum mechanics. In particle physics, the extremely short lived particles associated with such unstable states are called resonances. The Cauchy distribution has the probability density function

$$f(x; x_0, \gamma) = \frac{1}{\pi\gamma} \frac{1}{1 + \left(\frac{x - x_0}{\gamma}\right)^2}, \tag{A36}$$

where x_0 is the location parameter, specifying the location of the peak of the distribution, and γ is the scale parameter, which specifies the half-width at half-maximum. The amplitude of the above Lorentzian function is given by $1/\pi\gamma$.

Appendix B
The Basics of Lie Algebra, Lie Groups, and Their Representations

The theory of groups and Lie groups, Lie algebras, and their representations is widely known and many excellent books cover it, for instance, [74, 224, 225]. We recall here a few basic facts, mainly extracted from [11].

B.1
Group Transformations and Representations

A *transformation* of a set S is a one-to-one mapping of S onto itself. A group G is realized as a transformation group of a set S if with each $g \in G$ there is associated a transformation $s \mapsto g \cdot s$ of S where for any two elements g_1 and g_2 of G and $s \in S$ we have $(g_1 g_2) \cdot s = g_1 \cdot (g_2 \cdot s)$.

The set S is then called a G-space. A transformation group is *transitive* on S if, for each s_1 and s_2 in S, there is a $g \in G$ such that $s_2 = g \cdot s_1$. In that case, the set S is called a *homogeneous* G-space.

A (linear) *representation* of a group G is a continuous function $g \mapsto T(g)$ which takes values in the group of nonsingular continuous linear transformations of a vector space $\mathcal{V}$, and which satisfies the functional equation $T(g_1 g_2) = T(g_1)T(g_2)$ and $T(e) = I_d$, the identity operator in $\mathcal{V}$, where e is the identity element of G. It follows that $T(g^{-1}) = (T(g))^{-1}$. That is, $T(g)$ is a homomorphism of G into the group of nonsingular continuous linear transformations of $\mathcal{V}$.

A representation is *unitary* if the linear operators $T(g)$ are unitary with respect to the inner product $\langle \cdot | \cdot \rangle$ on $\mathcal{V}$. That is, $\langle T(g)\, v_1 | T(g)\, v_2 \rangle = \langle v_1 | v_2 \rangle$ for all vectors v_1, v_2 in $\mathcal{V}$. A representation is *irreducible* if there is no nontrivial subspace $\mathcal{V}_0 \subset \mathcal{V}$ such that for all vectors $v_o \in \mathcal{V}_0$, $T(g)\, v_o$ is in $\mathcal{V}_0$ for all $g \in G$. That is, there is no nontrivial subspace $\mathcal{V}_0$ of $\mathcal{V}$ which is invariant under the operators $T(g)$.

Let G be a transformation group of a set S. Let $\mathcal{V}$ be a linear space of functions $f(s)$ for $s \in S$. For each invariant subspace $\mathcal{V}_0 \subset \mathcal{V}$ we have a representation of the group G by shift operators: $(T(g)f)(s) = f(g^{-1} \cdot s)$. A *multiplier* representation is of the form $(T(g)f)(s) = A(g^{-1}, s) f(g^{-1} \cdot s)$, where $A(g, s)$ is an automorphic factor satisfying $A(g_1\, g_2, s) = A(g_1, g_2 \cdot s) A(g_2, s)$ and $A(e, s) = 1$.

Coherent States in Quantum Physics. Jean-Pierre Gazeau

ISBN: 978-3-527-40709-5

B.2
Lie Algebras

Let $\mathfrak{g}$ be a complex Lie algebra, that is, a complex vector space with an antisymmetric bracket $[\cdot, \cdot]$ that satisfies the Jacobi identity

$$[[X, Y], Z] + [[Y, Z], X] + [[Z, X], Y] = 0, \quad \forall X, Y, Z \in \mathfrak{g}. \tag{B1}$$

For $X, Y \in \mathfrak{g}$, the relation $(\text{ad}X)(Y) = [X, Y]$ gives a linear map ad: $\mathfrak{g} \to \text{End}\,\mathfrak{g}$ (endomorphisms of $\mathfrak{g}$), called the *adjoint representation* of $\mathfrak{g}$. Next, if dim $\mathfrak{g} < \infty$, it makes sense to define

$$B(X, Y) = \text{Tr}[(\text{ad}X)(\text{ad}Y)], \quad X, Y \in \mathfrak{g}. \tag{B2}$$

B is a symmetric bilinear form on $\mathfrak{g}$, called the *Killing form* of $\mathfrak{g}$. Alternatively, one may choose a basis $\{X_j, j = 1, \ldots, n\}$ in $\mathfrak{g}$, in terms of which the commutation relations read

$$[X_i, X_j] = \sum_{k=1}^{n} c_{ij}^k X_k, \quad i, j = 1, \ldots, n, \tag{B3}$$

where c_{ij}^k are called the structure constants and $n = \dim \mathfrak{g}$. Then it is easy to see that $g_{ij} = \sum_{k,m=1}^{n} c_{ik}^m c_{jm}^k = B(X_i, X_j)$ defines a metric on $\mathfrak{g}$, called the Cartan–Killing metric.

The Lie algebra $\mathfrak{g}$ is said to be *simple*, or *semisimple*, if it contains no nontrivial ideal, or Abelian ideal. A semisimple Lie algebra may be decomposed into a direct sum of simple ones. Furthermore, $\mathfrak{g}$ is semisimple if and only if the Killing form is nondegenerate (Cartan's criterion).

Let $\mathfrak{g}$ be semisimple. Choose in $\mathfrak{g}$ a Cartan subalgebra $\mathfrak{h}$, that is, a maximal nilpotent subalgebra (it is in fact maximal Abelian and unique up to conjugation). The dimension ℓ of $\mathfrak{h}$ is called the *rank* of $\mathfrak{g}$. A *root* of $\mathfrak{g}$ with respect to $\mathfrak{h}$ is a linear form on $\mathfrak{h}$, $\alpha \in \mathfrak{h}^*$, for which there exists $X \neq 0$ in $\mathfrak{g}$ such that $(\text{ad}H)X = \alpha(H)X, \forall H \in \mathfrak{h}$.

Then one can find (Cartan, Chevalley) a basis $\{H_i, E_\alpha\}$ of $\mathfrak{g}$, with the following properties. $\{H_j, j = 1, \ldots, \ell\}$ is a basis of $\mathfrak{h}$ and each generator E_α is indexed by a nonzero root α, in such a way that the commutation relations (B3) may be written in the following form:

$$[H_i, H_j] = 0, \quad j = 1, \ldots, \ell \tag{B4}$$

$$[H_i, E_\alpha] = \alpha(H_i) E_\alpha, \quad i = 1, \ldots, \ell \tag{B5}$$

$$[E_\alpha, E_{-\alpha}] = H_\alpha \equiv \sum_{i=1}^{\ell} \alpha^i H_i \in \mathfrak{h}, \tag{B6}$$

$$[E_\alpha, E_\beta] = N_{\alpha\beta}\, E_{\alpha+\beta}, \tag{B7}$$

where $N_{\alpha\beta} = 0$ if $\alpha + \beta$ is not a root. Let Δ denote the set of roots of $\mathfrak{g}$. Note that the nonzero roots come in pairs, $\alpha \in \Delta \Leftrightarrow -\alpha \in \Delta$, and no other nonzero multiple of a root is a root. Accordingly, the set of nonzero roots may be split into a subset Δ_+ of positive roots and the corresponding subset of negative roots $\Delta_- = \{-\alpha, \alpha \in \Delta_+\}$. The set Δ_+ is contained in a simplex (convex pyramid) in $\mathfrak{h}^*$, the edges of which are the so-called simple positive roots, that is, positive roots which cannot be decomposed as the sum of two other positive roots. Of course, the same holds for Δ_-. The consideration of root systems is the basis of the Cartan classification of simple Lie algebras into four infinite series $A_\ell, B_\ell, C_\ell, D_\ell$ and five exceptional algebras G_2, F_4, E_6, E_7, E_8 [225].

In addition to the Lie algebra $\mathfrak{g}$, one may also consider the universal enveloping algebra $\mathfrak{U}(\mathfrak{g})$, which consists of all polynomials in the elements of $\mathfrak{g}$, taking into account the commutation relations (B3). Of special interest are the so-called Casimir elements, that generate the center of $\mathfrak{U}(\mathfrak{g})$, and in particular the quadratic Casimir element $C_2 = \sum_{i,k=1}^{n} g_{ik} X^i X^k$, where $\{X^i\}$ is the basis of $\mathfrak{g}$ dual to $\{X_i\}$, that is, $B(X^i, X_j) = \delta^i_j$ and g_{ik} is the Cartan–Killing metric. The element C_2 does not depend on the choice of the basis $\{X_i\}$. In a Cartan–Chevalley basis $\{H_j, E_\alpha\}$, one gets

$$C_2 = \sum_{j=1}^{\ell} (H_j)^2 + \sum_{\alpha \in \Delta} E_\alpha E_{-\alpha} \,. \qquad \text{(B8)}$$

In the same way, one defines a real Lie algebra as a real vector space with an antisymmetric bracket $[\cdot,\cdot]$ that satisfies the Jacobi identity. The two concepts are closely related. If $\mathfrak{g}$ is a real Lie algebra, one can define its *complexification* $\mathfrak{g}^c$ by complexifying it as a vector space and extending the Lie bracket by linearity. If dim $\mathfrak{g} = n$, the complex dimension of $\mathfrak{g}^c$ is still n, but its real dimension is $2n$. Conversely, if $\mathfrak{g}$ is a complex Lie algebra, and one restricts its parameter space to real numbers, one obtains a real form $\mathfrak{g}_r$, that is, a real Lie algebra whose complexification is again $\mathfrak{g}$. A given complex Lie algebra has in general several nonisomorphic real forms (also classified by Cartan), among them a unique compact one, characterized by the fact that the Cartan–Killing metric g_{ij} is negative-definite. For instance, the complex Lie algebra A_2 yields two real forms, $\mathfrak{su}(3)$, the compact one, and $\mathfrak{su}(2,1)$.

For physical applications, it is not so much the Lie algebras themselves that matter, but their representations in Hilbert spaces. The key ingredient for building the latter, also due to Cartan, is the notion of a *weight vector*. Let T be a representation of the Lie algebra $\mathfrak{g}$ in the Hilbert space $\mathfrak{h}$, that is, a linear map of $\mathfrak{g}$ into the operators on $\mathfrak{h}$ such that

$$T([X,Y]) = [T(X), T(Y)] \,, \quad \forall X, Y \in \mathfrak{g} \,. \qquad \text{(B9)}$$

The representation T is called *Hermitian* if $T(X^*) = T(X)^*$, for every $X \in \mathfrak{g}$.

By Schur's lemma, it follows that the Casimir elements of $\mathfrak{U}(\mathfrak{g})$ are simultaneously diagonalizable in any irreducible Hermitian representation of $\mathfrak{g}$, and in fact their eigenvalues characterize the representation uniquely.

Let $\{H_j, E_\alpha\}$ be a Cartan–Chevalley basis of the complexified Lie algebra $\mathfrak{g}^c$ of $\mathfrak{g}$. Then a *weight* for T is a linear form $\alpha \in \mathfrak{h}^*$, for which there exists a nonzero vector

$|\lambda\rangle \in \mathfrak{h}$ such that $[T(H) - \alpha(H)]^n|\lambda\rangle = 0$, for all $H \in \mathfrak{h}$ and some n. Then $|\lambda\rangle$ is called a weight vector if $n = 1$, that is, $T(H)|\lambda\rangle = \alpha(H)|\lambda\rangle$, $\forall H \in \mathfrak{h}$. In particular, $|\lambda\rangle$ is called a *highest* (or *lowest*) *weight vector* if

$$T(E_{\alpha_+})|\lambda\rangle = 0\,, \quad \forall \alpha_+ \in \Delta_+\,, \quad (\text{or} \quad T(E_{\alpha_-})|\lambda\rangle = 0\,, \quad \forall \alpha_- \in \Delta_-)\,, \tag{B10}$$

where Δ_+, or Δ_-, is the set of the positive, or negative, roots. The interest of this notion is the fundamental result obtained by Cartan which says that the irreducible, finite-dimensional, Hermitian representations of simple Lie algebras are in one-to-one correspondence with highest-weight vectors for the geometrical construction of representations in those terms).

We illustrate this with the simplest example, namely, $\mathfrak{su}(2)$, with the familiar angular momentum basis $\{J_o, J_+, J_-\}$ and $\Delta_\pm = \{\pm\alpha\}$. In the unitary spin-$j$ irreducible representation $\mathcal{D}^j$ of $SU(2)$, of dimension $2j+1$, with standard basis $\{|j, m\rangle \mid m = -j, \ldots, j\}$, the highest (or lowest) weight vector $|j, j\rangle$ (or $|j, -j\rangle$) satisfies the relation $J_+|j, j\rangle = 0$ (or $J_-|j, -j\rangle = 0$) and has an isotropy subalgebra $\mathfrak{u}(1)$ (by this, we mean the subalgebra that annihilates the given vector). Notice that here $C_2 = J_o^2 + J_+J_- + J_-J_+$ is indeed diagonal in $\mathcal{D}^j$, with eigenvalue $j(j+1)$.

B.3
Lie Groups

A Lie group G may be defined in several ways, for instance, as a smooth manifold with a group structure such that the group operations $(g_1, g_2) \mapsto g_1 g_2$, $g \mapsto g^{-1}$ are C^k, for some $k \geq 2$. This actually implies that all group operations are in fact (real) analytical. A Lie group is said to be *simple*, or *semisimple*, if it has no nontrivial invariant subgroup, or Abelian invariant subgroup.

Let G be a Lie group. For $g \in G$ consider the map $L_g : G \to G$ with $L_g(g') = gg'$. The derivative of this map at $g' \in G$, denoted $T_{g'}(L_g)$, sets up an isomorphism $T_{g'}(L_g) : T_{g'}G \to T_{gg'}G$ between the tangent spaces at g and gg'. For any vector $X \in T_eG$ (the tangent space at the identity), let us define a vector field $\widehat{X}$ on G by $\widehat{X}_g = T_e(L_g)X$. Such a vector field is said to be left-invariant. It can be demonstrated that the usual Lie bracket $[\widehat{X}, \widehat{Y}]$ of two left-invariant vector fields is again a left-invariant vector field. Using this fact, we see that if $\widehat{Y}_g = T_e(L_g)Y$, $Y \in T_eG$, the relation $[X, Y] = [\widehat{X}, \widehat{Y}]$ defines a Lie bracket on the tangent space, T_eG, at the identity of the Lie group G. Thus, equipped with this bracket relation, T_eG becomes a Lie algebra which we denote by $\mathfrak{g}$.

Next, it can be shown that for any $X \in \mathfrak{g}$, there exists a unique analytical homomorphism, $t \mapsto \theta_X(t)$ of $\mathbb{R}$ into G such that

$$\dot{\theta}_X(0) = \frac{d}{dt}\theta_X(t)|_{t=0} = X\,,$$

and for each $X \in \mathfrak{g}$ we then write $\exp X = \theta_X(1)$. The map $X \mapsto \exp X$ is called the *exponential map* and it defines a homeomorphism between an open neighborhood

N_0 of the origin $0 \in \mathfrak{g}$ and an open neighborhood N_e of the identity element e of G. Each $\theta_X(t)$ defines a one-parameter subgroup of G, with infinitesimal generator X,

$$\theta_X(t) = \exp(tX),$$

and every one-parameter subgroup is obtained in this way. If G is a matrix group, the elements X of the Lie algebra are also matrices and the exponential map comes out in terms of matrix exponentials.

Using this tool, we may sketch the fundamental theorems of Lie as follows:

- Every Lie group G has a unique Lie algebra $\mathfrak{g}$, obtained as the vector space of infinitesimal generators of all one-parameter subgroups, in other words, the tangent space $T_e G$, at the identity element of G.
- Given a Lie algebra $\mathfrak{g}$, one may associate with it, by the exponential map $X \mapsto \exp X$, a unique connected and simply connected Lie group $\overline{G}$, with Lie algebra $\mathfrak{g}$ ($\overline{G}$ is called the universal covering of G). Any other connected Lie group G with the same Lie algebra $\mathfrak{g}$ is of the form $G = \overline{G}/D$, where D is an invariant discrete subgroup of $\overline{G}$.

Furthermore, a Lie group G is simple, or semisimple, if and only if its Lie algebra $\mathfrak{g}$ is simple, or semisimple.
Here again, one may start with a *real* Lie group G, build its complexification G^c, and find all real forms of G^c. One, and only one, of them is compact (it corresponds, of course, to the compact real form of the Lie algebra). For instance, the complex Lie group $SL(2, \mathbb{C})$ has two real forms, $SU(2)$ and $SU(1, 1)$, the former being compact.

A Lie group has natural actions on its Lie algebra and its dual. These are the *adjoint* and *coadjoint* actions, respectively, and may be understood in terms of the exponential map. For $g \in G$, $g' \mapsto g g' g^{-1}$ defines a differentiable map from G to itself. The derivative of this map at $g' = e$ is an invertible linear transformation, Ad_g, of $T_e G$ (or equivalently, of $\mathfrak{g}$) onto itself, giving the adjoint action. Thus, for $t \in (-\varepsilon, \varepsilon)$, for some $\varepsilon > 0$, such that $\exp(tX) \in G$ and $X \in \mathfrak{g}$,

$$Y = \frac{d}{dt}[g \exp(tX)\, g^{-1}]|_{t=0} := \mathrm{Ad}_g(X) \tag{B11}$$

is a tangent vector in $T_e G = \mathfrak{g}$. If G is a matrix group, the adjoint action is simply

$$\mathrm{Ad}_g(X) = gXg^{-1}. \tag{B12}$$

Now considering $g \mapsto \mathrm{Ad}_g$ as a function on G with values in $\mathrm{End}\,\mathfrak{g}$, its derivative at the identity, $g = e$, defines a linear map $\mathrm{ad} : \mathfrak{g} \to \mathrm{End}\,\mathfrak{g}$. Thus, if $g = \exp X$, then $\mathrm{Ad}_g = \exp(\mathrm{ad}X)$, and it can be verified that $(\mathrm{ad}X)(Y) = [X, Y]$.

The corresponding coadjoint actions are now obtained by dualization: the coadjoint action $\mathrm{Ad}^\#_g$ of $g \in G$ on the dual $\mathfrak{g}^*$, of the Lie algebra, is given by

$$\langle \mathrm{Ad}^\#_g(X^*); X \rangle = \langle X^*; \mathrm{Ad}_{g^{-1}}(X) \rangle, \quad X^* \in \mathfrak{g}^*, \quad X \in \mathfrak{g}, \tag{B13}$$

where $\langle \cdot; \cdot \rangle \equiv \langle \cdot; \cdot \rangle_{\mathfrak{g}^*, \mathfrak{g}}$ denotes the dual pairing between $\mathfrak{g}^*$ and $\mathfrak{g}$. Once again, the (negative of the) derivative of the map $g \mapsto \mathrm{Ad}^\#_g$ at $g = e$ is a linear transformation

$\mathrm{ad}^{\#} : \mathfrak{g} \to \mathrm{End}\,\mathfrak{g}^*$, such that for any $X \in \mathfrak{g}$, $(\mathrm{ad}^{\#})(X)$ is the map

$$\langle (\mathrm{ad}^{\#})(X)(X^*); Y \rangle = \langle X^*; (\mathrm{ad})(X)(Y) \rangle , \quad X^* \in \mathfrak{g}^* , \quad Y \in \mathfrak{g} . \tag{B14}$$

Clearly, $\mathrm{Ad}^{\#}_g = \exp(-\mathrm{ad}^{\#} X)$. If we introduce a basis in $\mathfrak{g}$ and represent Ad_g by a matrix in this basis, then, in terms of the dual basis in $\mathfrak{g}^*$, $\mathrm{Ad}^{\#}_g$ is represented by the transposed inverse of this matrix.

Under the coadjoint action, the vector space $\mathfrak{g}^*$ splits into a union of disjoint *coadjoint orbits*

$$\mathcal{O}_{X^*} = \{\mathrm{Ad}^{\#}_g(X^*) | g \in G\} . \tag{B15}$$

According to the Kirillov–Souriau–Kostant theory (see [226]), each coadjoint orbit carries a natural symplectic structure. In addition, Ω is G-invariant. This implies in particular that the orbit is of even dimension and carries a natural G-invariant (Liouville) measure. Therefore, a coadjoint orbit is a natural candidate for realizing the *phase space* of a classical system and hence a starting point for a quantization procedure.

Semisimple Lie groups have several interesting decompositions. In the sequel, we present three of them, the Cartan, the Iwasawa, and the Gauss decompositions.

(1) Cartan decomposition

This is the simplest case. Given a semisimple Lie group G, its real Lie algebra $\mathfrak{g}$ always possesses a *Cartan involution*, that is, an automorphism $\theta : \mathfrak{g} \to \mathfrak{g}$, with square equal to the identity,

$$\theta[X, Y] = [\theta(X), \theta(Y)] , \quad \forall X, Y \in \mathfrak{g} , \quad \theta^2 = I_d , \tag{B16}$$

and such that the symmetric bilinear form $B_\theta(X, Y) = -B(X, \theta Y)$ is positive-definite, where B is the Cartan–Killing form. Then the Cartan involution θ yields an eigenspace decomposition

$$\mathfrak{g} = \mathfrak{k} \oplus \mathfrak{p} \tag{B17}$$

of $\mathfrak{g}$ into +1 and –1 eigenspaces. It follows that

$$[\mathfrak{k}, \mathfrak{k}] \subseteq \mathfrak{k} , \quad [\mathfrak{k}, \mathfrak{p}] \subseteq \mathfrak{p} , \quad [\mathfrak{p}, \mathfrak{p}] \subseteq \mathfrak{k} . \tag{B18}$$

Assume for simplicity that the center of G is finite, and let K denote the analytical subgroup of G with Lie algebra $\mathfrak{k}$. Then:

- K is closed and maximal-compact,
- there exists a Lie group automorphism Θ of G, with differential θ, such that $\Theta^2 = I_d$ and the subgroup fixed by Θ is K,
- the mapping $K \times \mathfrak{p} \to G$ given by $(k, X) \mapsto k \exp X$ is a diffeomorphism.

One may also write the diffeomorphism as $(X, k) \mapsto \exp X\, k \equiv p\, k$.

(2) Iwasawa decomposition

Any connected semisimple Lie group G has an Iwasawa decomposition into three closed subgroups, namely, $G = KAN$, where K is a maximal compact subgroup, A is Abelian, and N nilpotent, and the last two are simply connected. This means that every element $g \in G$ admits a *unique* factorization $g = kan$, $k \in K, a \in A, n \in N$, and the multiplication map $K \times A \times N \to G$ given by $(k, a, n) \mapsto kan$ is a diffeomorphism.

Assume that G has a finite center. Let M be the centralizer of A in K, that is, $M = \{k \in K : ka = ak, \forall a \in A\}$ (if the center of G is not finite, the definition of M is slightly more involved). Then $P = MAN$ is a closed subgroup of G, called the minimal parabolic subgroup. The interest in this subgroup is that the quotient manifold $\mathcal{X} = G/P \sim K/M$ carries the unitary irreducible representations of the principal series of G (which are induced representations), in the sense that these representations are realized in the Hilbert space $L^2(\mathcal{X}, \nu)$, with ν the natural G-invariant measure. To give a concrete example, take $G = SO_o(3,1)$, the Lorentz group. Then the Iwasawa decomposition reads

$$SO_o(3,1) = SO(3) \cdot A \cdot N\,, \tag{B19}$$

where $A \sim SO_o(1,1) \sim \mathbb{R}$ is the subgroup of Lorentz boosts in the z-direction and $N \sim \mathbb{C}$ is two-dimensional and Abelian. Then $M = SO(2)$, the subgroup of rotations around the z-axis, so that $\mathcal{X} = G/P \sim SO(3)/SO(2) \sim S^2$, the 2-sphere.

A closely related decomposition is the so-called KAK decomposition. Let G again be a semisimple Lie group with a finite center, K a maximal compact subgroup, and $G = KAN$ the corresponding Iwasawa decomposition. Then every element in G has a decomposition as $k_1 a k_2$ with $k_1, k_2 \in K$ and $a \in A$. This decomposition is in general *not* unique, but a is unique up to conjugation. A familiar example of a KAK decomposition is the expression of a general rotation $\gamma \in SO(3)$ as the product of three rotations, parameterized by the Euler angles:

$$\gamma = m(\psi)\, u(\theta)\, m(\varphi)\,, \tag{B20}$$

where m and u denote rotations around the z-axis and the y-axis, respectively.

(3) Gauss decomposition

Let G again be a semisimple Lie group and $G^c = \exp \mathfrak{g}^c$ the corresponding complexified group. If $\mathfrak{b}$ is a subalgebra of $\mathfrak{g}^c$, we call it *maximal* (in the sense of Perelomov [10]) if $\mathfrak{b} \oplus \overline{\mathfrak{b}} = \mathfrak{g}^c$, where $\overline{\mathfrak{b}}$ is the conjugate of $\mathfrak{b}$ in $\mathfrak{g}^c$. Let $\{H_j, E_\alpha\}$ be a Cartan–Chevalley basis of the complexified Lie algebra $\mathfrak{g}^c$.

Then G^c possesses remarkable subgroups:

- H^c, the Cartan subgroup generated by $\{H_j\}$.
- $B_\pm$, the Borel subgroups, which are maximal connected solvable subgroups, corresponding to the subalgebras $\mathfrak{b}_\pm$, generated by $\{H_j, E_\alpha \,|\, \alpha \in \Delta_\pm\}$; if $\mathfrak{b}$ is maximal, then $\mathfrak{b}_+ = \mathfrak{b}$ and $\mathfrak{b}_- = \overline{\mathfrak{b}}$ generate Borel subgroups.

- $Z_\pm$, the connected nilpotent subgroups generated by $\{E_{\alpha_\pm} \,|\, \alpha_\pm \in \Delta_\pm\}$.

The interest in these subgroups is that almost all elements of G^c admit a *Gauss decomposition:*

$$g = z_+ h z_- = b_+ z_- = z_+ b_- \,, \quad z_\pm \in Z_\pm \,, \quad h \in H^c \,, \quad b_\pm \in B_\pm \,. \tag{B21}$$

It follows that the quotients $X_+ = G^c/B_-$ and $X_- = B_+\backslash G^c$ are compact complex homogeneous manifolds, on which G^c acts by holomorphic transformations.

B.3.1
Extensions of Lie algebras and Lie groups

It is useful to have a method for constructing a group G from two smaller ones, one of them at least becoming a closed subgroup of G. Several possibilities are available. Here, we describe the two simplest ones.

(1) Direct product

This is the most trivial solution, which consists in glueing the two groups together, without interaction. Given two (topological or Lie) groups G_1, G_2, their direct product $G = G_1 \times G_2$ is simply their Cartesian product, endowed with the group law:

$$(g_1, g_2)(g_1', g_2') = (g_1 g_1', g_2 g_2') \,, \quad g_1, g_1' \in G_1 ,\; g_2, g_2' \in G_2 \,. \tag{B22}$$

With the obvious identifications $g_1 \sim (g_1, e_2)$, $g_2 \sim (e_1, g_2)$, where e_j denotes the neutral element of G_j, $j = 1, 2$, it is clear that both G_1 and G_2 are invariant subgroups of $G_1 \times G_2$. In the case of Lie groups, the notion of direct product corresponds to that of direct sum of the corresponding Lie algebras, $\mathfrak{g} = \mathfrak{g}_1 \oplus \mathfrak{g}_2$, and again both $\mathfrak{g}_1$ and $\mathfrak{g}_2$ are ideals of $\mathfrak{g}$.

(2) Semidirect product

A more interesting construction arises when one of the groups, say, G_2, acts on the other one, G_1, by automorphisms. More precisely, there is given a homomorphism α from G_2 into the group $\operatorname{Aut} G_1$ of automorphisms of G_1. Although the general definition may be given as in the first case, we consider only the case where $G_1 \equiv V$ is Abelian, in fact a vector space (hence group operations are noted additively), and $G_2 \equiv S$ is a subgroup of $\operatorname{Aut} V$. Then we define the semidirect product $G = V \rtimes S$ as the Cartesian product, endowed with the group law:

$$(v, s)(v', s') = (v + \alpha_s(v'), s s') \,, \quad v, v' \in V \,, \; s, s' \in S \,. \tag{B23}$$

The law (B23) entails that the neutral element of G is $(0, e_S)$ and the inverse of (v, s) is $(v, s)^{-1} = (-\alpha_s^{-1}(v), s^{-1}) = (-\alpha_{s^{-1}}(v), s^{-1})$. It is easy to check that V is an invariant subgroup of G, while S is not in general. As a matter of fact, S is invariant if and only if the automorphism α is trivial, that is, the product is direct. Indeed one has readily

$$(v, s)(v', e_S)(v, s)^{-1} = (\alpha_s(v'), e_S) \in V ,$$

and

$$(v, s)(0, s')(v, s)^{-1} = (v, ss')(-\alpha_s^{-1}(v), s^{-1}) = (v - \alpha_{ss's^{-1}}(v), ss's^{-1}) .$$

In addition to the Weyl–Heisenberg group $G_{WH} = \mathbb{R} \rtimes \mathbb{R}^2$ that we discussed in Chapter 3, the following groups are examples of semidirect products of this type are:

- The Euclidean group $E(n) = \mathbb{R}^n \rtimes SO(n)$;
- The Poincaré group $\mathcal{P}_+^{\uparrow}(1, 3) = \mathbb{R}^n \rtimes SO_o(1, 3)$, where the second factor is the Lorentz group;
- The similitude group $SIM(n) = \mathbb{R}^n \rtimes (\mathbb{R}_*^+ \times SO(n))$, where $\mathbb{R}_*^+$ is the group of dilations, whereas $SO(n)$ denotes the rotations, as in the first example (since these two operations commute, one gets here a direct product).

Appendix C
SU(2) Material

In this appendix, we just give a list of formulas concerning the group $SU(2)$ and its representations, material necessary for the construction of sigma-spin coherent states and the resulting fuzzy sphere. They are essentially extracted from Talman [74] and Edmonds [75], and also from [212].

C.1
SU(2) Parameterization

$$SU(2) \ni \xi = \begin{pmatrix} \xi_0 + i\xi_3 & -\xi_2 + i\xi_1 \\ \xi_2 + i\xi_1 & \xi_0 - i\xi_3 \end{pmatrix} . \tag{C1}$$

In bicomplex angular coordinates,

$$\xi_0 + i\xi_3 = \cos\omega e^{i\psi_1} , \quad \xi_1 + i\xi_2 = \sin\omega e^{i\psi_2} \tag{C2}$$

$$0 \leqslant \omega \leqslant \frac{\pi}{2} , \quad 0 \leqslant \psi_1, \psi_2 < 2\pi \tag{C3}$$

and so

$$SU(2) \ni \xi = \begin{pmatrix} \cos\omega e^{i\psi_1} & i\sin\omega e^{i\psi_2} \\ i\sin\omega e^{-i\psi_2} & \cos\omega e^{-i\psi_1} \end{pmatrix} , \tag{C4}$$

in agreement with Talman [74].

C.2
Matrix Elements of *SU*(2) Unitary Irreducible Representation

$$\begin{aligned} D^j_{m_1 m_2}(\xi) = {} & (-1)^{m_1 - m_2} \left[(j+m_1)!(j-m_1)!(j+m_2)!(j-m_2)! \right]^{1/2} \\ & \times \sum_t \frac{(\xi_0 + i\xi_3)^{j-m_2-t}}{(j-m_2-t)!} \frac{(\xi_0 - i\xi_3)^{j+m_1-t}}{(j+m_1-t)!} \\ & \times \frac{(-\xi_2 + i\xi_1)^{t+m_2-m_1}}{(t+m_2-m_1)!} \frac{(\xi_2 + i\xi_1)^t}{t!} , \end{aligned} \tag{C5}$$

Coherent States in Quantum Physics. Jean-Pierre Gazeau

ISBN: 978-3-527-40709-5

in agreement with Talman. With angular parameters the matrix elements of the unitary irreducible representation of $SU(2)$ are given in terms of Jacobi polynomials [18] by

$$\begin{aligned} D^j_{m_1 m_2}(\xi) =& e^{-i m_1(\psi_1+\psi_2)} e^{-i m_2(\psi_1-\psi_2)} i^{m_2-m_1} \sqrt{\frac{(j-m_1)!(j+m_1)!}{(j-m_2)!(j+m_2)!}} \\ & \times \frac{1}{2^{m_1}} (1+\cos 2\omega)^{\frac{m_1+m_2}{2}} (1-\cos 2\omega)^{\frac{m_1-m_2}{2}} \times \\ & \times P^{(m_1-m_2, m_1+m_2)}_{j-m_1}(\cos 2\omega) , \end{aligned} \tag{C6}$$

in agreement with Edmonds [75] (up to an irrelevant phase factor).

C.3
Orthogonality Relations and 3*j* Symbols

Let us equip the $SU(2)$ group with its Haar measure:

$$\mu(d\xi) = \sin 2\omega \, d\omega \, d\psi_1 \, d\psi_2 , \tag{C7}$$

in terms of the bicomplex angular parameterization. Note that the volume of $SU(2)$ with this choice of normalization is $8\pi^2$. The orthogonality relations satisfied by the matrix elements $D^j_{m_1 m_2}(\xi)$ read as

$$\int_{SU(2)} D^j_{m_1 m_2}(\xi) \, \overline{D^{j'}_{m'_1 m'_2}(\xi)} \, \mu(d\xi) = \frac{8\pi^2}{2j+1} \delta_{j j'} \delta_{m_1 m'_1} \delta_{m_2 m'_2} . \tag{C8}$$

In connection with the reduction of the tensor product of two unitary irreducible representations of $SU(2)$, we have the following equivalent formula involving the so-called 3*j* symbols (proportional to Clebsch–Gordan coefficients), in the Talman notation:

$$\begin{aligned} D^j_{m_1 m_2}(\xi) \, D^{j'}_{m'_1 m'_2}(\xi) = & \sum_{j'' m''_1 m''_2} (2j''+1) \begin{pmatrix} j & j' & j'' \\ m_1 & m'_1 & m''_1 \end{pmatrix} \\ & \times \begin{pmatrix} j & j' & j'' \\ m_2 & m'_2 & m''_2 \end{pmatrix} \overline{D^{j''}_{m''_1 m''_2}(\xi)} , \end{aligned} \tag{C9}$$

$$\begin{aligned} & \int_{SU(2)} D^j_{m_1 m_2}(\xi) \, D^{j'}_{m'_1 m'_2}(\xi) \, D^{j''}_{m''_1 m''_2}(\xi) \, \mu(d\xi) = \\ & \qquad 8\pi^2 \begin{pmatrix} j & j' & j'' \\ m_1 & m'_1 & m''_1 \end{pmatrix} \begin{pmatrix} j & j' & j'' \\ m_2 & m'_2 & m''_2 \end{pmatrix} . \end{aligned} \tag{C10}$$

One of the multiple expressions of the $3j$ symbols (in the convention that they are all real) is given by

$$\begin{pmatrix} j & j' & j'' \\ m & m' & m'' \end{pmatrix} = (-1)^{j-j'-m''} \left[\frac{(j+j'-j'')!(j-j'+j'')!(-j+j'+j'')!}{(j+j'+j''+1)!} \right]^{1/2}$$
$$\times \left[(j+m)!(j-m)!(j'+m')!(j'-m')!(j''+m'')!(j''-m'')! \right]^{1/2}$$
$$\times \sum_s (-1)^s \frac{1}{s!(j'+m'-s)!(j-m-s)!(j''-j'+m+s)!}$$
$$\times \frac{1}{(j''-j-m'+s)!(j+j'-j''-s)!} . \quad \text{(C11)}$$

C.4 Spin Spherical Harmonics

The spin spherical harmonics, as functions on the 2-sphere S^2, are defined as follows:

$$ {}_\sigma Y_{j\mu}(\hat{r}) = \sqrt{\frac{2j+1}{4\pi}} \left[D^j_{\mu\sigma}(\xi(\mathcal{R}_{\hat{r}})) \right]^* = (-1)^{\mu-\sigma} \sqrt{\frac{2j+1}{4\pi}} D^j_{-\mu-\sigma}(\xi(\mathcal{R}_{\hat{r}}))$$
$$= \sqrt{\frac{2j+1}{4\pi}} D^j_{\sigma\mu}\left(\xi^\dagger(\mathcal{R}_{\hat{r}})\right) , \quad \text{(C12)}$$

where $\xi(\mathcal{R}_{\hat{r}})$ is a (nonunique) element of $SU(2)$ which corresponds to the space rotation $\mathcal{R}_{\hat{r}}$ which brings the unit vector $\widehat{e_3}$ to the unit vector $\hat{r}$ with polar coordinates

$$\hat{r} = \begin{cases} x^1 = & \sin\theta\cos\phi , \\ x^2 = & \sin\theta\sin\phi , \\ x^3 = & \cos\theta . \end{cases} \quad \text{(C13)}$$

We immediately infer from the definition (6.55) the following properties:

$$\left({}_\sigma Y_{j\mu}(\hat{r})\right)^* = (-1)^{\sigma-\mu} {}_{-\sigma}Y_{j-\mu}(\hat{r}) , \quad \text{(C14)}$$

$$\sum_{\mu=-j}^{\mu=j} \left| {}_\sigma Y_{j\mu}(\hat{r}) \right|^2 = \frac{2j+1}{4\pi} . \quad \text{(C15)}$$

Let us recall here the correspondence (homomorphism) $\xi = \xi(\mathcal{R}) \in SU(2) \leftrightarrow \mathcal{R} \in SO(3) \simeq SU(2)/\mathbb{Z}_2$:

$$\hat{r}' = (x'_1, x'_2, x'_3) = \mathcal{R} \cdot \hat{r} \longleftrightarrow \quad \text{(C16)}$$

$$\begin{pmatrix} ix'_3 & -x'_2 + ix'_1 \\ x'_2 + ix'_1 & -ix'_3 \end{pmatrix} = \xi \begin{pmatrix} ix_3 & -x_2 + ix_1 \\ x_2 + ix_1 & -ix_3 \end{pmatrix} \xi^\dagger . \quad \text{(C17)}$$

In the particular case (C12), the angular coordinates ω, ψ_1, ψ_2 of the $SU(2)$-element $\xi(\mathcal{R}_{\hat{r}})$ are constrained by

$$\cos 2\omega = \cos\theta\,, \quad \sin 2\omega = \sin\theta\,, \quad \text{so} \quad 2\omega = \theta\,, \tag{C18}$$

$$e^{i(\psi_1+\psi_2)} = i e^{i\phi} \quad \text{so} \quad \psi_1 + \psi_2 = \phi + \frac{\pi}{2}\,. \tag{C19}$$

Here we should pay special attention to the range of values for the angle ϕ, depending on whether j and consequently σ and m are half-integer or not. If j is half-integer, then angle ϕ should be defined $\mod (4\pi)$, whereas if j is integer, it should be defined $\mod (2\pi)$.

We still have one degree of freedom concerning the pair of angles ψ_1, ψ_2. We leave open the option concerning the σ-dependent phase factor by putting

$$i^{-\sigma} e^{i\sigma(\psi_1-\psi_2)} \stackrel{\text{def}}{=} e^{i\sigma\psi}\,, \tag{C20}$$

where ψ is arbitrary. With this choice and considering (C5), we get the expressions for the spin spherical harmonics in terms of ϕ, $\theta/2$, and ψ:

$$\begin{aligned} {}_\sigma Y_{j\mu}(\hat{r}) &= (-1)^\sigma e^{i\sigma\psi} e^{i\mu\phi} \sqrt{\frac{2j+1}{4\pi}} \sqrt{\frac{(j+\mu)!(j-\mu)!}{(j+\sigma)!(j-\sigma)!}} \\ &\quad \times \left(\cos\frac{\theta}{2}\right)^{2j} \sum_t (-1)^t \binom{j-\sigma}{t} \binom{j+\sigma}{t+\sigma-\mu} \left(\tan\frac{\theta}{2}\right)^{2t+\sigma-\mu}\,, \end{aligned} \tag{C21}$$

$$\begin{aligned} &= (-1)^\sigma e^{i\sigma\psi} e^{i\mu\phi} \sqrt{\frac{2j+1}{4\pi}} \sqrt{\frac{(j+\mu)!(j-\mu)!}{(j+\sigma)!(j-\sigma)!}} \\ &\quad \times \left(\sin\frac{\theta}{2}\right)^{2j} \sum_t (-1)^{j-t+\mu-\sigma} \binom{j-\sigma}{t-\mu} \binom{j+\sigma}{t+\sigma} \left(\cot\frac{\theta}{2}\right)^{2t+\sigma-\mu}\,, \end{aligned} \tag{C22}$$

which are not in agreement with the definitions of Newman and Penrose [77], Campbell [76] (note that there is a mistake in the expression given by Campbell, in which a $\cos\frac{\theta}{2}$ should read $\cot\frac{\theta}{2}$), and Hu and White [78]. Besides the presence of different phase factors, the disagreement is certainly due to a different relation between the polar angle θ and the Euler angle.

Now, considering (C6), we get the expression for the spin spherical harmonics in terms of the Jacobi polynomials, valid in the case in which $\mu \pm \sigma > -1$:

$$\begin{aligned} {}_\sigma Y_{j\mu}(\hat{r}) &= (-1)^\mu e^{i\sigma\psi} \sqrt{\frac{2j+1}{4\pi}} \sqrt{\frac{(j-\mu)!(j+\mu)!}{(j-\sigma)!(j+\sigma)!}} \\ &\quad \times \frac{1}{2^\mu} (1+\cos\theta)^{\frac{\mu+\sigma}{2}} (1-\cos\theta)^{\frac{\mu-\sigma}{2}} P_{j-\mu}^{(\mu-\sigma,\mu+\sigma)}(\cos\theta)\, e^{i\mu\phi}\,. \end{aligned} \tag{C23}$$

For other cases, it is necessary to use alternative expressions based on the relations [18]

$$P_n^{(-l,\beta)}(x) = \frac{\binom{n+\beta}{l}}{\binom{n}{l}} \left(\frac{x-1}{2}\right)^l P_{n-l}^{(l,\beta)}(x) , \quad P_0^{(\alpha,\beta)}(x) = 1 . \tag{C24}$$

Note that with $\sigma = 0$ we recover the expression of the normalized spherical harmonics

$$\begin{aligned} {}_0Y_{jm}(\hat{r}) &= Y_{jm}(\hat{r}) \\ &= (-1)^m \sqrt{\frac{2j+1}{4\pi}} \sqrt{(j-m)!(j+m)!} \frac{1}{j!\,2^m} (\sin\theta)^m P_{j-m}^{(m,m)}(\cos\theta)\, e^{im\phi} \\ &= \sqrt{\frac{2j+1}{4\pi}} \sqrt{\frac{(j-m)!}{(j+m)!}} P_j^m(\cos\theta) e^{im\phi} , \end{aligned} \tag{C25}$$

since we have the following relation between associated Legendre polynomials and Jacobi polynomials:

$$P_{j-m}^{(m,m)}(z) = (-1)^m 2^m (1-z^2)^{-\frac{m}{2}} \frac{j!}{(j+m)!} P_j^m(z) , \tag{C26}$$

for $m > 0$. We recall also the symmetry formula

$$P_j^{-m}(z) = (-1)^m \frac{(j-m)!}{(j+m)!} P_j^m(z) . \tag{C27}$$

Our expression for spherical harmonics is rather standard, in agreement with Arfken [227, 228].[19]

C.5 Transformation Laws

We consider here the transformation law of the spin spherical harmonics under the rotation group. From the relation

$$\mathcal{R}\mathcal{R}_{{}^t\mathcal{R}\hat{r}} = \mathcal{R}_{\hat{r}} \tag{C28}$$

for any $\mathcal{R} \in SO(3)$, and from the homomorphism $\xi(\mathcal{R}\mathcal{R}') = \xi(\mathcal{R})\xi(\mathcal{R}')$ between $SO(3)$ and $SU(2)$, we deduce from the definition (6.55) of the spin spherical har-

19) Sometimes (e.g., Arfken [227]), the Condon–Shortley phase $(-1)^m$ is prepended to the definition of the spherical harmonics. Talman adopted this convention.

monics the transformation law

$$
\begin{aligned}
{}_\sigma Y_{j\mu}({}^t\mathcal{R}\cdot\hat{r}) &= \sqrt{\frac{2j+1}{4\pi}}\, D^j_{\sigma\mu}\left(\xi^\dagger\left(\mathcal{R}_{{}^t\mathcal{R}\cdot\hat{r}}\right)\right) = \sqrt{\frac{2j+1}{4\pi}}\, D^j_{\sigma\mu}\left(\xi^\dagger\left({}^t\mathcal{R}\mathcal{R}_{\hat{r}}\right)\right) \\
&= \sqrt{\frac{2j+1}{4\pi}}\, D^j_{\sigma\mu}\left(\xi^\dagger\left(\mathcal{R}_{\hat{r}}\right)\xi\left(\mathcal{R}\right)\right) \\
&= \sqrt{\frac{2j+1}{4\pi}} \sum_\nu D^j_{\sigma\nu}\left(\xi^\dagger\left(\mathcal{R}_{\hat{r}}\right)\right) D^j_{\nu\mu}\left(\xi\left(\mathcal{R}\right)\right) \\
&= \sum_\nu {}_\sigma Y_{j\nu}(\hat{r}) D^j_{\nu\mu}\left(\xi\left(\mathcal{R}\right)\right) ,
\end{aligned}
\tag{C29}
$$

as expected if we think of the special case ($\sigma = 0$) of the spherical harmonics.

Given a function $f(\hat{r})$ on the sphere S^2 belonging to the $(2j+1)$-dimensional Hilbert space $\mathcal{H}^{\sigma j}$ and a rotation $\mathcal{R} \in SO(3)$, we define the representation operator $\mathcal{D}^{\sigma j}(\mathcal{R})$ by

$$
\left(\mathcal{D}^{\sigma j}(\mathcal{R}) f\right)(\hat{r}) = f(\mathcal{R}^{-1}\cdot\hat{r}) = f({}^t\mathcal{R}\cdot\hat{r}) . \tag{C30}
$$

Thus, in particular,

$$
\left(\mathcal{D}^{\sigma j}(\mathcal{R}) \sigma Y_{j\mu}\right)(\hat{r}) =_\sigma Y_{j\mu}({}^t\mathcal{R}\cdot\hat{r}) . \tag{C31}
$$

The generators of the three rotations $\mathcal{R}^{(a)}$, $a = 1, 2, 3$, around the three usual axes, are the angular momentum operator in the representation. When $\sigma = 0$, we recover the spherical harmonics, and these generators are the usual angular momentum operators $J_a^{(j)}$ (or simply J_a), $a = 1, 2, 3$, for that representation. In the general case $\sigma \neq 0$, we call them $\Lambda_a^{(\sigma j)}$. We study their properties below.

C.6
Infinitesimal Transformation Laws

Recalling that the components $J_a = -i\varepsilon_{abc}x^b\partial_c$ of the ordinary angular momentum operator are given in spherical coordinates by

$$
\begin{aligned}
J_3 &= -i\partial_\phi , \\
J_+ &= J_1 + iJ_2 = e^{i\phi}\left(\partial_\theta + i\cot\theta\partial_\phi\right) , \\
J_- &= J_1 - iJ_2 = -e^{-i\phi}\left(\partial_\theta - i\cot\theta\partial_\phi\right) ,
\end{aligned}
\tag{C32}
$$

we have for the "sigma-spin" angular momentum operators:

$$
\Lambda_3^{\sigma j} = J_3 = -i\partial_\phi , \tag{C33}
$$

$$
\Lambda_+^{\sigma j} = \Lambda_1^{\sigma j} + i\Lambda_2^{\sigma j} = J_+ + \sigma\csc\theta e^{i\phi} , \tag{C34}
$$

$$\Lambda_{-}^{\sigma j} = \Lambda_{1}^{\sigma j} - i\Lambda_{2}^{\sigma j} = J_{-} + \sigma \csc\theta e^{-i\phi} . \tag{C35}$$

They obey the expected commutation rules,

$$\left[\Lambda_{3}^{\sigma j}, \Lambda_{\pm}^{\sigma j}\right] = \pm\Lambda_{\pm}^{\sigma j}, \quad \left[\Lambda_{+}^{\sigma j}, \Lambda_{-}^{\sigma j}\right] = 2\Lambda_{3}^{\sigma j} . \tag{C36}$$

These operators are the infinitesimal generators of the action of $SU(2)$ on the spin spherical harmonics:

$$\Lambda_{3}^{\sigma j} \, {}_{\sigma}Y_{j\mu} = \mu \, {}_{\sigma}Y_{j\mu} \tag{C37}$$

$$\Lambda_{+}^{\sigma j} \, {}_{\sigma}Y_{j\mu} = \sqrt{(j-\mu)(j+\mu+1)} \, {}_{\sigma}Y_{j\mu+1} \tag{C38}$$

$$\Lambda_{-}^{\sigma j} \, {}_{\sigma}Y_{j\mu} = \sqrt{(j+\mu)(j-\mu+1)} \, {}_{\sigma}Y_{j\mu-1} . \tag{C39}$$

C.7 Integrals and 3j Symbols

Specifying (C8) for the spin spherical harmonics leads to the following orthogonality relations which are valid for integer j (and consequently integer σ):

$$\int_{S^2} {}_{\sigma}Y_{j\mu}(\hat{r}) \, \overline{{}_{\sigma}Y_{j'\nu}(\hat{r})} \, \mu(d\hat{r}) = \delta_{jj'}\delta_{\mu\nu} . \tag{C40}$$

We recall that in the integer case, the range of values assumed by the angle ϕ is $0 \leqslant \phi < 2\pi$. Now, if we consider half-integer j (and consequently σ), the range of values assumed by the angle ϕ becomes $0 \leqslant \phi < 4\pi$. The integration above has to be carried out on the "doubled" sphere $\widetilde{S}^2$ and an extra normalization factor equal to $\frac{1}{\sqrt{2}}$ is needed in the expression of the spin spherical harmonics.

For a given integer σ the set $\left\{{}_{\sigma}Y_{j\mu}, \, -\infty \leqslant \mu \leqslant \infty, \, j \geqslant \max(0, \sigma, m)\right\}$ forms an orthonormal basis of the Hilbert space $L^2(S^2)$. Indeed, at μ fixed so that $\mu \pm \sigma \geqslant 0$, the set

$$\left\{\sqrt{\frac{2j+1}{4\pi}} \sqrt{\frac{(j-\mu)!(j+\mu)!}{(j-\sigma)!(j+\sigma)!}} \frac{1}{2^{\mu}} (1+\cos\theta)^{\frac{\mu+\sigma}{2}} (1-\cos\theta)^{\frac{\mu-\sigma}{2}} \right.$$
$$\left. \times P_{j-\mu}^{(\mu-\sigma,\mu+\sigma)}(\cos\theta) , \quad j \geqslant \mu \right\} \tag{C41}$$

is an orthonormal basis of the Hilbert space $L^2([-\pi, \pi], \sin\theta \, d\theta)$. The same holds for other ranges of values of μ by using alternative expressions such as (C24) for Jacobi polynomials. Then it suffices to view $L^2(S^2)$ as the tensor product $L^2([-\pi, \pi], \sin\theta \, d\theta) \bigotimes L^2(S^1)$. Similar reasoning is valid for half-integer σ. Then, the Hilbert space to be considered is the space of "fermionic" functions on the doubled sphere $\widetilde{S}^2$, that is, such that $f(\theta, \phi + 2\pi) = -f(\theta, \phi)$.

Specifying (C9) for the spin spherical harmonics leads to

$$ {}_{\sigma}Y_{j\mu}(\hat{r})\,{}_{\sigma'}Y_{j'\mu'}(\hat{r}) = \sum_{j''\mu''\sigma''} \sqrt{\frac{(2j+1)(2j'+1)(2j''+1)}{4\pi}} \times \begin{pmatrix} j & j' & j'' \\ \mu & \mu' & \mu'' \end{pmatrix} \begin{pmatrix} j & j' & j'' \\ \sigma & \sigma' & \sigma'' \end{pmatrix} \overline{{}_{\sigma''}Y_{j''\mu''}(\hat{r})}\,. \tag{C42} $$

We easily deduce from (C42) and *with the constraint that* $\sigma+\sigma'+\sigma''=0$ the following integral involving the product of three spherical spin harmonics (in the integer case, but analogous formula exists in the half-integer case):

$$ \int_{S^2} {}_{\sigma}Y_{j\mu}(\hat{r})\,{}_{\sigma'}Y_{j'\mu'}(\hat{r})\,{}_{\sigma''}Y_{j''\mu''}(\hat{r})\,\mu(d\hat{r}) = \sqrt{\frac{(2j+1)(2j'+1)(2j''+1)}{4\pi}} \times \begin{pmatrix} j & j' & j'' \\ \mu & \mu' & \mu'' \end{pmatrix} \begin{pmatrix} j & j' & j'' \\ \sigma & \sigma' & \sigma'' \end{pmatrix}. \tag{C43} $$

Note that this formula is independent of the presence of a constant phase factor of the type $e^{i\sigma\psi}$ in the definition of the spin spherical harmonics because of the a priori constraint $\sigma+\sigma'+\sigma''=0$. On the other hand, we have to be careful in applying (C43) because of this constraint, that is, since it has been derived from (C42) on the ground that σ'' was already *fixed* at the value $\sigma'' = -\sigma-\sigma'$. Therefore, the computation of

$$ \int_{S^2} {}_{\sigma}Y_{j\mu}(\hat{r})\,{}_{\sigma'}Y_{j'\mu'}(\hat{r})\,{}_{\sigma''}Y_{j''\mu''}(\hat{r})\,\mu(d\hat{r}) \tag{C44} $$

for an arbitrary triplet $(\sigma, \sigma', \sigma'')$ should be carried out independently.

C.8
Important Particular Case: $j = 1$

In the particular case $j = 1$, we get the following expressions for the spin spherical harmonics:

$$ {}_{\sigma}Y_{10}(\hat{r}) = e^{i\sigma\psi}\sqrt{\frac{3}{4\pi}}\frac{1}{\sqrt{(1+\sigma)!(1-\sigma)!}}\left(\cot\frac{\theta}{2}\right)^{\sigma}\cos\theta\,, \tag{C45} $$

$$ {}_{\sigma}Y_{11}(\hat{r}) = -e^{i\sigma\psi}\sqrt{\frac{3}{4\pi}}\frac{1}{\sqrt{2(1+\sigma)!(1-\sigma)!}}\left(\cot\frac{\theta}{2}\right)^{\sigma}\sin\theta\, e^{i\phi}\,, \tag{C46} $$

$$ {}_{\sigma}Y_{1-1}(\hat{r}) = (-1)^{\sigma}e^{-i\sigma\psi}\sqrt{\frac{3}{4\pi}}\frac{1}{\sqrt{2(1+\sigma)!(1-\sigma)!}}\left(\tan\frac{\theta}{2}\right)^{\sigma}\sin\theta\, e^{-i\phi}\,. \tag{C47} $$

For $\sigma = 0$, we recover familiar formulas connecting spherical harmonics to components of the vector on the unit sphere:

$$ Y_{10}(\hat{r}) = \sqrt{\frac{3}{4\pi}}\cos\theta = \sqrt{\frac{3}{4\pi}}x^3\,, \tag{C48} $$

$$Y_{11}(\hat{r}) = -\sqrt{\frac{3}{4\pi}}\frac{1}{\sqrt{2}}\sin\theta e^{i\phi} = -\sqrt{\frac{3}{4\pi}}\frac{x^1 + ix^2}{\sqrt{2}}, \tag{C49}$$

$$Y_{1-1}(\hat{r}) = \sqrt{\frac{3}{4\pi}}\frac{1}{\sqrt{2}}\sin\theta e^{-i\phi} = \sqrt{\frac{3}{4\pi}}\frac{x^1 - ix^2}{\sqrt{2}}. \tag{C50}$$

C.9
Another Important Case: $\sigma = j$

For $\sigma = j$, owing to the relations (6.63), the spin spherical harmonics reduce to their simplest expressions:

$$_jY_{j\mu}(\hat{r}) = (-1)^j e^{ij\psi}\sqrt{\frac{2j+1}{4\pi}}\sqrt{\binom{2j}{j+\mu}}\left(\cos\frac{\theta}{2}\right)^{j+\mu}\left(\sin\frac{\theta}{2}\right)^{j-\mu} e^{i\mu\phi}. \tag{C51}$$

They are precisely the states which appear in the construction of the Gilmore–Radcliffe coherent states. Otherwise said, the latter and related quantization are just particular cases of our approach.

Appendix D
Wigner–Eckart Theorem for Coherent State Quantized Spin Harmonics

We give more precision on the rotational covariance properties of some operators obtained in Chapter 17 from the coherent state quantization of spin spherical harmonics.

By construction, the operators $A_{_\nu Y_{kn}}$ acting on $\mathcal{H}^{\sigma j}$ are tensorial-irreducible. Indeed, under the action of the representation operator $\mathcal{D}^{\sigma j}(\mathcal{R})$ in $\mathcal{H}^{\sigma j}$, due to (6.83), the rotational invariance of the measure and $\mathcal{N}(\hat{r})$, and (C29), they transform as

$$\begin{aligned}\mathcal{D}^{\sigma j}(\mathcal{R})A_{_\nu Y_{kn}}\mathcal{D}^{\sigma j}(\mathcal{R}^{-1}) &= \int_{S^2} {_\nu Y_{kn}}(\hat{r})\,|\mathcal{R}\cdot\hat{r};\sigma\rangle\langle\mathcal{R}\cdot\hat{r};\sigma|\,\mathcal{N}(\hat{r})\,\mu(d\hat{r})\\ &= \int_X {_\nu Y_{kn}}(\mathcal{R}^{-1}\cdot\hat{r})\,|\hat{r};\sigma\rangle\langle\hat{r};\sigma|\,\mathcal{N}(\hat{r})\,\mu(d\hat{r})\\ &= \sum_{n'} D^k_{n'n}(\xi(\mathcal{R}))\int_{S^2} {_\nu Y_{kn'}}(\hat{r})\,|\hat{r};\sigma\rangle\langle\hat{r};\sigma|\,\mathcal{N}(\hat{r})\,\mu(d\hat{r})\\ &= \sum_{n'} A_{_\nu Y_{kn'}} D^k_{n'n}(\xi(\mathcal{R}))\,.\end{aligned} \tag{D1}$$

Therefore, the Wigner–Eckart theorem [75] tells us that the matrix elements of the operator $A_{_\nu Y_{kn}}$ with respect to the spin spherical harmonic basis $\left\{{_\sigma Y_{jm}}\right\}$ are given by

$$\left[A_{_\nu Y_{kn}}\right]_{mm'} = (-1)^{j-m}\begin{pmatrix} j & j & k\\ -m & m' & n\end{pmatrix}\mathcal{K}(\nu,\sigma,j,k)\,. \tag{D2}$$

Note that the presence of the $3j$ symbol in (D2) implies the selection rules $n + m' = m$ and the triangular rule $0 \leqslant k \leqslant 2j$. The proportionality coefficient $\mathcal{K}$ can be computed directly from (17.6) by choosing therein suitable values of m, m'.

On the other hand, we have by definition (C29,C31)

$$\sum_{n'} {_\nu Y_{kn'}} D^k_{n'n}(\xi(\mathcal{R})) = \mathcal{D}^{\nu k}(\mathcal{R})_\nu Y_{kn}\,.$$

Thus, from the formula above,

$$\mathcal{D}^{\sigma j}(\mathcal{R})A_{_\nu Y_{kn}}\mathcal{D}^{j}(\mathcal{R}^{-1}) = A_{\mathcal{D}^{\nu k}(\mathcal{R})_\nu Y_{kn}}\,.$$

Coherent States in Quantum Physics. Jean-Pierre Gazeau

ISBN: 978-3-527-40709-5

In the special case $\nu = 0$,

$$\mathcal{D}^{\sigma j}(\mathcal{R}) A_{Y_{kn}} \mathcal{D}^{j}(\mathcal{R}^{-1}) = A_{\mathcal{D}^{0k}(\mathcal{R}) Y_{kn}} \, . \tag{D3}$$

Its infinitesimal version for each of the three rotations $\mathcal{R}_i$ reads as

$$\left[\Lambda_i^{(\sigma j)}, A_{Y_{kn}} \right] = A_{J_i^{(k)} Y_{kn}} \, . \tag{D4}$$

Appendix E
Symmetrization of the Commutator

We want to prove that

$$S\left(\left[J_3, J_1^{\alpha_1} J_2^{\alpha_2} J_3^{\alpha_3}\right]\right) = \left[J_3, S\left(J_1^{\alpha_1} J_2^{\alpha_2} J_3^{\alpha_3}\right)\right],$$

where J_i is a representation of so(3).

Let us first comment on the symmetrization:

$$S\left(J_1^{\alpha_1} J_2^{\alpha_2} J_3^{\alpha_3}\right) = \frac{1}{l!} \sum_{\sigma \in S_l} J_{i_{\sigma(1)}} \cdots J_{i_{\sigma(l)}},$$

where $l = \alpha_1 + \alpha_2 + \alpha_3$. The terms of the sum are not all distinct, since the exchange of, say, two J_1 gives the same term: each term appears in fact $\alpha_1!\alpha_2!\alpha_3!$ times, so there are $l!/(\alpha_1!\alpha_2!\alpha_3!)$ distinct terms. This is the number of sequences of length l, with values in $\{1, 2, 3\}$, where there are α_i occurrences of the value i (for $i = 1, 2, 3$). One denotes this set as $U_{\alpha_1,\alpha_2,\alpha_3}$. After grouping of identical terms, one obtains

$$S\left(J_1^{\alpha_1} J_2^{\alpha_2} J_3^{\alpha_3}\right) = \frac{\alpha_1!\alpha_2!\alpha_3!}{l!} \sum_{u \in U_{\alpha_1,\alpha_2,\alpha_3}} J_{u_1} \cdots J_{u_l},$$

where all the terms of the summation are now different.

Let us now calculate $S\left(\left[J_3, J_1^{\alpha_1} J_2^{\alpha_2} J_3^{\alpha_3}\right]\right)$. First, we write

$$\left[J_3, J_1^{\alpha_1} J_2^{\alpha_2} J_3^{\alpha_3}\right] = \underbrace{\left[J_3, J_1^{\alpha_1}\right] J_2^{\alpha_2} J_3^{\alpha_3}}_{A} + \underbrace{J_1^{\alpha_1}\left[J_3, J_2^{\alpha_2}\right] J_3^{\alpha_3}}_{B},$$

with

$$A = \sum_{k=1}^{\alpha_1} \underbrace{J_1 \cdots J_1}_{k-1 \text{ terms}} J_2 \underbrace{J_1 \cdots J_1}_{\alpha_1 - k \text{ terms}} J_2^{\alpha_2} J_3^{\alpha_3}.$$

The different terms in A give the same symmetrized operator. Thus,

$$S(A) = \alpha_1 S\left(J_1^{\alpha_1 - 1} J_2^{\alpha_2 + 1} J_3^{\alpha_3}\right) \tag{E1}$$

Coherent States in Quantum Physics. Jean-Pierre Gazeau

ISBN: 978-3-527-40709-5

$$= \alpha_1 \frac{(\alpha_1 - 1)!(\alpha_2 + 1)!\alpha_3!}{l!} \sum_{u \in U_{\alpha_1 - 1, \alpha_2 + 1, \alpha_3}} J_{u_1} \cdots J_{u_l} . \tag{E2}$$

Similarly, for B,

$$S(B) = -\alpha_2 \frac{(\alpha_1 + 1)!(\alpha_2 - 1)!\alpha_3!}{l!} \sum_{u \in U_{\alpha_1 + 1, \alpha_2 - 1, \alpha_3}} J_{u_1} \cdots J_{u_l} .$$

Now we calculate

$$I = \left[J_3, S(J_1^{\alpha_1} J_2^{\alpha_2} J_3^{\alpha_3}) \right]$$
$$= \frac{\alpha_1!\alpha_2!\alpha_3!}{l!} \sum_{u \in U_{\alpha_1,\alpha_2,\alpha_3}} \sum_{k=1}^{l} J_{u_1} \cdots J_{u_{k-1}} [J_3, J_{u_k}] J_{u_{k+1}} \cdots J_{u_l} .$$

The sum splits into two parts, according to the value of $u_k = 1$ or 2:

$$I = A' + B' ,$$

with

$$A' = \frac{\alpha_1!\alpha_2!\alpha_3!}{l!} \sum_{u \in U_{\alpha_1,\alpha_2,\alpha_3}} \sum_{k|u_k=1} J_{u_1} \cdots J_{u_{k-1}} J_2 J_{u_{k+1}} \cdots J_{u_l}$$

and

$$B' = -\frac{\alpha_1!\alpha_2!\alpha_3!}{l!} \sum_{u \in U_{\alpha_1,\alpha_2,\alpha_3}} \sum_{k|u_k=2} J_{u_1} \cdots J_{u_{k-1}} J_1 J_{u_{k+1}} \cdots J_{u_l} .$$

Let us examine the constituents of A'. There are of the form $J_{u_1} \cdots J_{u_l}$, with $u \in U_{\alpha_1-1,\alpha_2+1,\alpha_3}$. Their number is $l!/(\alpha_1!\alpha_2!\alpha_3!) \times \alpha_1$, but they are not all different. Each monomial emerges from a term where a J_1 has been transformed into a J_2. Since there are α_2+1 occurrences of J_2 in each term, each monomial appears α_2+1 times. We now group these identical terms:

$$A' = \frac{\alpha_1!\alpha_2!\alpha_3!}{l!} (\alpha_2 + 1) \sum_{?} J_{u_1} \cdots J_{u_l} .$$

It remains to determine the definition set of the summation. Let us first estimate the number of its terms, namely,

$$N = \frac{l!}{\alpha_1!\alpha_2!\alpha_3!} \frac{\alpha_1}{\alpha_2 + 1} = \frac{l!}{(\alpha_1 - 1)!(\alpha_2 + 1)!\alpha_3!} .$$

This is the number of elements in $U_{\alpha_1-1,\alpha_2+1,\alpha_3}$. On the other hand, all the elements of $U_{\alpha_1-1,\alpha_2+1,\alpha_3}$ appear. In the contrary case, the retransformation of a J_2 into a J_1 would provide some elements not appearing in I, which cannot be true. It re-

sults that the sum comprises exactly all symmetrized expressions of $J_1^{\alpha_1-1} J_2^{\alpha_2+1} J_3^{\alpha_3}$.
Thus,

$$\begin{aligned}
A' &= \frac{\alpha_1!\alpha_2!\alpha_3!}{l!}(\alpha_2+1) \sum_{u \in U_{\alpha_1-1,\alpha_2+1,\alpha_3}} J_{u_1} \cdots J_{u_l} \\
&= \alpha_1 \frac{(\alpha_1-1)!(\alpha_2+1)!\alpha_3!}{l!} \sum_{u \in U_{\alpha_1-1,\alpha_2+1,\alpha_3}} J_{u_1} \cdots J_{u_l} \\
&= S(A)\,.
\end{aligned}$$

The application of the same method to B' leads to the proof.

References

1 Schrödinger, E. (1926) Der stetige Übergang von der Mikro- zur Makromechanik, *Naturwiss.*, **14**, 664.

2 Klauder, J.R. (1960) The action option and the Feynman quantization of spinor fields in terms of ordinary c-numbers, *Ann. Phys.*, **11**, 123.

3 Klauder, J.R. (1963) Continuous–Representation theory I. Postulates of continuous-representation theory, *J. Math. Phys.*, **4**, 1055.

4 Klauder, J.R. (1963) Continuous–Representation theory II. Generalized relation between quantum and classical dynamics, *J. Math. Phys.*, **4**, 1058.

5 Glauber, R.J. (1963) Photons correlations, *Phys. Rev. Lett.*, **10**, 84.

6 Glauber, R.J. (1963) The quantum theory of optical coherence, *Phys. Rev.*, **130**, 2529.

7 Glauber, R.J. (1963) Coherent and incoherent states of radiation field, *Phys. Rev.*, **131**, 2766.

8 Sudarshan, E.C.G. (1963) Equivalence of semiclassical and quantum mechanical descriptions of statistical light beams, *Phys. Rev. Lett.*, **10**, 277.

9 Zhang, W.-M., Feng, D.H., and Gilmore, R. (1990) Coherent states: Theory and some applications, *Rev. Mod. Phys.*, **26**, 867.

10 Perelomov, A.M. (1986) *Generalized Coherent States and their Applications*, Springer-Verlag, Berlin.

11 Syad Twareque Ali, Antoine, J.P., and Gazeau, J.P. (2000) *Coherent states, wavelets and their generalizations. Graduate Texts in Contemporary Physics*, Springer-Verlag, New York.

12 Dodonov, V.V. (2002) Nonclassical states in quantum optics: a "squeezed" review of the first 75 years, *J. Opt. B: Quantum Semiclass. Opt.*, **4**, R1.

13 Dodonov, V.V. and Man'ko, V.I. (eds) (2003) *Theory of Nonclassical States of Light*, Taylor & Francis, London, New York.

14 Vourdas, A. (2006) Analytic representations in quantum mechanics, *J. Phys. A*, **39**, R65.

15 Klauder, J.R. and Skagerstam, B.S. (eds) (1985) *Coherent states. Applications in physics and mathematical physics*, World Scientific Publishing Co., Singapore.

16 Feng, D.H., Klauder, J.R. and Strayer, M. (eds) (1994) *Coherent States: Past, Present and Future*, Proceedings of the 1993 Oak Ridge Conference. World Scientific, Singapore.

17 Landau, L.D. and Lifshitz, E.M. (1958) *Quantum Mechanics. Non-relativistic Theory*, Pergamon Press, Oxford.

18 Magnus, W., Oberhettinger, F. and Soni, R.P. (1966) *Formulas and Theorems for the Special Functions of Mathematical Physics*, Springer-Verlag, Berlin, Heidelberg and New York.

19 Lieb, E.H. (1973) The classical limit of quantum spin systems, *Commun. Math. Phys.*, **31**, 327.

Coherent States in Quantum Physics. Jean-Pierre Gazeau

ISBN: 978-3-527-40709-5

20 Berezin, F.A. (1975) General concept of quantization, *Commun. Math. Phys.*, **40**, 153.

21 von Neumann, J. (1955) *Mathematical Foundations of Quantum Mechanics*, Princeton University Press, Princeton, NJ, English translation by R.T. Byer.

22 Torrésani, B. (1995) *Analyse continue par ondelettes*, EDP Sciences, Paris.

23 Cohen, L. (1994) *Time-frequency analysis: theory and applications*, Prentice Hall Signal Processing Series, NJ.

24 Mallat, S.G. (1998) *A Wavelet Tour of Signal Processing*, Academic Press, New York, NY.

25 Scully, M.O. and Suhail Zubairy, M. (1997) *Quantum Optics*, Cambridge Univ. Press, Cambridge.

26 Schleich, W.P. (2001) *Quantum optics in Phase space* Wiley-VCH Verlag GmbH.

27 Feynman, R.P. (1948) Space-time approach to nonrelativistic quantum mechanics, *Rev. Mod. Phys.*, **20**, 367.

28 Klauder, J.R. (2003) The Feynman Path Integral: An Historical Slice, in *A Garden of Quanta* (eds Arafune, J. *et al.*), World Scientific, Singapore, pp. 55–76.

29 Feynman, R.P. (1951) An operator calculus having applications in quantum electrodynamics, *Phys. Rev.*, **84**, 108.

30 Daubechies, I., and Klauder, J.R. (1985) Quantum mechanical path integrals with Wiener measures for all polynomial Hamiltonians. II, *J. Math. Phys.*, **26**, 2239.

31 dos Santos, L.C. and de Aguiar, M.A.M. (2006) Coherent state path integrals in the Weyl representation, *J. Phys. A: Math. Gen.*, **39**, 13465.

32 Preskill, J. (2008) *Quantum Computation and Information*, Caltech PH-229 Lecture Notes. http://www.theory.caltech.edu/~preskill/ph219/ph219_2008-09.

33 Nielsen, M.A. and Chuang, I.L. (2000) *Quantum Computation and Quantum Information*, Cambridge University Press, Cambridge.

34 Audenaert, K.M.R. (2007) Mathematical Aspects of Quantum Information Theory, in *Physics and Theoretical Computer Science*, (eds Gazeau, J.-P. *et al.*), IOS Press, pp. 3–24.

35 Holevo, A.S. (2001) *Statistical Structure of Quantum Theory*, Springer-Verlag, Berlin.

36 Fuchs, C.A. (1996) Distinguishability and Accessible Information in Quantum Theory, Ph.D. thesis, University of New Mexico.

37 Peres, A. (1995) *Quantum Theory: Concepts and Methods*, Kluwer Academic Publishers, Dordrecht.

38 Helstrom, C.W. (1976) *Quantum Detection and Estimation Theory*, Academic Press, New York.

39 Bennett, C.H. and Brassard, G. (1984) in *Proceedings of the IEEE International Conference on Computers, Systems and Signal Processing*, IEEE, New York, p. 175.

40 Cook, R.L., Martin, P.J. and Geremia, J.M. (2007) Optical coherent state discrimination using a closed-loop quantum measurement, *Nature*, **446**, 774.

41 Dolinar, S. (1973) Quaterly Progress Report. Tech. Rep. 111, Research Laboratory of Electronics, MIT, p. 115, unpublished.

42 Geremia, J.M. (2004) Distinguishing between optical coherent states with imperfect detection, *Phys. Rev. A*, **70**, 062303-1.

43 Loudon, R. (1973) *The Quantum Theory of Light*, Oxford Univ. Press, Oxford.

44 Kennedy, R.S. (1972) Quaterly Progress Report. Tech. Rep. 110, Research Laboratory of Electronics, MIT, p. 219, unpublished.

45 Sasaki, M. and Hirota, O. (1996) Optimal decision scheme with a unitary control process for binary quantum-state signals, *Phys. Rev. A*, **54**, 2728.

46 Dolinar, S. (1976) Ph.D. thesis, Massachussets Institute of Technology.

47 Bertsekas, D.P. (2000) *Dynamic Programming and Optimal Control*, vol. 1, Athena Scientific, Belmont, MA.

48 Wittman, C., Takeoka, M., Cassemiro, K.N., Sasaki, M., Leuchs, G. and Andersen, U.L. (2008) Demonstration of near-optimal discrimination of optical coherent states, arXiv:0809.4953v1 [quant-ph].

49 Takeoka, M., and Sasaki, M. (2008) Discrimination of the binary coherent signal: Gaussian-operation limit and simple non-Gaussian near optimal receivers, *Phys. Rev. A*, **78**, 022320-1.

50 Kotz, S., Balakrishnan, N., Read, C.B. and Vidakovic, B. (2006) editors-in-chief: under the entries "Bayesian Inference" and

"Conjugate Families of Distributions". *Encyclopaedia of Statistical Sciences*, 2nd edn, vol. 1, John Wiley & Sons, Inc., Hoboken, NJ.
51 Box, G.E.P. and Tiao, G.C. (1973) *Bayesian Inference in Statistical Analysis*, Wiley Classics Library, Wiley-Interscience.
52 Heller, B. and Wang, M. (2004) Posterior distributions on certain parameter spaces by using group theoretic methods adopted from quantum mechanics, Univ. of Chicago, Dept. of Statistics Technical Report Series, Tech. Rep. No. 546, p. 26.
53 Heller, B. and Wang, M. (2007) Group invariant inferred distributions via noncommutative probability, in *Recent Developments in Nonparametric Inference and Probability*, IMS Lecture Notes-Monograph Series, **50**, p. 1.
54 Heller, B. and Wang, M. (2007) Posterior distribution for negative binomial parameter *p* using a group invariant prior, *Stat. Probab. Lett.*, **77**, 1542.
55 Ali, S.T., Gazeau, J.-P. and Heller, B. (2008) Coherent states and Bayesian duality, *J. Phys. A: Math. Theor.*, **41**, 365302.
56 Daubechies, I. (1992) *Ten Lectures on Wavelets*, SIAM, Philadelphia.
57 Klauder, J.R. (1996) Coherent states for the hydrogen atom, *J. Phys. A: Math. Gen.*, **29**, L293.
58 Gazeau, J.-P. and Klauder, J.R. (1999) Coherent states for systems with discrete and continuous spectrum, *J. Phys. A: Math. Gen.*, **32**, 123.
59 Gazeau, J.-P. and Monceau, P. (2000) Generalized coherent states for arbitrary quantum systems. in *Conférence Moshé Flato 1999 – Quantization, Deformations, and Symmetries*, vol. II (eds Dito, G. and Sternheimer, D.), Kluwer, Dordrecht, pp. 131–144.
60 Busch, P., Lahti, P. and Mittelstaedt, P. (1991) *The Quantum Theory of Measurement*, LNP vol. m2, Springer-Verlag, Berlin, second revised edition 1996.
61 Ali, S.T., Antoine, J.-P., and Gazeau, J.-P. (1993) Continuous frames in Hilbert spaces, *Ann. Phys.*, **222**, 1.
62 De Bièvre, S. and González, J.A. (1993) Semiclassical behaviour of coherent states on the circle, in *Quantization and Coherent States Methods in Physics* (eds Odzijewicz, A. *et al.*), World Scientific, Singapore.
63 Kowalski, K., Rembieliński, J. and Papaloucas, L.C. (1996) Coherent states for a quantum particle on a circle, *J. Phys. A: Math. Gen.*, **29**, 4149.
64 Kowalski, K. and Rembielinski, J. (2002) Exotic behaviour of a quantum particle on a circle, *Phys. Lett. A*, **293**, 109.
65 Kowalski, K. and Rembielinski, J. (2002) On the uncertainty relations and squeezed states for the quantum mechanics on a circle, *J. Phys. A: Math. Gen.*, **35**, 1405.
66 Kowalski, K. and Rembielinski, J. (2003) Reply to the Comment on "On the uncertainty relations and squeezed states for the quantum mechanics on a circle", *J. Phys. A: Math. Gen.*, **36**, 5695.
67 González, J.A., and del Olmo, M.A. (1998) Coherent states on the circle, *J. Phys. A: Math. Gen.*, **31**, 8841.
68 Hall, B.C., and Mitchell, J.J. (2002) Coherent states on spheres, *J. Math. Phys.*, **43**, 1211.
69 Radcliffe, J.M. (1971) Some properties of spin coherent states, *J. Phys. A*, **4**, 313.
70 Gilmore, R. (1972) Geometry of symmetrized states, *Ann. Phys. (NY)*, **74**, 391.
71 Gilmore, R. (1974) On properties of coherent states, *Rev. Mex. Fis.*, **23**, 143.
72 Perelomov, A.M. (1972) Coherent states for arbitrary Lie group, *Commun. Math. Phys.*, **26**, 222.
73 Hamilton, W.R. (1844) On a new species of imaginary quantities connected with a theory of quaternions, *Proc. R. Ir. Acad.*, **2**, 424.
74 Talman, J.D. (1968) *Special Functions, A Group Theoretical Approach*, W.A. Benjamin, New York, Amsterdam.
75 Edmonds, A.R. (1968) *Angular Momentum in Quantum Mechanics*, 2nd ed., Princeton University Press, Princeton, NJ, rev. printing.
76 Campbell, W.B. (1971) Tensor and spinor spherical harmonics and the spins harmonics ${}_sY_{lm}(\theta, \phi)$, *J. Math. Phys.*, **12**, 1763.
77 Newman, E.T. and Penrose, R. (1966) Note on the Bondi-Metzner-Sachs Group, *J. Math. Phys.*, **7**, 863.

78 Hu, W. and White, M. (1997) CMB anisotropies: total angular momentum method, *Phys. Rev. D*, **56**, 596.

79 Carruthers, P. and Nieto, M.M. (1965) Coherent states and the forced quantum oscillator, *Am. J. Phys.*, **33**, 537.

80 Wang, Y.K. and Hioe, F.T. (1973) Phase transition in the Dicke maser model, *Phys. Rev. A*, **7**, 831.

81 Hepp, K. and Lieb, E.H. (1973) Equilibrium statistical mechanics of matter interacting with the quantized radiation field, *Phys. Rev. A*, **8**, 2517.

82 Erdélyi, A. (1987) *Asymptotic Expansions*, Dover, New York.

83 Fuller, W. and Lenard, A. (1979) Generalized quantum spins, coherent states, and Lieb inequalities, *Commun. Math. Phys.*, **67**, 69.

84 Biskup, M., Chayes, L. and Starr, S. (2007) Quantum spin systems at positive temperature, *Commun. Math. Phys.*, **269**, 611.

85 Encyclopedia of Laser Physics and Technology, http://www.rp-photonics.com/encyclopedia.html.

86 Dicke, R.H. (1954) Coherence in Spontaneous Radiation Processes, *Phys. Rev.*, **93**, 99.

87 Rehler, N.E. and Eberly, J.H. (1971) Superradiance, *Phys. Rev. A*, **3**, 1735.

88 Bonifacio, R., Schwendimann, P. and Haake, F. (1971) Quantum statistical theory of superradiance. I, *Phys. Rev. A*, **4**, 302.

89 Bonifacio, R., Schwendimann, P. and Haake, F. (1971) Quantum statistical theory of superradiance. II, *Phys. Rev. A*, **4**, 854.

90 Berezin, F.A. (1971) Wick and anti-Wick symbols of operators, *Mat. Sb.*, **86**, 578, (english transl. in (1971) *Math. USSR-Sb.*, **15**, 577).

91 Berezin, F.A. (1972) Covariant and contravariant symbols of operators, *SSSR Ser. Mat.*, **36**, 1134, (english transl. in (1973) *Math. USSR-Izv.*, **6**, 1117).

92 Thirring, W.E. and Harrell, E.M. (2004) *Quantum Mathematical Physics*, Springer, Berlin.

93 Gazeau, J.-P. and Hussin, V. (1992) Poincaré contraction of $SU(1,1)$ Fock–Bargmann structure, *J. Phys. A: Math. Gen.*, **25**, 1549.

94 Gazeau, J.-P. and Renaud, J. (1993) Lie algorithm for an interacting $SU(1,1)$ elementary system and its contraction, *Ann. Phys. (NY)*, **222**, 89.

95 Gazeau, J.-P. and Renaud, J. (1993) Relativistic harmonic oscillator and space curvature, *Phys. Lett. A*, **179**, 67.

96 Doubrovine, B., Novikov, S. and Famenko, A. (1982) *Géométrie Contemporaine, Méthodes et Applications*, (1ère Partie) Mir, Moscow.

97 Bargmann, V. (1947) Irreducible unitary representations of the Lorentz group, *Ann. Math.*, **48**, 568.

98 Gelfand, I. and Neumark, M. (1947) Unitary representations of the Lorentz group, *Acad. Sci. USSR. J. Phys.*, **10**, 93.

99 Vilenkin, N.J. and Klimyk, A.U. (1991) *Representations of Lie Groups and Special Functions*, Kluwer Academic, Boston.

100 Pukánszky, L. (1964) The Plancherel formula for the universal covering group of $SL(\mathbb{R},2)$, *Math. Ann.*, **156**, 96.

101 Basu, D. (2007) The Plancherel formula for the universal covering group of $SL(2,R)$ revisited, arXiv:0710.2224v3 [hep-th].

102 Miller Jr., W. (1968) *Lie Theory and Special Functions*, Academic, New York, Chap. I and V.

103 Gazeau, J.-P. and Maquet, A. (1979) Bound states in a Yukawa potential: A Sturmian group-theoretical approach, *Phys. Rev. A*, **20**, 727.

104 Castañeda, J. A., Hernández, M. A., and Jáuregui, R. (2008) Continuum effects on the temporal evolution of anharmonic coherent molecular vibrations, *Phys. Rev. A*, **78**, 78.

105 Benedict, M.G, and Molnar, B. (1999) Algebraic construction of the coherent states of the Morse potential based on supersymmetric quantum mechanics, *Phys. Rev. A*, **60**, R1737.

106 Molnar, B., Benedict, M.G, and Bertrand, J. (2001) Coherent states and the role of the affine group in the quantum mechanics of the Morse potential, *J. Phys. A: Math. Gen.*, **34**, 3139.

107 Ferapontov, E.V. and Veselov, A.P. (2001) Integrable Schrödinger operators with

magnetic fields: Factorization method on curved surfaces, *J. Math. Phys.*, **42**, 590.
108 Mouayn, Z. (2003) Characterization of hyperbolic Landau states by coherent state transforms, *J. Phys. A: Math. Gen.*, **36**, 8071.
109 Aslaksen, E.W. and Klauder, J.R. (1968) Unitary representations of the affine group, *J. Math. Phys.*, **9**, 206.
110 Aslaksen, E.W. and Klauder, J.R. (1969) Continuous representation theory using the affine group, *J. Math. Phys.*, **10**, 2267.
111 Pöschl, G. and Teller, E. (1933) Bemerkungen zur Quantenmechanik des anharmonischen Oszillators, *Z. Phys.*, **83**, 143.
112 Antoine, J.-P., Gazeau, J.-P., Monceau, P., Klauder, J.R., Penson, K.A. (2001) Temporally stable coherent states for infinite well, *J. Math. Phys.*, **42**, 2349.
113 Barut, A.O. and Girardello, L. (1971) New "coherent" states associated with non compact groups, *Commun. Math. Phys.*, **21**, 41.
114 Flügge, S. (1971) *Practical Quantum Mechanics I*, Springer-Verlag, Berlin, Heidelberg and New York.
115 Akhiezer, N.I. (1965) *The classical moment problem* Oliver & Boyd LTD, Edinburgh and London.
116 Simon, B. (1998) The classical moment problem as a self-adjoint finite difference operator, *Adv. Math.*, **137**, 82.
117 Besicovitch, A.S. (1932) *Almost Periodic Functions*, Cambridge University Press, Cambridge.
118 Nieto, M.M. and Simmons, Jr., L.M. (1987) Coherent states for general potentials, *Phys. Rev. Lett.*, **41**, 207.
119 Aronstein, D.L. and Stroud, C.R. (1997) Fractional wave-function revivals in the infinite square well, *Phys. Rev. A*, **55**, 4526.
120 Averbuch, I.S. and Perelman, N.F. (1989) Fractional revivals: Universality in the long-term evolution of quantum wave packets beyond the correspondence principle dynamics, *Phys. Lett. A*, **139**, 449.
121 Kinzel, W. (1995) Bilder elementarer Quantenmechanik, *Phys. Bl.*, **51**, 1190.
122 Matos Filho, R.L. and Vogel, W. (1996) Nonlinear coherent states, *Phys. Rev. A*, **54**, 4560.
123 Großmann, F., Rost, J.-M. and Schleich, W.P. (1997) Spacetime structures in simple quantum systems, *J. Phys. A: Math. Gen.*, **30**, L277.
124 Stifter, P., Lamb, Jr., W.E. and Schleich, W.P. (1997) The particle in the box revisited, in *Proceedings of the Conference on Quantum Optics and Laser Physics* (eds Jin, L. and Zhu, Y.S.) World Scientific, Singapore.
125 Marzoli, I., Friesch, O.M. and Schleich, W.P. (1998) Quantum carpets and Wigner functions, in *Proceedings of the 5th Wigner Symposium (Vienna,1997)* (eds Kasperkovitz, P. and Grau, D.) World Scientific, Singapore, pp. 323–329.
126 Bluhm, R., Kostelecky, V.A. and Porter, J.A. (1996) The evolution and revival structure of localized quantum wave packets, *Am. J. Phys.*, **64**, 944.
127 Robinett, R.W. (2000) Visualizing the collapse and revival of wave packets in the infinite square well using expectation values, *Am. J. Phys.*, **68**, 410.
128 Robinett, R.W. (2004) Quantum wave packet revivals, *Phys. Rep.*, **392**, 1.
129 Mandel, L. (1979) Sub-Poissonian photon statistics in resonance fluorescence, *Opt. Lett.*, **4**, 205.
130 Peřina, J. (1984) *Quantum Statistics of Linear and Nonlinear Optical Phenomena*, Reidel, Dordrecht.
131 Solomon, A.I. (1994) A characteristic functional for deformed photon phenomenology, *Phys. Lett. A*, **196**, 29.
132 Katriel, J. and Solomon, A.I. (1994) Nonideal lasers, nonclassical light, and deformed photon states, *Phys. Rev. A*, **49**, 5149.
133 Peřina, J. (ed.) (2001) *Coherence and Statistics of Photons and Atoms*, Wiley–Interscience.
134 Bonneau, G., Faraut, J., and Valent, G. (2001) Self-adjoint extensions of operators and the teaching of quantum mechanics, *Am. J. Phys.*, **69**, 322.
135 Voronov, B.L., Gitman, D.M. and Tyutin, I.V. (2007) Constructing quantum observables and self-adjoint extensions of symmetric operators I, *Russ. Phys. J.*, **50**(1), 1.
136 Voronov, B.L., Gitman, D.M. and Tyutin, I.V. (2007) Constructing quantum

observables and self-adjoint extensions of symmetric operators II, *Russ. Phys. J.*, **50**(9), 3.
137 Voronov, B.L., Gitman, D.M. and Tyutin, I.V. (2001) Constructing quantum observables and self-adjoint extensions of symmetric operators III, *Russ. Phys. J.*, **51**(2), 645.
138 Fox, R. and Choi, M.H. (2000) Generalized coherent states and quantum-classical correspondence, *Phys. Rev. A*, **61**, 032107-1.
139 Hollenhorst, J.N. (1979) Quantum limits on resonant-mass gravitational-radiation detectors, *Phys. Rev. D*, **19**, 1669.
140 Yuen, H.P. (1976) Two-photon coherent states of the radiation field, *Phys. Rev. A*, **13**, 2226.
141 Stoler, D. (1970) Equivalence classes of minimum uncertainty packets, *Phys. Rev. D*, **1**, 3217.
142 Stoler, D. (1971) Equivalence classes of minimum uncertainty packets II, *Phys. Rev. D*, **4**, 1925.
143 Lu, E.Y.C. (1971) New coherent states of the electromagnetic field, *Nuovo Cim. Lett.*, **2**, 1241.
144 Slusher, R.E., Hollberg, L.W., Yurke, B., Mertz, J.C. and Valley, J.F. (1985) Observation of squeezed states generated by four-wave mixing in an optical cavity, *Phys. Rev. Lett.*, **55**, 2409.
145 Caves, C.V. (1981) Quantum-mechanical noise in an interferometer, *Phys. Rev. D*, **23**, 1693.
146 Walls, D.F. (1983) Squeezed states of light, *Nature*, **306**, 141.
147 Nieto, M.M (1997) The Discovery of Squeezed States – In 1927, in *Proceedings of the 5th International Conference on Squeezed States and Uncertainty Relations, Balatonfured, 1997* (eds Han, D. *et al.*) NASA/CP-1998-206855.
148 Dell-Annoa, F., De Siena, S., and Illuminatia, F. (2006) Multiphoton quantum optics and quantum state engineering, *Phys. Rep.*, **428**, 53.
149 Gilmore, R. and Yuan, J.M. (1987) Group theoretical approach to semiclassical dynamics: Single mode case, *J. Chem. Phys.*, **86**, 130.
150 Gilmore, R. and Yuan, J.M. (1989) Group theoretical approach to semiclassical dynamics: Multimode case, *J. Chem. Phys.*, **91**, 917.
151 Gazdy, B. and Micha, D.A. (1985) The linearly driven parametric oscillator: Application to collisional energy transfer, *J. Chem. Phys.*, **82**, 4926.
152 Gazdy, B. and Micha, D.A (1985) The linearly driven parametric oscillator: Its collisional time-correlation function, *J. Chem. Phys.*, **82**, 4937.
153 Helgason, S. (1962) *Differential Geometry and Symmetric Spaces*, Academic Press, New York.
154 Klauder, J.R. (2000) *Beyond Conventional Quantization*, Cambridge University Press, Cambridge.
155 Ali, S.T., Engliš, M. (2005) Quantization methods: a guide for physicists and analysts, *Rev. Math. Phys.*, **17**, 391.
156 Hall, B.C. (2005) Holomorphic methods in analysis and mathematical physics, *First Summer School in Analysis and Mathematical Physics (Cuernavaca Morelos, 1998), Contemp. Math.***260** 1, (Am. Math. Soc., Providence, RI, 2000).
157 Cahill, K.E. (1965) Coherent-State Representations for the Photon Density, *Phys. Rev.*, **138**, 1566.
158 Miller, M.M. (1968) Convergence of the Sudarshan expansion for the diagonal coherent-state, *J. Math. Phys.*, **9**, 1270.
159 Klauder, J.R. and Sudarshan, E.C.G. (1968) *Fundamentals of Quantum Optics*, Dover Publications.
160 Chakraborty, B., Gazeau, J.-P. and Youssef, A. (2008) Coherent state quantization of angle, time, and more irregular functions and distributions, submitted, arXiv:0805.1847v2 [quant-ph].
161 Van Hove, L. (1961) Sur le problème des relations entre les transformations unitaires de la Mécanique quantique et les transformations canoniques de la Mécanique classique, *Bull. Acad. R. Belg., Cl. Sci.*, **37**, 610.
162 Kastrup, H.A. (2007) A new look at the quantum mechanics of the harmonic oscillator, *Ann. Phys. (Leipzig)*, **7–8**, 439.
163 Dirac, P.A.M. (1927) The Quantum Theory of Emission and Absorption of Radiation, *Proc. R. Soc. Lond. Ser. A*, **114**, 243.

164 Schwartz, L. (2008) *Mathematics for the Physical Sciences*, Dover Publications.

165 Bužek, V., Wilson-Gordon, A.D., Knight, P.L. and Lai, W.K. (1992) Coherent states in a finite-dimensional basis: Their phase properties and relationship to coherent states of light, *Phys. Rev. A*, **45**, 8079.

166 Kuang, L.M., Wang, F.B. and Zhou, Y.G. (1993) Dynamics of a harmonic oscillator in a finite-dimensional Hilbert space, *Phys. Lett. A*, **183**, 1.

167 Kuang, L.M., Wang, F.B. and Zhou, Y.G. (1994) Coherent states of a harmonic oscillator in a finite-dimensional Hilbert space and their squeezing properties, *J. Mod. Opt.*, **41**, 1307.

168 Kehagias, A.A. and Zoupanos, G. (1994) Finiteness due to cellular structure of $\mathbb{R}^N$ I. Quantum Mechanics, *Z. Phys. C*, **62**, 121.

169 Polychronakos, A.P. (2001) Quantum Hall states as matrix Chern-Simons theory, *JHEP*, **04**, 011.

170 Miranowicz, A., Ski, W. and Imoto, N.: Quantum-Optical States in Finite-Dimensional Hilbert Spaces. 1. General Formalism, *in Modern Nonlinear Optics* (ed. Evans, M.W.) *Adv. Chem. Phys.*, **119** (I) (2001), 155, John Wiley & Sons, New York; quant-ph/0108080; *ibidem* p. 195; quant-ph/0110146.

171 Gazeau, J.-P., Josse-Michaux, F.X. and Monceau, P. (2006) Finite dimensional quantizations of the (q, p) plane: new space and momentum inequalities, *Int. J. Mod. Phys. B*, **20**, 1778.

172 Lubinsky, D.S. (1987) A survey of general orthogonal polynomials for weights on finite and infinite intervals, *Act. Appl. Math.*, **10**, 237.

173 Lubinsky, D.S. (1993) An update on orthogonal polynomials and weighted approximation on the real line, *Act. Appl. Math.*, **33**, 121.

174 Landsman, N.P. (2006) Between classical and quantum, in *Handbook of the Philosophy of Science*, vol. 2: Philosophy of Physics (eds Earman, J. and Butterfield, J.), Elsevier, Amsterdam.

175 Woodhouse, N.J.M. (1992) *Geometric Quantization* 2nd edn, Clarendon Press, Oxford.

176 Dito, G. and Sternheimer, D. (2002) Deformation quantization: genesis, developments and metamorphoses, in *Deformation quantization* (ed. Halbout, G.), IRMA Lectures in Math. Theor. Phys., Walter de Gruyter, p. 9.

177 Karaali, G.: Deformation quantization – a brief survey, http://pages.pomona.edu/~gk014747/research/deformationquantization.pdf

178 Madore, J. (1995) *An Introduction to Noncommutative Differential Geometry and its Physical Applications*, Cambridge University Press, Cambridge.

179 Connes, A. and Marcolli, M. (2006) A walk in the noncommutative garden, arXiv:math/0601054v1. Connes, A. and Marcolli, M. (2006) A walk in the noncommutative garden, arXiv:math/0601054v1.

180 Taylor, W. (2001) M(atrix) Theory, *Rev. Mod. Phys.*, **73**, 419.

181 Deltheil, R. (1926) *Probabilités géométriques*, Traité de Calcul des Probabilités et de ses Applications par Émile Borel, Tome II. Gauthiers-Villars, Paris.

182 Filippov, A.T., Isaev, A.P. and Kurdikov, A.B. (1996) Paragrassmann Algebras, Discrete Systems and Quantum Groups, *Problems in Modern Theoretical Physics, dedicated to the 60th anniversary of the birthday of A.T. Filippov*, Dubna 96–212, p. 83.

183 Isaev, A.P. (1996) Paragrassmann integral, discrete systems and quantum groups (arXiv:q-alg/9609030).

184 Daoud, M. and Kibler M. (2002) A fractional supersymmetric oscillator and its coherent states, in *Proceedings of the International Wigner Symposium, Istanbul, Aout 1999* (eds Arik, M. *et al.*), Bogazici University Press, Istanbul.

185 Majid, S. and Rodriguez-Plaza, M. (1994) Random walk and the heat equation on superspace and anyspace, *J. Math. Phys.*, **35**, 3753.

186 Cotfas, N. and Gazeau, J.-P. (2008) Probabilistic aspects of finite frame quantization, arXiv:0803.0077v2 [math-ph].

187 García de Léon, P. and Gazeau, J.-P. (2007) Coherent state quantization and phase operator, *Phys. Lett. A*, **361**, 301.

188 Carruthers, P. and Nieto, M.M. (1968) Phase and Angle Variables in Quantum Mechanics, *Rev. Mod. Phys.*, **40**, 411.

189 Louisell, W.H. (1963) Amplitude and phase uncertainty relations, *Phys. Lett.*, **7**, 60.

190 Susskind, L. and Glogower, J. (1964) Quantum mechanical phase and time operator, *Physics*, **1**, 49.

191 Popov, V.N. and Yarunin, V.S. (1973) Quantum and quasi-classical states of the photon phase operator, *Vestnik Leningrad University*, **22**, 7.

192 Popov, V.N. and Yarunin, V.S. (1992) Quantum and Quasi-classical States of the Photon Phase Operator, *J. Mod. Opt.*, **39**, 1525.

193 Barnett, S.M. and Pegg, D.T. (1989) Phase in quantum optics, *J. Mod. Opt.*, **36**, 7.

194 Busch, P., Grabowski, M. and Lahti, P.J. (1995) Who is afraid of POV measures? Unified approach to quantum phase observables, *Ann. Phys. (N.Y.)*, **237**, 1.

195 Busch, P., Lahti, P., Pellonpa, J.-P. and Ylinen, K. (2001) Are number and phase complementary observables? *J. Phys. A: Math. Gen.*, **34**, 5923.

196 Dubin, D.A., Hennings, M.A. and Smith, T.B. (1994) Quantization in polar coordinates and the phase operator, *Publ. Res. Inst. Math. Sci.*, **30**, 479.

197 Busch, P., Hennings, M.A. and Smith, T.B. (2000) *Mathematical Aspects of Weyl Quantization and Phase*, World Scientific, Singapore.

198 Gazeau, J.-P., Piechocki, W. (2004) Coherent state quantization of a particle in de Sitter space, *J. Phys. A: Math. Gen.*, **37**, 6977.

199 Garcia de Leon, P., Gazeau, J.-P. and Quéva, J. (2008) The infinite well revisited: coherent states and quantization, *Phys. Lett. A*, **372**, 3597.

200 Garcia de Leon, P., Gazeau, J.-P., Gitman, D., and Quéva, J. (2009) Infinite quantum well: on the quantization problem, *Quantum Wells: Theory, Fabrication and Applications*. Nova Science Publishers, Inc.

201 Ashtekar, A., Fairhurst, S. and Willis, J.L. (2003) Quantum gravity, shadow states, and quantum mechanics, *Class. Quantum Grav.*, **20**, 1031.

202 Lévy-Leblond, J.M. (2003) Who is afraid of Nonhermitian Operators? A quantum description of angle and phase, *Ann. Phys.*, **101**, 319.

203 Rabeie, A., Huguet, E. and Renaud, J. (2007) Wick ordering for coherent state quantization in $1+1$ de Sitter space, *Phys. Lett. A*, **370**, 123.

204 Znojil, M. (2001) $\mathcal{PT}$-symmetric square well, *Phys. Lett. A*, **285**, 7.

205 Ali, S.T., Engliš, M. and Gazeau, J.-P (2004) Vector coherent states from Plancherel's theorem, Clifford algebras and matrix domains, *J. Phys. A: Math. Gen.*, **37**, 6067.

206 Reed, M. and Simon, B. (1980) *Methods of Modern Mathematical Physics, Functional Analysis vol. I*, Academic Press, New York.

207 Lagarias, J. (2000) Mathematical Quasicrystals and the Problem of Diffraction, in *Directions in Mathematical Quasicrystals* (eds Baake, M. and Moody, R.V.), CRM monograph Series, Am. Math. Soc., Providence, RI.

208 Bouzouina, A., and De Bièvre, S. (1996) Equipartition of the eigenfunctions of quantized ergodic maps on the torus, *Commun. Math. Phys.*, **178**, 83.

209 Abramowitz, M. and Stegun, I.A. (1972) *Handbook of Mathematical Functions*, National Bureau of Standards Applied Mathematics Series – 55.

210 Calderon, A.P. and Vaillancourt, R. (1972) A class of bounded pseudodifferential operators, *Proc. Natl. Acad. Sci. USA*, **69**, 1185.

211 Bouzouina, A. and De Bièvre, S. (1998) Equidistribution des valeurs propres et ergodicité semi-classique de symplectomorphismes du tore quantifiés., *C. R. Acad. Sci., Série I*, **326**, 1021.

212 Gazeau, J.-P, Huguet, E., Lachièze Rey, M. and Renaud, J. (2007) Fuzzy spheres from inequivalent coherent states quantizations, *J. Phys. A: Math. Theor.*, **40**, 10225.

213 Grosse, H. and Prešnajder, P. (1993) The construction of non-commutative manifolds using coherent states, *Lett. Math. Phys.*, **28**, 239.

214 Kowalski, K. and Rembielinski, J. (2000) Quantum mechanics on a sphere and coherent states, *J. Phys. A: Math. Gen.*, **33**, 6035.

215 Kowalski, K. and Rembielinski, J. (2001) The Bargmann representation for the quantum mechanics on a sphere, *J. Math. Phys.*, **42**, 4138.
216 Hall, B. and Mitchell, J.J. (2002) Coherent states on spheres, *J. Math. Phys.*, **43**, 1211.
217 Freidel, L. and Krasnov, K. (2002) The fuzzy sphere $*$-product and spin networks, *J. Math. Phys.*, **43**, 1737.
218 Gazeau, J.-P., Mourad, J. and Quéva, J. (2009) Fuzzy de Sitter space-times via coherent states quantization, in *Proceedings of the XXVIth Colloquium on Group Theoretical Methods in Physics, New York, USA, 2006* (eds Birman, J. and Catto, S.) to appear.
219 Gazeau, J.-P. (2007) An Introduction to Quantum Field Theory in de Sitter space-time, in *Cosmology and Gravitation: XIIth Brazilian School of Cosmology and Gravitation* (eds Novello, M. and Perez-Bergliaffa, S.E.), *AIP Conference Proceedings* 910, 218.
220 Sinai, Y.G. (1992) *Probability Theory. An introductory Course*, Springer Textbook, Springer-Verlag, Berlin.
221 Johnson, N.L. and Kotz, S. (1969) *Discrete Distributions*, John Wiley & Sons, Inc., New York.
222 Ross, S.M. (1985) *Introduction to Probability Models* Academic Press, New York.
223 Douglas, J.B. (1980) *Analysis with Standard Contagious Distributions*, International Co-operative Publishing House, Fairland, Maryland.
224 Vilenkin, N.Y. (1968) *Special Functions and the Theory of Group Representations*, Am. Math. Soc., Providence, RI.
225 Barut, A.O. and Rączka, R. (1977) *Theory of Group Representations and Applications*, PWN, Warszawa.
226 Kirillov, A.A. (1976) *Elements of the Theory of Representations*, Springer-Verlag, Berlin.
227 Arfken, G. (1985) *Spherical Harmonics and Integrals of the Products of Three Spherical Harmonics*. 12.6 and 12.9 in *Mathematical Methods for Physicists*, p. 680 and p. 698, 3rd ed., Academic Press, Orlando, FL.
228 Weisstein, E.W. *Spherical Harmonic*, MathWorld – A Wolfram Web Resource. http://mathworld.wolfram.com/SphericalHarmonic.html.

Index

Coherent States in Quantum Physics. Jean-Pierre Gazeau

ISBN: 978-3-527-40709-5